MARGARET FELL AND THE RISE OF QUAKERISM

Also by Bonnelyn Young Kunze

COURT, COUNTRY AND CULTURE: Essays on Early Modern British History in Honor of Perez Zagorin
(*edited with Dwight D. Brautigam*)

Margaret Fell and the Rise of Quakerism

Bonnelyn Young Kunze
Adjunct Assistant Professor in History
LeMoyne College, Syracuse
New York

© Bonnelyn Young Kunze 1994

All rights reserved. No reproduction, copy or transmission of this publication may be made without written permission.

No paragraph of this publication may be reproduced, copied or transmitted save with written permission or in accordance with the provisions of the Copyright, Designs and Patents Act 1988, or under the terms of any licence permitting limited copying issued by the Copyright Licensing Agency, 90 Tottenham Court Road, London W1P 9HE.

Any person who does any unauthorised act in relation to this publication may be liable to criminal prosecution and civil claims for damages.

First published 1994 by
THE MACMILLAN PRESS LTD
Houndmills, Basingstoke, Hampshire RG21 2XS
and London
Companies and representatives
throughout the world

ISBN 0-333-59389-8

A catalogue record for this book is available
from the British Library.

Copy-edited and typeset by Grahame & Grahame Editorial, Brighton

Printed in Hong Kong

To Bill, Karl, Kaaren and George

Contents

List of Abbreviations	viii
Chronology of Margaret Fell's Life	x
Preface and Acknowledgements	xviii
Introduction	1

Part I Life in a Seventeenth-Century Quaker Family — 11

1. "Weak in Body but Alive to God": Margaret Fell's Religious Self-Reflections — 13
2. A Family History — 27

Part II The Domestic and Economic World of the Fells — 63

3. The Swarthmoor Farm — 65
4. "Poore and in Necessity": Margaret Fell's Charitable Activities — 83
5. Feuding Friends — 101

Part III Political and Religious World of Margaret Fell — 129

6. "We Have Been a Suffering People under Every Power and Change": Margaret Fell and Politics, 1659–61 — 131
7. "Walk as Becomes Truth": Margaret Fell and Women's Meetings — 143
8. Gender, Religion, and Class: A Seventeenth-Century Friendship — 169

Part IV The Mental World of Margaret Fell — 185

9. Fell's Worldview — 187
10. "Let Us Be of One Spirit . . . ": Fell's Spiritualist Theology — 197
11. Fell's Work to Convert the Jews — 211
12. Conclusion — 229

Notes	235
Appendix	290
Bibliography	292
Index	318

List of Abbreviations

ABSF	Norman Penney, editor, *The Household Account Book of Sarah Fell* (1920)
BQ	William C. Braithwaite, *Beginnings of Quakerism* (1981)
CSPD	Calendar of State Papers Domestic
DQB	*Dictionary of Quaker Biography* in the FHL
FHL	Friends' House Library, London (Library of the Society of Friends)
Journal	George Fox's *Journal*, in three editions: Ellwood (1694, reprints 1831, 1975) Cambridge University Press edition (1911) Nickalls (1975)
JFHS	The *Journal of the Friends Historical Society*
Kendal WMMMB	Kendal Women's Monthly Meeting Minute Book
LBC	Lewis Benson, *Concordance of George Fox's Works*
LQM	Lancaster Quarterly Meeting
LWQM	Lancaster Women's Quarterly Meeting
PRO	Public Record Offices in Preston, Lancashire, and Barrow-in-Furness
PWP	Papers of William Penn, Pennsylvania Historical Society, Philadelphia
SMMMM	Swarthmoor Men's Monthly Meeting Minutes
SWM	Swarthmore Manuscripts
SWMMM	Swarthmoor Women's Monthly Meeting Minutes
SDMM	Second Day Morning Meeting – acted as a Board editing Quaker publications. It preserved numerous early Quaker documents.
SPQ	William C. Braithwaite, *Second Period of Quakerism* (1979)
TCWAAS	Transactions of the Cumberland and Westmorland Archeological and Antiquarian Society
VCHL	*Victoria County History of Lancashire*
WTWM	Women's Two Week Meeting (London)

WWP	Richard S. Dunn and Mary Maples Dunn, editors, *The World of William Penn* (1986)
Works	*A Brief Collection of Remarkable Passages and Occurrences Relating to the Birth, Education, Life, Conversion, Travels, Services and Deep Sufferings of that Ancient and Faithful Servant of the Lord, Margaret Fell, but by her Second Marriage M. Fox* (1712)

Chronology of Margaret Fell's Life**

1598–9	Birth of (Judge) Thomas Fell.
1603	**Death of Queen Elizabeth I. Accession of James I.**
1614	Birth of Margaret Askew at Marsh Grange, near Dalton.
1616	**Death of William Shakespeare.**
1624	Birth of George Fox.
1625	**Accession of Charles I.**
1632	Marriage of Thomas Fell to Margaret Askew.
c. 1633	Birth of Margaret Fell, Junior.
c. 1635	Birth of Bridget Fell.
c. 1637	Birth of Isabel Fell.
c. 1638	Birth of George Fell.
1640	**Long Parliament first met.**
1642	**Outbreak of the Civil War.**
* 1642	Thomas Fell, estate is valued at £150 per annum.
1642	Birth of Sarah Fell.
1647	Birth of Mary Fell.
1649 (Jan.)	**Execution of Charles I.**
1650	Birth of Susannah Fell.
1651	**Battle of Worcester.**
1652 (June) *	George Fox first comes to Swarthmoor. Margaret Fell's conversion to Quakerism. Fox visits St. Mary's Church, Ulverston.
1653 (April)	**Cromwell dissolves Long Parliament.**

** With the permission of the William Sessions Book Trust, this chronology has been reproduced from *Margaret Fell, Mother of Quakerism*, by Isabel Ross (William Sessions Book Trust, The Ebor Press, York, England, 1984). All items designated with an asterisk are this author's own additions.

Chronology of Margaret Fell's Life

1653 (Oct.)	Birth of Rachel Fell.
* 1653	Margaret Fell, with the help of George Taylor and Thomas Willan of Kendal, establishes the Kendal Fund to aid traveling ministers and their families.
* 1655	Margaret Fell writes two letters to Oliver Cromwell described in her *Works*.
1656	Margaret Fell writes *To Manasseth-ben-Israel*.
* 1656	Margaret Fell writes a third letter to Oliver Cromwell reminding him of his promise of "liberty of conscience."
* 1657 (May)	John Lilburne Leveller writes to Margaret Fell asking for her legal advice and help. Lilburne allegedly became a Quaker before his death in 1657.
1657	Margaret Fell writes *A Loving Salutation*.
* 1657	Margaret Fell writes her fourth letter to Oliver Cromwell concerning persecutions of Quakers.
1658 (Sept.)	Death of Oliver Cromwell.
1658 (Oct.)	Death of Judge Thomas Fell.
* 1658–63	Thomas Rawlinson, steward of Force Forge, in the employ of Sarah, Mary, Susannah, and Rachel Fell.
* 1659 (July)	Margaret Fell's and daughters' signatures lead a petition of 7,000 Quaker women presented to Parliament protesting the tithe.
* 1659 (Oct.)	Margaret Fell writes an epistle to the General Council officers of the English army exhorting them to be just to Quakers.
1660 (April)	**Declaration of Breda.**
1660 (May)	**Restoration of Charles II.**
1660 (May)	Arrest of George Fox at Swarthmoor and imprisonment in Lancaster Castle until September.
1660 (June)	Margaret Fell's first visit to London with Margaret Fell, Jr.
1660 (June)	**Margaret Fell's declaration against violence and war.**
* 1660 (July)	Margaret Fell's broadsheet *To the Citie of London*.
1660 (Dec.)	Marriage of George Fell and Hannah Potter.
* 1660	Margaret Fell writes to Major General Harrison

Chronology of Margaret Fell's Life

	urging him to convert to the Light of Christ.
1661 (Jan.)	**Rising of Fifth Monarchy Men. Nationwide persecution of Friends.**
1661 (Sept.)	Margaret Fell and Margaret, Jr. return to Swarthmoor Hall, after freedom granted to Friends.
1662 (Jan.)	Marriage of Margaret Fell, Jr., to John Rous.
1662 (March)	Marriage of Bridget Fell to John Draper.
1662 (May)	**Quaker Act passed.**
1662 (May)	Margaret Fell's second visit to London.
1662 (Aug.)	Quakers freed from prisons.
1662 (Sept.)	Margaret Fell returns to Swarthmoor.
1662–3 (winter)	Margaret Fell goes to be with Bridget Draper at Headlam, Durham.
1663 (March)	Death of Bridget Draper and her baby, Isaac.
1663 (May–Aug.)	1,000-mile journey through England – Margaret Fell, Sarah Fell, Mary Fell, Leonard Fell, Thomas Salthouse, William Caton in S.W. (George Fox and Thomas Lower with them there), Yorks, Bishoprick, Nurthumberland, Westmorland.
1663 (Aug.)	**Kaber Rigg Plot in the North. Renewed persecution of Quakers.**
* 1663 (Feb.)	Thomas Rawlinson writes to the SMMM concerning the Force Forge dispute.
1664 (winter)	George Fox, at Swarthmoor, arrested; examination of George Fox and Margaret Fell at Holker Hall.
1664 (Jan.)	George Fox tried and imprisoned at Lancaster.
1664 (Feb.)	Margaret Fell arrested and imprisoned at Lancaster.
1664 (March)	Trial and imprisonment of both George Fox and Margaret Fell.
1664 (July)	Marriage of Isabel Fell to William Yeamans at Bristol.
1664 (July)	**First Conventicle Act.**
1664 (Aug.)	Margaret Fell praemunired (writes *A Call to the Universal Seed of God*).
1664–7	War between England and Holland.
1664 (Nov.–Dec.)	Mary Fell ill, probably plague, in London.
1665 (Jan.)	Swarthmoor estate of Margaret Fell granted to George Fell by Charles II.

Chronology of Margaret Fell's Life

1665 (March)	George Fox praemunired.
1665 (April)	George Fox moved from Lancaster to Scarborough Castle. John and Margaret Rous, Mary Fell return to Swarthmoor.
1665	**The plague in London.**
1665 (Nov.)	Death of William Caton in Holland.
1666 (Sept.)	George Fox released from Scarborough.
1666 (Sept.)	**The Great Fire of London.**
1666	Margaret Fell writes *Women's Speaking Justified*.
1666 (Aug.)	Margaret Fell writes *Epistle to Charles II*.
1666 (Sept.)	Margaret Fell writes *A Touch-Stone*.
1666	Margaret Fell writes *Standard of the Lord Revealed*.
1666–7 (winter)	Mary Fell in Bristol with Isabel Yeamans.
* 1667 (May)	Thomas Rawlinson writes to the SMMM protesting Margaret Fell's ill usage of him.
1667	William Penn and Robert Barclay join the Quakers.
1668	Margaret Fell writes *A Call unto the Seed of Israel*.
1668 (June)	Margaret Fell released from Lancaster Castle.
1668 (Aug.)	Marriage of Mary Fell to Thomas Lower. Margaret Fell travels in north and western England with Rachel, and to Bristol, Devon, and Somerset.
* 1668	Margaret Fell writes *M[argaret] ff[ell]s Answer to Allan Smallwood Dr. priest of Grastock in Cumberland*.
1668–9 (winter)	Margaret Fell's third visit to London, with Rachel.
1669 (spring)	Margaret Fell and Rachel visit Friends in Kent.
1669 (summer)	Margaret Fell and Rachel in London.
1669 (Oct.)	Marriage of Margaret Fell to George Fox at Bristol. George Fox travels in the south and Midlands. Margaret Fox returns to Swarthmoor.
1669 (Dec.)	George Fell scheming in London to get his mother's estate and Swarthmoor. Susannah and Rachel at Swarthmoor; Sarah in London. Thomas and Mary Lower, John and Margaret Rous living in London.
1670 (April)	Margaret Fox's second imprisonment, Lancaster.
1670 (May)	**Second Conventicle Act.**

* 1670 (May)	Thomas Rawlinson writes to the Lancaster Quarterly Meeting concerning his dispute with Margaret Fell.
1670 (summer)	Sarah Fell returns home.
1670 (Oct.)	Death of George Fell.
1670 (Sept. to spring '71)	George Fox very ill in London, nursed by Margaret Rous.
1671 (April)	Margaret Fox released by discharge under the Broad Seal.
1671 (May–Aug.)	Margaret Fox's fourth visit to London.
* 1671 (August)	George Fox sets sail for America.
1671–3	George Fox in America.
* 1671 (Oct.)	Swarthmoor Women's Monthly Meeting formed at Swarthmoor, the first continuous women's meeting outside London.
1672	Charles II's 'Great Pardon' releases 500 Friends and other Nonconformists.
1672 (April)	Margaret Fox and Rachel Fell travel in Yorkshire (268 miles).
* 1672 (June)	Thomas Rawlinson's "Ten Queries to Margaret Fell and her children."
1672 (late)	Margaret Fox travels again – district not known.
1673 (June)	Margaret Fox, with Sarah and Rachel Fell and Thomas Lower, go to Bristol to meet George Fox on his return from America. Margaret Fox's fifth visit to London.
1673	**Test Act passed.**
1673–85	Continuous and severe persecution of Quakers.
1673 (Dec.)	George Fox arrested at Armscott and imprisoned at Worcester. Margaret Fox and Rachel Fell return to Swarthmoor. Thomas Lower stays at Worcester.
1674 (Aug.)	Margaret Fox with Thomas Lower and Susannah Fell go to Worcester.
1674 (Oct.)	Margaret Fox's sixth visit to London, interviews King.
* 1674–91	Extant correspondence of Margaret Fell, George Fox, and William Penn spans these years.
1675 (Feb.)	George Fox is discharged freely from Worcester.
1675 (June)	George Fox, Margaret Fox, and Susannah Fell return

Chronology of Margaret Fell's Life

	to Swarthmoor. George Fox stays at Swarthmoor for twenty-one months.
* 1676	A representative year in the Swarthmoor Account Book of Sarah Fell (ABSF).
* 1676	Swarthmoor estate valued at £450, according to ABSF.
1676 (April)	William Penn at Swarthmoor.
1676	Thomas and Mary Lower move to Marsh Grange.
1677 (March)	George Fox leaves Swarthmoor for London.
1677 (summer)	Journey to Holland – George Fox, William Penn, Robert Barclay, George Keith, and wife, and Isabel Yeamans.
1678	Isabel Yeamans in Scotland with Robert Barclay and wife.
1678 (Sept. to March 1680)	George Fox at Swarthmoor (last visit).
* 1680 (April)	Thomas Rawlinson's "Salme of Praise" acknowledging the winning of his suit against Margaret Fell.
1680–90	Isabel Yeamans lives in various places, including London and Stockton-on-Tees.
1681–91	George Fox in or near London.
* 1681	Force Forge signed over by the Fells to Thomas Rawlinson.
1681 (June)	Marriage of Sarah Fell to William Meade in London.
1681 (May–July)	Margaret Fox's seventh visit to London.
1682	William Penn founds Pennsylvania.
1683–6	Thomas Lower in prison at Launceston, Cornwall.
1683 (March)	Marriage of Rachel Fell to Daniel Abraham.
1683 (Sept.)	Margaret Fox, Rachel and Daniel Abraham arrested and tried by Roger and William Kirkby in Lancaster.
1684 (June)	These three fined heavily and imprisoned.
1684 (Nov. to March 1685)	Margaret Fox's eighth visit to London – Mary Lower and Leonard Fell with her, and Susannah Fell. Margaret Fox nearly 70 years old
1685 (Feb.)	**Death of Charles II.** Margaret Fox interviews James II.
1685 (May or June)	Robert Barclay visits Swarthmoor on his way from London to Scotland.
1686 (March)	**James II's General Pardon and Royal Warrant**

	releases Friends from prison.
1687 (spring)	**Declaration of Indulgence for all Nonconformists.**
1688	Swarthmoor Meeting House given by George Fox.
1689	**Fall of James II. Accession of William and Mary.**
1689 (May)	**Toleration Act gives limited religious freedom.**
1690	Marriage of Isabel Yeamans to Abraham Morrice of Lincoln.
1690 (April–June)	Margaret Fox's ninth visit to London; now 76 years old.
* 1690	Margaret Fell writes "A Relation of Margaret Fell, Her Birth, Life, Testimony and Sufferings for the Lord's Everlasting Truth in her Generation."
1690	Death of Robert Barclay.
1691 (Jan.)	Death of George Fox in London.
1691	Marriage of Susannah Fell to William Ingram.
1694	Death of Gulielma Penn.
* 1694	Margaret Fell writes "The Testimony of Margaret Fox concerning her late husband."
1696 (March)	Death of John Rous at sea on returning from Barbados.
1697–8	Margaret Fox's tenth and last visit to London; now 83 years old.
1698	Margaret Fox's *Epistle to Friends Concerning Oaths*. Trouble in Society of Friends concerning Whigs and Tories.
1698 (May)	Margaret Fox attends Yearly Meeting.
1698 (June)	Margaret Fox sends a letter to William III by Susannah Ingram. Death of William Yeamans.
1698 (late June)	Margaret Fox's First Epistle against regimentation.
1699	Second visit of William Penn to Pennsylvania.
1700 (April)	Margaret Fox's Second Epistle against regimentation.
1701 (Nov.)	Margaret Fox's last Testimony.
1702 (March)	**Death of William III. Accession of Anne.**
1702 (April)	Death of Margaret Fox, aged nearly eighty-eight, at Swarthmoor Hall.
* 1712	Margaret Fell's *Works* published under the direction

of her daughters and sons-in-laws under the title, *A Brief Collection of Remarkable Passages and Occurrances Relating to the Birth, Education, Life, Conversion, Travels, Services and Deep Sufferings of that ancient Eminent and Faithful Servant of the Lord, Margaret Fell, but by her Second Marriage M. Fox.*

Preface and Acknowledgements

When a historian encounters substantial written and behavioral evidence of a seventeenth-century Englishwoman's workaday and mental world and recovers some entirely new evidence to support the existing material, it is a comparatively rare and happy occurrence. Such is the reason for this study of Margaret Askew Fell Fox (1615–1702) and radical religion in early modern England. Margaret Fell was a literate woman who left in print her thoughts relating to her family, religion, and political milieu. Fell was a religious sectarian, specifically a Quaker, in a world that deplored radical sectarianism. She was an outspoken woman in public life, in a world that looked askance at women in public roles. Fell was a gentry woman of some wealth who transgressed class categories in marrying a man of lower social status. A controversial figure in her own day, Margaret Fell came to be designated "mother of Quakerism" by those who remembered her. This label has been given to Fell in much of the secondary literature on the formative phase of Quakerism, yet has failed to define her roles or give us a picture of what life was like for a nonconformist woman of old gentry background in early modern England. This study is an attempt to understand and recreate the life, times, and thoughts of a woman and a religious movement that had influence far beyond Quakerism's place of origin in county Lancaster, England, in the mid-seventeenth century.

In the process of writing this book, I have been supported by numerous people. The University of Rochester Department of History sheltered me with a fellowship while I completed my original research and writing. The librarians at Friends House Library, London, especially Malcolm J. Thomas and Josef Keith, have given me considerable assistance with my research. I received courteous and efficient help from the library staffs of the Friends Historical Library of Swarthmoor College, Swarthmoor, Pennsylvania, and the Quaker collection of Haverford College, Haverford, Pennsylvania. The British Library manuscripts department was especially considerate in giving me early access to a newly acquired and yet uncatalogued document written by Fell in 1668. Likewise, Craig Horle, of the Pennsylvania Historical Society, graciously put me in touch with another important and unique document relating to Fell. I remember the kindness of the late Ralph Randles, custodian of the early records of the Lancaster Quaker Meeting House, Lancaster, England, who gave me access to the original Lancashire quarterly meeting minutes and sufferings.

During the past few years, portions of this book have been presented as papers at the Quaker Historians and Archivists Meeting (1986); the Berkshire Conference on the History of Women (1987) at Wellesley, Massachusetts; and the North American Conference of British Historians (1987) in Portland, Oregon. I benefited substantially from the comments and criticisms of Carol Stoneburner, Esther Cope, and Paul Seaver.

I am grateful to Bonnie Smith, Donald R. Kelley, and David Nicholas, who read earlier versions of this text and gave me helpful suggestions. I also thank my mother, Esther Hill Young, for her careful reading of my final manuscript. Early in my research, Winthrop S. Hudson gave me insightful ideas for my work. I wish to convey special thanks to Perez Zagorin, who oversaw this work from its inception. His scholarly assistance, penetrating questions, and moral support, I have appreciated deeply. Finally, I thank my family who have observed my progress and encouraged me along the way, and so this book is for them.

BONNELYN YOUNG KUNZE

Introduction

Margaret Askew Fell Fox (1614–1702) was an authoritative female public minister in a radical religious movement of the seventeenth-century, Quakerism. This book is dedicated to reconsidering Margaret Fell's life and work in the rise of the English Quaker movement.[1] It is dedicated to recreating the wider world in which Fell worked, thought, and moved as a mother, wife, and literate sectarian leader in the second half of the seventeenth century. Further, this study is a reassessment of Fell's role in relation to George Fox and other early leaders of one of the few, if not the only, radical sectarian movements of the Interregnum that has endured, although in significantly altered form, until today.

The use of the term radical to describe Margaret Fell requires some explanation. Although the word "radical" is anachronistic, historians have often noted that in studying the past, it is frequently necessary to draw on modern terminology to help the reader comprehend past experience and convey that past experience as accurately as possible to the present. Richard L. Greaves has observed that an opponent of seventeenth-century Quakerism would very likely have used the label "fanatic" to describe in a highly pejorative manner a sectarian group like the early Quakers. A contemporary co-religionist would most probably have employed a word like "godly" to describe the same dissenting persons. Therefore a nuanced and careful use of the modern term radical is both logical and justified to convey as precisely as possible the immediacy and essence of the historical situation to the modern reader. I am indebted to Greaves for his definition of the term for my use of it in this study. Radicals were people "who sought fundamental change by striking at the very root of contemporary assumptions and institutions." Further, radicals sought "replacement of the status quo by something new." I will add only one restriction, that it specifically applies to the religious, political, and social context in which Fell lived, wrote, and worked.[2]

Margaret Fell's life unfolds most compellingly through the use of the tools of social and family historiography, as well as the tools of intellectual and topical biography. Historians of the social and family history of early modern England are well acquainted with two significant studies that expertly reconstruct the family life and wider milieux of two seventeenth-century religious men. One is Alan Macfarlane's work on the Essex priest, Ralph Josselin. This work is still considered by some social historians as one of the most thorough studies of family relationships. It creates, in opposition to some other interpretations, a "picture of great parental care and warm emotional bonding." Another more recent study in the field of English social history is Paul Seaver's recreation of the world of the London Puritan artisan, Nehemiah Wallington. Seaver has avoided straightforward biography by reconstructing the larger Puritan "context," not merely the experiences of Wallington. He has thereby fashioned a fine book that, although approximating a biography, avoids the pitfall of reductionist overemphasis on the role of the single individual in the historical process.[3]

Like the Josselin and Wallington books, this work is a microstudy of Fell's religious, social, and family life. It investigates the gender roles of Fell in terms of the family, the household unit, and the neighborhood network, in order to understand her activities from a woman's point of view. Further, it probes her public role as authoritative first-generation Quaker co-leader with George Fox. Both facets of Fell's life and worldview illustrate that the hypothesis of patriarchy as an explanatory model of seventeenth-century social order is inadequate, for it fails to draw out the complexity of gender roles in religious movements in early modern England.

To understand Margaret Fell and the world in which she lived, a broader and highly useful mode of thinking for me has been one of dual perspective. Combined with the orientation of social history, I have included an intellectual and topical biographical focus. This approach has the advantage of seeing a seventeenth-century nonconformist woman at close range in her public and domestic life, extending from her conversion in 1652 to her death in 1702. In telling her story within the larger framework of her English sectarian community, I admittedly attempt to avoid a stereotypical biography. I resist considering this a traditional biographical work in part because three sympathetic biographies of Fell already exist. These biographies, although having some merit, idealize her life in an attempt to uphold the ideals of a religious persuasion. Further, I wish to avoid straightforward biography because the sources on Fell, which amount to 600 to 700 pages of published and unpublished works, letters, and contemporary materials about her activities and ideas allow for a more inclusive social and intellectual history of Fell and her sectarian circle.[4]

My approach is to place Fell's life in several contexts. This is, I do not attempt to cover the entire chronology of her life. Rather, departing from chronology, I discuss several types of evidence on Fell, namely, economic, personal, political, and religious, which allow for what Alan Macfarlane has called a "total life" view of the historical figure at hand.

To counteract this retreat from biographical chronology, and in order to stress the influential activities and pivotal episodes in her experience, I have included a list of important dates of her life, borrowed from an earlier biography and supplemented with my own dates (see Chronology). Although Fell's life has been reasonably well documented, there is much that still remains in shadow. This study not only constitutes a new reading of existing evidence but also incorporates new evidence in the form of a new-found religious treatise and critical tract of Fell's leadership written by a disaffected Quaker. Both pieces of evidence were unavailable to earlier biographers and to historians of the primitive period of English Quakerism. This evidence brings to light one woman's work in a revolutionary sectarian English movement in the second half of the 1600s that was both ordinary and exceptional. Her way of life was, in part, similar to many religiously motivated landed women of her era. At the same time her experiences as a literate, independent, wealthy, and widely traveled religious leader place her among a tiny minority of women who may claim some degree of historical influence.[5]

Although not a traditional biography, this study does benefit from some useful comparisons with Lois Schwoerer's biography of *Lady Rachel Russell*, a younger and more aristocratic contemporary of Margaret Fell. Schwoerer's careful study illuminates the female experience of an aristocratic woman of the late Stuart period, and thus invites a close look at how the advantages of class freed aristocratic women, to a degree, from the seventeenth-century patriarchal model of womanhood. Margaret Fell and Lady Rachel Russell differed in social status and wealth. However, falling within the broadly defined structure of greater and lesser aristocracy, Fell and Russell enjoyed certain privileges and held some values that were class rather than gender defined. Both women, during their widowhoods, oversaw their estates and family money matters, and were matriarchs in their households. Both women shared the privileges of wealth, education, social connections, and leisure time to write. Leisure allowed Margaret and Rachel to become voluminous writers and this in turn gave them high visibility and recognition in the public sphere. Additionally, the two women possessed a similar self-image and self-assurance derived from their gentle origins that helped them transcend the traditional gender restraints of their day. Lady Rachel Russell and Margaret Fell (for different reasons), used their class advantages for their own personal advantage. In so doing their behavior and attitudes did not conform to the submissive model. They were not reticent in using their connections to influence people in high places. Beyond this, however, the comparison ceases to be useful, for Lady Rachel Russell was far more conventional for her class than

Margaret Fell was for hers. After 1652, when Margaret Fell converted to Quakerism, and especially after her first husband's death in 1658, Margaret Fell became radicalized in her religious and political views. Thus, while Schwoerer found Lady Russell "unusual rather than unique among contemporary upper-class women," Fell was very exceptional to the point of being probably unique in her public role as a powerful spokeswoman for a revolutionary religious ideology. What is significant about the comparison of Lady Rachel Russell and Margaret Fell is that both women serve as models that help establish some of the complexities within seventeenth-century patriarchy.[6]

Margaret Fell is recognized as significant in virtually all the histories of the early movement. This notwithstanding, a reading of the standard histories of English Quakerism leaves unanswered several questions about Fell and her influence: What was Fell's religious worldview and what predisposed her to it? What was the significance of her religious thought and activities especially to women? How well did she coordinate her religious ideology, domesticity, and charitable activities from a feminine perspective? Did she conform to the predominant "weaker vessel" image of her day or attempt to overthrow it or ignore it? What was her role *vis-à-vis* George Fox, and to what extent was he influenced by her ideas? Admittedly, the answers to these questions must remain partial. Thus, while much has been written about Fell, she remains, aside from her traditional reputation, a figure only imperfectly understood. In the larger accounts of early Quakerism where the Fox-as-founder tradition has been largely accepted without question by later historians, she is cast only in a secondary role.

A reconsideration of Margaret Fell requires, therefore, a close look at the existing secondary literature on Quakerism. It is now apparent that an older historiographical approach to this radical religious movement has, by and large, yielded to the revisionism of recent scholarship. For example, Fox is no longer unquestioningly considered the "Moses" of the early movement. Rather he is seen as one charismatic leader among several important early apostles of Quakerism, including, among others, Francis Howgill, Richard Farnsworth, James Nayler, and Edward Burrough. His longevity and extensive writings, especially his famous *Journal*, have contributed at least in part to his acknowledged leadership.[7] In contrast to this, the enduring hagiographic approach to Margaret Fell has not been seriously reappraised in light of new methods and recent research.

The standard portrayal of Fell depicts her as the intrepid mother-overseer of the struggling first two generations of Quakers, wife of George Fox, and faithful supporter of his work in a secondary role. This standard account is still fettered by a near "ancestor worship" approach that inadvertently serves to control the traditions of the early movement. The majority of historians of early Quakerism and Fell's three biographers have been Quakers themselves. When historians write from within a religious tradition, there is a constant temptation to emphasize the victories and to minimize defeats and internal dissensions. Founders are frequently presented as plaster saints whose human dimensions are obscured. Evidence is sometimes suppressed or selectively used to support a standard history that incorporates, to some degree, a "myth of beginnings". By and large the story of Margaret Fell has been circumscribed by this internalist approach.[8] For example, William C. Braithwaite's two-volume work stands out as the most notable and comprehensive account of the early movement and its leaders.[9] However, Braithwaite can be criticized for his appraisal of Fell's position in the first fifty years of Quakerism. Despite his frequent references to Fell's extensive work and authority, Braithwaite does not sufficiently analyze her position in relation to Fox and other early leaders. Likewise, he fails to rank her among the significant leaders of seventeenth-century Quakerism. Rather he follows the Quaker historiographical tradition that consistently stresses Fox's primary leadership position flanked by other first-generation Quaker notables cited above, as well as by William Penn and Robert Barclay, who joined the movement in the late 1660s. Further, in the matter of setting up women's monthly meetings, Braithwaite hands down the traditional belief that George Fox "set up a system which should allow every man and woman with spiritual gifts to have an office and be serviceable." According to Braithwaite, Fox was successful in this venture of establishing women's meetings, even though it "taxed seventeenth century feminine capacity to the utmost." Braithwaite fails to cite any evidence on how it taxed Quaker females. This explanation of the establishment of church order in the 1650s to 1670s is inadequate in light of the evidence.[10] Fell's role in establishing the earliest Quaker women's meeting outside London was pivotal in offering a model of sectarian female leadership and a model for subsequent Quaker women's meetings. This subject will receive detailed treatment in this study.

Margaret Fell is remembered in Quaker tradition for having written several important statements. As far as is known, she was the first radical sectarian to publish, in June 1660, a statement on the peaceful principle of Quakerism. Fell in 1666 was the first Quaker woman to write in favor of a female public ministry. She was the first among Quaker leaders to write to the Jewish leader Menassah Ben Israel and to the Jews in Holland when Jewish reentry into England was a subject of interest and debate in the Cromwellian government. In an important statement written in the last two years of her life, she was, it appears, the first to speak out against the growing popularity of Quaker plain dress. Her public work for Quakerism in and of itself calls for a fresh appraisal of Fell, but there are other aspects of Fell's role in nonconformity that call for reappraisal.

Recent historiography has yet to reappraise critically Fell's position in relation to Fox and other early nonconformist leaders. For instance, Luella Wright's work on early Quaker literature refers to Fell as George Fox's "unofficial secretary in the northern counties." Melvin Endy's fine work on William Penn speaks of the Fell-Fox-Penn friendship peripherally. His references to Fell only hint at her influence in that triumvirate of leadership. Barry Reay's recent book overlooks Fell's activist role during the feverish pre- and post-Restoration political activities in London between 1659 and 1661. For instance, he fails to acknowledge that Fell wrote the first peace testimony instead of George Fox. Michael Watts indicates that Fell was, along with Fox, a chief organizer of the Quaker church administration, but he does not develop this point. Richard Vann discusses the social aspects of early Quaker church development and membership, but gives mainly an account of the masculine perspective. A short article by Milton D. Speizman and Jane C. Kronick about the Fell women and the Swarthmoor Women's Monthly Meeting (SWMM) asserts simply that Fell was a "co-founder" with George Fox. The problem of Fell's true position in the rise of Quakerism has been aptly summarized by Hugh Barbour, who maintains that in the organization of the early movement, Fell's roles "must always be read between the lines."[11]

Historians of Stuart women have noted how seventeenth-century English Quaker nonconformity was peculiar in its relationship to the traditional construction of gender. Quakerism appealed to women in a way that few other religious groups did. An important older study of Quaker women by Mabel Brailsford points this out. It still stands as a good descriptive source of early Quaker female ministers who were contemporaries of Margaret Fell. Brailsford's book includes short biographical sketches of important and sometimes infamous early ministers such as Mary Fisher, Barbara Blaugdone, and Elizabeth Hooten, along with Margaret Fell. More recently, Keith Thomas's groundbreaking article, "Women and the Civil War Sects" (1958) stimulated interest in female activity in English sectarianism. Thomas found that the emergence of the concept of the spiritual equality of men and women in the Interregnum sects reached its apogee in Quakerism, where the doctrine of the Inner Light knew no gender boundaries. Thanks in part to the Thomas article, the question of women's roles in English nonconformity has received sustained attention since.

Recent works touching on this area of female religious sectarian thought include those of Joyce Irwin and Hilda Smith. They focus generally on Fell's "bold" assertiveness or "feminism" concerning the issue of equality of women in religious activities. Both Barbara Kanner and Mary Prior allude indirectly to Fell in referring to Quaker women writers. A recent dictionary of English women writers attributes more than one-third of its entries to Quaker women writers. If Fell's literary output is compared, not only to other Quaker female writers, but to other literate Englishwomen of the period, she stands out as one of the exceptional women of the seventeenth century. While Antonia Fraser gives a lively descriptive treatment of Margaret Fell and her daughters, other recent works interpret the prophetic stance of female ministers. Phyllis Mack ably addresses the gender and knowledge issue in the emergence of the female prophet and her audience in mid-century. In a careful exploration of the model of the female prophet, Mack makes us aware that a woman like Fell must be placed in a different category of sectarian female public ministry. It is evident that Fell was more than the rank-and-file pious Quaker woman prophet who was authoritative only in the prophetic role. Fell's religious authority derived from a more deeply rooted social and economic foundation that will become apparent as her story unfolds. She not only wrote prophetic tracts and important pastoral epistles to Friends throughout England and beyond the seas, she was also one of the chief controversialists for early Quakerism. Although she was less prolific than Fox and Penn, Fell's polemic tracts rank her near their level. Aside from these recent studies, the secondary literature is largely consistent in its approach to Fell: her public and private roles *vis-à-vis* Fox, other Quaker leaders, and women cry out for reappraisal. The problem of Fell remains.[12]

Studies in the history of the early modern family have progressed to the point that we are able to see that Margaret Fell's family life did not conform to that of the typical early modern patriarchal family. Recently, historians have noted that spousal relations of the period were far more complicated and varying in power dynamics than the conduct literature of early modern England would have us believe. The Fell family is a case in point. We know that by the time of Thomas Fell's death in 1658, and perhaps much earlier, a powerful woman had charge of Swarthmoor Hall. The Fell family first emerges from the shadows in 1652, the year of Margaret Fell's conversion to Quakerism. From then until her death in 1702, Fell's family displayed characteristics of conjugal mutuality, high affect, incessant female household and external business-oriented activities, and strong leadership. Fell's relationship to her first husband, Thomas, was a companionate one, with definite distaff authority and power in matters relating to household and children. Subsequently, as a widow she exercised overlordship of the Swarthmoor estate. She also made her own choice in marrying George Fox, despite oppositional pressure from her only son. Her second marriage was, seemingly, a relationship of equality between spouses. The union was based largely on companionship and affection as well as on a sharing of the same religious ideals. But of greater significance, Fell remained in full control of the resources of the Swarthmoor estate throughout her second marriage. Fox was atypical in that he renounced all interest in and right to Fell's property prior to their wedding in October 1669. Margaret Fell was no patriarch's wife, rather she was a model of matriarchy. The renunciation of patriarchal control by Fox may well have created the foundation for the depth of affection and closeness that existed between him and his stepdaughters.[13]

To reconstruct the separate but often overlapping public and domestic worlds of Margaret Fell, I have divided the book into four sections. Part I includes a brief survey of Fell's life written in her own words late in her lifetime. Thus, the book begins with her private world and autobiographical statements which demonstrate the potent force of sectarian religion in her life. Following this, the discussion is extended to Fell's family. Fell's position in the domestic sphere *vis-à-vis* her two husbands, her seven daughters, and one son is the focus of another chapter. The Swarthmoor household accounts of the 1670s and the family correspondence between 1652 and 1702 give us some snapshot images of domestic life that enable us to form a picture of her roles as wife and mother, and to draw out her family relationships. Through the Swarthmoor Hall household inventory, we may view Fell's role as overseer of the household servants and her of daughters' and grandchildren's comings and goings.

Part II establishes a fuller picture of Fell's world by incorporating the local community and Quaker community in which she lived and worked. The Swarthmoor home economy and the dynamic interaction between the Fell women and the circle of neighbors who lived in the area and who frequently had business and friendly dealings with them, is the focus of one chapter. Another chapter probes the philanthropic neighborly activities of the Fells and the SWMM during two decades. Additionally, I have drawn on both known and new evidence to retell the case of the long-term business dispute of Margaret Fell and Thomas Rawlinson, who was her neighbor, co-religionist, and employee. In 1663 a disagreement arose between these two friends concerning Rawlinson's stewardship at Force Forge, an iron smelting forge owned by the Fell family. This case illuminates Quaker business philosophy and Fell's role in some internal church discipline at the local level in the earliest phase of Quakerism.

Part III considers Fell's world beyond the local community by examining her public persona both within and outside of the Quaker movement. Her political activities in the important political years 1659–61 receive attention. It is evident that Fox's influence on Fell was deep and lasting. Too little has been said about that dynamic in the reverse. Her activities in the chaotic years at the close of the Interregnum demonstrate that her influence on Fox was likewise very great. Additionally, her friendship with William Penn was a significant one. In probing the long-term friendship of Fell, Penn, and Fox, this book will suggest that social status and religious authority were significant factors in early Quaker church order despite Quakerism's appeal to radical egalitarian principles.

Part IV expands Fell's horizons still further by considering her sectarian premodern worldview and her spiritualist opposition to Puritan and Anglican theology. A new-found source *M[argaret] F[ell]s Answer to Allan Smallwood Dr. priest of Grastock in Cumberland* (1668), reopens this subject of Quaker and Anglican antipathy with its causes and effects. This is followed by an analysis of some of her more significant written work directed beyond her own circle of co-religionists. Christopher Hill recently considered undertaking the project of analyzing Fell's written works. Hill attests that Fell is one of the more intriguing women who survived and came to terms with the defeat of the ideals put forth in the English Revolution. However, Hill concluded that Fell was mainly an organizer and not a contributor of ideas, so he abandoned the project.[14] In contrast to this point of view, I suggest that Fell's role as an ideologue has been seriously underestimated in recent studies of radical religious movements in early modern England. Moreover, her public letters and theological writings, although not singularly creative, are very useful in defining primitive Quaker thought and practice from a woman's perspective, not only within the seventeenth-century movement, but within the wider range of Protestant hermeneutics. For instance, a chapter presents an interpretation of four works with millennial overtones that Fell wrote to the Jews. Her tracts, translated into Latin, Dutch, and Hebrew, were carried to the Netherlands by Quaker missionaries. Fell's and Fox's religious ideas relating to the Jews carried over into their Quaker theology of female ministry. Both Fell's and Fox's theology was radical in an age when female stereotypes created a contrary image of women. In their Quaker conception of universal salvation, Fell and Fox believed that those who were to go forth to preach the gospel included ministers of both sexes. The fact that her religious ideas allowed her to transcend the ordinary religious presuppositions and attitudes toward gender and ethnicity deserves further attention.

A study of Fell's female experience reveals to us the expectations, social tensions, and political situation for a woman of old gentry background in early modern England. As such, it is one stone in the foundation of the history of seventeenth-century English nonconformity. Her family and public life illustrate the enduring strength of radical religion in the lives of men and women. Her life expresses a gender difference that was unusual for her time. She changed the rules for gender in her domestic and public life, and this in turn allowed her to change the rules for race as well, at least in the spiritual sphere, for she took a universalist approach in her theology. This study yields an example of gender constructions at work on a local and individual level, as well as on a wider national and even an international level, in that period. Her portrait is a composite that, of necessity, includes paradoxes. Nonetheless, a study of Fell as an authoritative daughter of seventeenth-century radical nonconformity sheds new light on the wider pictures of seventeenth-century gender constructions and English sectarianism.[15]

Part I
Life in a Seventeenth-Century Quaker Family

1 "Weak in Body but Alive to God": Margaret Fell's Religious Self-Reflections

Late in her life, Margaret Fell looked back on her fifty years of involvement in the Quaker movement and summed up her early conversion to Quakerism in laconic fashion: "I received the truth in the love of it; and it was opened to me so clear, that I never had a tittle in my heart against it; but I desired the Lord that I might be kept in it, and then I desired no greater portion."[1] Fell recorded her personal reminiscences drawn from a full and long life in two brief autobiographies that survive among her published tracts. In these writings Fell's self-fashioned image emerges as one of intense commitment to her newfound faith, having rejected traditional Protestant female piety, of either Anglican or the more intense Puritan form. Her personal account contains very little information about the non-religious aspects of her life. In fact, her account fails to describe in detail her earlier religious experiences as if they were unworthy of note. Fell's account of her conversion to Quakerism reveals none of the traditional Puritan obsession with self-doubt and self-introspection, while going through a gradual faith-awakening process. Instead, we learn only that from the moment she met the itinerant minister George Fox, she changed abruptly her religious allegiance and she made her decision without her husband's knowledge or approval. Additionally, she conveyed no sense of guilt in making her sudden unilateral decision. After her Quaker conversion, Margaret ceased attending the local Anglican parish of St. Marys in the nearby town of Ulverston, although her husband, Judge Thomas Fell, continued to attend regularly without her. For Margaret, her earlier religious commitment paled in the light of her immediate spiritual experience under the influence of Fox. Thus Margaret relates her own story.[2]

When George Fox first came to Swarthmoor and convinced Margaret and her household in June 1652, her husband Thomas was not at home. Upon his return home, the Judge found his family no longer attending the local Anglican church but rather deeply involved in a new "Principle and Persuasion" through the efforts of a total stranger, George Fox. Judge Fell was "much troubled . . . and surpriz'd" at their "sudden change." He was also predisposed against Fox because while he was enroute home, several "great ones of the country went to meet him" as he was crossing the sands of Morecambe Bay. They told Judge Fell "that a great disaster was befallen amongst his family, and that they [Fox and his circle] were witches," and that his family had forsaken their religion as a result. Margaret described her husband's negative demeanor upon arriving home after an extended absence on the assize court circuit: "So my husband came home greatly offended; and any may think what a condition I was like to be in, that either I must displease my husband or offend God; for he was very much troubled with us all in the house and family, they had so prepossessed him against us."[3]

Upon accepting her new faith, Margaret was confronted by the conflict between self-assertion in religious matters and deference to her spouse as head of the household. After expressing his displeasure to his wife, the Judge went silently to dinner. During dinner Margaret shared her newfound religious faith: "His dinner being ready he went to it, and I . . . sat down by him. And whilst I was sitting the power of the Lord seized upon me, and he was struck with amazement, and knew not what to think; but was still and quiet. And the children were all quiet and . . . could not play on their music." Judge Fell was apparently a moderate man, for he was willing to hear George Fox, and very soon thereafter Margaret invited Fox to speak to the Judge. She recorded that Fox spoke "so powerfully and convincingly, that the witness of God in his [Thomas Fell's] conscience answer'd that [Fox] spake Truth." Fox's charismatic personality and Margaret Fell's domestic diplomacy accomplished an apparent change in Thomas Fell's opinion of Fox and of his traveling companions, James Nayler and Richard Farnsworth. She wrote of her husband's next encounter with their local Anglican priest: "The next morning came [William] Lampit, priest of Ulverston, and got my husband into the garden, and spoke much to him there; but my husband had seen so much the night before, that the priest got little entrance upon him."[4] Margaret added that the Judge from thenceforth allowed a meeting of Friends at Swarthmoor, which met continuously for thirty-eight years, beginning in the summer of 1652. Thus Thomas Fell "went . . . to the steeplehouse [St. Marys], and none with him but his clerk and groom." Although Judge Fell never became a Quaker, in his position as judge he protected Fox and other Quakers from persecution in his lifetime, and so "Truth increased in the Counties . . . and many came in, and were convinced." After his death in 1658, Quakers missed his kindness at the local bench.[5]

Fox's momentous impression on Fell is revealed in her description of his "steeplehouse" visitation that occurred during his first visit to Swarthmoor. In her spouse's absence she invited Fox to attend her church of St. Marys. Fox made a dramatic late entrance into the church, arriving just prior to the sermon. He stood up on a pew and requested "liberty to speak; and he that was in the pulpit said he might." Fell recalled that his first words were: "He is not a Jew that is one outward, neither is that circumcision which is outward; but he is a Jew who is one inward, and that is circumcision which is of the heart." This statement, which she reiterated many times in the course of her life, became a theological watchword for her and other Quaker preachers. Fox was referring to the interior religious experience of the Inner Light. To know Christ's spiritual presence inwardly was to experience circumcision of the heart in early Quaker parlance. Upon hearing Fox's words, Fell was deeply moved, and she wrote, "I stood up in my pew, and wondered at his doctrine, for I had never heard such before." Fox continued to speak from his pew, "opening the Scriptures," and describing the Inner Light of Christ given inwardly by God. He declared that neither the priest of St. Mary's, William Lampitt, a man of Puritan sympathies, nor any of his ilk possessed the Inner Light. Margaret was considerably affected by Fox's preaching. She recalled: "This opened me so, that it cut me to the heart; and then I saw clearly, we were all wrong. So I sat down in my pew again, and cried bitterly; and I cried in my spirit to the Lord."[6]

That same evening Fox returned to Swarthmoor Hall and preached to Fell, her children, and household. She recalled, "I was struck into such a sadness, I knew not what to do, my husband being from home." This equivocating phrase appears inconsistent with her demeanor after 1652. She may have been giving mere lip service to the system of patriarchy or she may have accepted it at face value then. However, it was probably the last time she felt any wavering on religious issues in deference to her husband's sensibilities. Thereafter she seemingly shed the "weaker vessel" role, only giving it lipservice when convention required. Fell remained firm in her conversion to Quakerism, as expressed in letters and actions to her spouse and others.

Fell reported that her husband Thomas lived for six more years after her conversion. Then according to Margaret, it pleased the Lord to visit him with sickness, wherein he became more than usually loving . . . to our Friends . . . having been a merciful Man to the Lord's People. I, and many other Friends were well satisfy'd the Lord in mercy receiv'd him to himself. It was in the beginning of the 8th Month, 1658, that he died."[7]

Margaret Fell witnessed enormous political and religious change over the course of her lifetime. She witnessed the English Civil War in the

1640s, which culminated with the execution of Charles I in 1649. The public execution of the only anointed king in English history brought with it a complete loosening of social, political, and religious restraints in the Interregnum. Anglicanism and prelacy were abolished and in their wake a rash of sectarian groups emerged. The Familists, Seekers, Muggletonians, Ranters, and Quakers, among others, became visible in this unusually free-thinking period of English history. Besides participating in this rising wave of English sectarianism, Margaret viewed with great interest from distant Swarthmoor Hall the political events under Cromwell's leadership. The 1650s were marked by gradual political disillusionment as people watched the failure of Puritan-style government under the Rump Parliament and Cromwell's leadership. With Cromwell's passing in 1658, she and others watched the godly revolution, the great hope of English Puritans, grind to a halt.

Fell was forty-six years old when she witnessed the Restoration of the Stuart monarchy. Along with the restoration of kingship came that of the Anglican Church and its bishops. In spite of the Breda Accord, Charles's arrival in England struck fear into Quaker hearts, and not without reason. Although George Fox had been arrested and imprisoned during the Interregnum, he was arrested again while visiting Swarthmoor in 1660 and was imprisoned in Lancaster Castle on the warrant that "he drew away the king's liege people, to the endangering [and] the imbruing [of] the nation in blood." With Fox's arrest, Margaret Fell wasted no time in traveling to Whitehall to gain an audience with Charles II. She recorded that her position compelled her to go for, "I having a great family, and he being taken in my house, I was moved of the Lord to go to the King . . . and took with me a declaration and an information of our principles." She asked the king for his protection of Quakers and for the release of George Fox. She recalled, "with much ado, after [Fox] had been kept prisoner nearly half a year at Lancaster, we got a habeas corpus, and removed him to the king's bench where he was released."[8] Fell's devotion to Fox in her lifetime was singularly unswerving. Considering Fox's lesser social stature this is the more remarkable. Her motivations were more than simply devotion to a charismatic religious leader. Of course, love was a significant factor here, but she also was capable of influencing him in his work for Quakerism, as will be seen in succeeding chapters.

Margaret Fell was swept into active participation in the politics of religious dissent in 1660. Her autobiography and letters to her family at Swarthmoor give a woman's-eye view of contemporary politics in London between 1660 and 1662. Fell expressed it thus:

> I spake often with the King, and writ many Letters and Papers unto him, and many Books were given by our Friends to Parliament . . . and they were fully inform'd of our peaceable Principles and Practices . . . And I writ and gave Papers and Letters to everyone of the Family several times . . . to the Duke of York, to the Duke of Gloucester . . . the Queen Mother . . . the Prince of Orange, and to the Queen of Bohemia. [I] did lay our Principles and Doctrines before them, and desired that they would let us have Discourse with their Priests, Preachers and Teachers, and if they would prove us Erroneous, then let them manifest it: But if our Principles and Doctrines be found according to the Doctrine of Christ, and the Apostles and Saintes in the Primitive Times, then let us have our Liberty.[9]

The trappings of royal power and style did not intimidate her, for she returned several times to remind Charles to allow religious liberty to tender consciences. That she was from a "great family" gave her an entrée to court, that George Fox did not have, and which she was prepared to use for her religious purposes.

When the Fifth Monarchy uprising of January 1661 caused renewed hostility toward Friends, Margaret's return to Swarthmoor and her family became delayed. Her visits to the king and her papers for release of Quakers from prison helped in some degree to bring about the general Proclamation of 1662 to end Quaker imprisonments. Once she had accomplished that, she found the "freedom in Spirit to Return Home" to her children. However, the mistress of Swarthmoor remained at home only nine months, after which she commenced her travels again to visit Quakers in prison. Fell returned also to London to protest the act of Parliament against Quakers for refusing to take oaths of allegiance. Margaret recalled:

> At this time Friends Meetings at London were much troubled with Soldiers, pulling Friends out of their Meetings, and beating them with their Muskets and Swords; many were cast into Prison, through which many lost their Lives; and all this being done to a Peaceable People, only for Worshipping God as they in conscience were perswaded.[10]

The king, although promising Margaret to give imprisoned Quakers liberty, was apparently powerless to do so against the will of his Privy Council and a strongly Anglican Parliament that feared further civil-religious resistance from sectarian groups such as the Quakers. From 1664 until 1689 conventicle worship was illegal in England. Nevertheless, the Quakers continued to meet publicly for worship, a policy that incurred sporadic and sometimes severe persecution from authorities.

In the winter of 1663–64 George Fox again came to Swarthmoor and, while a guest of Margaret Fell, was arrested by the local constables. Shortly thereafter, in March 1663–64, Fell too was arrested, and, like Fox, was tendered the oath of allegiance before the local justices at Lancaster. Refusing the oath, she was imprisoned at Lancaster Castle until the following court session. At her trial in August 1664, Fell was found guilty of not taking the oath to the king and of refusing to stop holding Quaker meetings in her home. The judge passed a sentence of praemunire against her, a serious sentence for property holders like Fell. Upon hearing her sentence, Margaret responded that she must first serve the King of Kings before the King of England. In later years she reflected, "the great God of Heaven and Earth supported my Spirit under the severe Sentence, [so] that I was not terrified."[11] Fortunately for Margaret the king did not seize Swarthmoor, and the Fell daughters continued to live there while their mother was in prison. Her oath trial became a well-remembered show trial for nonconformity. Margaret Fell was a widely-known sectarian gentlewoman when she entered Lancaster Castle for an incarceration that began in August 1664 and lasted until June 1668.

By 1666 Fell was not reticent about warning Charles of his broken promises toward sectarians, saying, "These six years . . . the Laws you have made . . . have laid Oppression and Bondage on the consciences of God's People." Writing from Lancaster Prison, she referred to her earlier warnings, "I also writ to thee to beware how thee Rulest in this Nation, for the People of this Nation was a brittle People generally . . . and the Lord had a People here that was dear unto him. And I desired thee not to touch them, nor hurt them." Referring to her nearly four-year imprisonment, she added:

> Thou kept me in Prison three long Winters, in a place not fit for People to lie in; sometime for Wind, and Storm, and Rain, and sometime for Smoke; so that it is much that I am alive, but that the Power and Goodness of God hath been with me. I was kept a Year and Seven Months in this Prison, before I was suffered to see the House that was mine, or Children or Family, except they came to me over the dangerous Sands [of Morecambe Bay] in the cold Winter, when they came with much danger of their Lives: But since the last Assizes I have had a little more respect from this Sheriff, than formerly from others. And in all this I am very well satisfied; and praise the Lord, who counts me worthy to suffer for his sake.

She closed her epistle to Charles with the admonition, "I desire thee . . . to fear the Lord God, by whom Kings rule, and Princes decree Justice; who sets up one and pulls down another, at his pleasure. And let not the Guilt of the Burthen of the Breach of that Word that passed from thee at Breda lie upon thy Conscience."[12]

While Margaret was in Lancaster prison, she received visits from her young daughters who were living alone at Swarthmoor and running the farm under the guidance of the middle daughter, Sarah. Fell's family worries (undoubtedly they existed) seemingly never superseded her work for Quakerism. In retelling her story so many years later she still wished to emphasize that she was single-minded in her religious motivation.

During her imprisonment she filled her time with writing. She penned several tracts in defense of Quaker principles, among which was her most famous published piece entitled *Women's Speaking Justified* (1666). In this tract Fell defended the right of women to be public preachers, arguing that the Quaker ministry was a truly "called" ministry and those persons, whether male or female, who experienced the divine leading or call of the Holy Spirit were the true ministers of Christ. As Christ had not received money or education during his own ministry, the real ministers of the gospel were to emulate Christ and the original apostles.

This concept of female ministry stirred up a major debate between the Quakers and other Protestants. The idea of women inspired by the Holy Spirit and acting in an authoritative public role was anathema to most Protestants, except for some Baptists and Ranters and a few other sectarians of the period. The importance of Fell's defense of women preachers is that it was the first defense written by a Quaker woman. Fell, following earlier arguments, claimed that women who were in the Spirit and not in carnal wisdom could speak as men did, for it was the Spirit speaking through them, not natural men or women per se speaking. She garnered from Scripture examples of twenty-four female biblical figures to support her argument that women should preach, teach, and minister. She dismissed the Pauline injunction on women keeping silent by using the argument that the gospel purified women and the power of the Inner Light released them from the prohibitions under the law. Fell made the distinction that foolish gossiping, carnal women were to be barred from speaking for they were not of the "Seed of the Promise," that is, part of the converted community who experienced the Inner Light. For Fell, those who were given the Seed of the Promise entered a new life and became new women and men in Christ. Once Quakers received the Inner Light, they were in the Seed of the Promise, and nothing else mattered – neither gender, class, nor race. In theory at least, the Seed democratized the Quaker fellowship and smashed through the political and religious restraints placed upon women. Her use of a logical and sustained argument to define women's authority role in ministry places Fell among the significant seventeenth-century sectarian religious apologists.[13]

In conjunction with her writing *Women Speaking Justified*, one of Margaret's greatest single accomplishments in her Quaker career was her pivotal work in the establishment of separate women's meeting which she commenced shortly after she was released from Lancaster prison. This institution was essentially unique in seventeenth-century Protestantism. But this is a subject to which we will return in a later chapter.[14] Although she does not dwell on her women's meetings' activities in her recollections, it is, nonetheless, a telling example of her proselytizing and organizational concerns and how she combined these activities in her work for Quakerism.

Shortly after her release from Lancaster Castle, Margaret carried on a lively and acidic debate with an Anglican clergyman, Allen Smallwood of the parish of Carlisle, in Westmorland. Smallwood preached a sermon on oath-taking in August 1668, and it received some acclaim because of his casuistic treatment of the Matthean text, "swear not at all" (Matt: 5.34). He gave an appealing argument for the Anglican position that supported oaths, arguing that Christ's words, "swear not at all," did not necessarily forbid all swearing in all situations. Fell wrote a fiery reply to Smallwood, calling him not only a blind, ignorant priest, but a man who, through his clever words and distortions of Scripture, was leading poor, illiterate people astray. She stated why Quakers rejected ordained ministers, infant baptisms, and the eucharist. She presented the Quaker view against tithes and stated her conviction that women should be allowed to preach. As a representative of hierarchical Anglicanism, Smallwood, in Fell's eyes, was nothing less than the Antichrist. But she also included among the Antichrists at work in the world all other Protestants, Catholics, and Antimoninians.[15]

Her rejection of Anglicanism is expressed in her prison writing about clergy and bishops who, in her eyes, were debauched swearmongers: "And what is your spirituall men and clergymen as thou calls them whether Bishops or whatsoeuer titles thou gives them, else what are they butt a heape of evill doers and workers off iniquity." She deplored the symbiotic relationship between Anglican bishops and the Crown: "Why may there not be a king in Greate Brittan without bishops. . . . And what doth the bishops acknowledge him in words, but robs him both by there power and there greate benefits and liveings that they keep from him to maintaine there prid and haughtiness."[16]

In October 1669, eleven years after her first husband's death, Margaret met Fox in Bristol, and they were married. Margaret's autobiography describes the event in a matter-of-fact way: "And then it was Eleven Years after my former Husband's Decease; and G. Fox being then returned from visiting Friends in Ireland. At Bristol he declared his Intentions of Marriage with me; and there was also our Marriage solemnized, in a publick Meeting of many Friends, who were our Witnesses."[17] The wedding took place in the presence of her daughters and their husbands. Her only son, George Fell, who strongly disapproved of his mother marrying a man of socially inferior status, was absent. From that point until George Fell's premature death exactly one year later, mother and son were estranged. A few days after their marriage, Margaret and George parted ways, Fox continuing on his missionary travels to different parts of England while she returned to Swarthmoor.

Of the several imprisonments experienced by both Fell and Fox, probably the most trying one was Fox's incarceration in Worcester Gaol from December 1673 until February 1674–5. Fox was arrested while visiting Friends in Worcestershire. During his imprisonment he became seriously ill, and local Quakers attending Fox at Worcester feared he would die. They wrote Fell at Swarthmoor that if she were to see him alive she should go to him, "which," Margaret recorded, "accordingly I did." Margaret described her husband's long imprisonment:

> And after I tarried seventeen weeks with him at Worcester, and no discharge like to be obtained for him, I went up to London, and wrote to the King an account of his long imprisonment. . . . And I went with it to Whitehall myself; and I met with the King, and gave him the paper. And he said, I must go to the Chancellor, he could do nothing in it. Then I writ also to the Lord Chancellor . . . a very tender man, and spoke to the judge; who gave out an Habeas Corpus presently. And when we got it, we sent it down to Worcester; and they would not part with him at first, but said, he was praemunired, and was not to go out on that manner. And then we were forced to go [back to London] to Judge North, and to the attorney general, and we got another order, and sent [it] down from them; and with much ado and great labour and industry of William Mead and other Friends, we got him up to London, where he appeared in Westminister Hall at the King's Bench, before Judge Hales, who was a very honest, tender man. And he knew they had imprisoned him but in envy. So that which they had against him was read; and our Counsel pleaded, that he was taken up in his travel and journey. And there was but a little said, till he was quitted. And this

was the last prison that he was in, being freed by the Court of King's Bench.[18]

It was during this imprisonment that Fox dictated to his son-in-law, Thomas Lower, portions of his travels and experiences which were subsequently included in his published *Journal* in 1694.

When Fox was released from Worcester in February 1674–5, he was so sickly that he could hardly ride a horse. Husband and wife traveled slowly back to Swarthmoor where he convalesced for nearly two years. Fell wrote of his prolonged sojourn at the Hall: "And this was the first time that he came to Swarthmoor after we were married, and he staid here about two years, and then went to London again . . . and after awhile went to Holland, and . . . Germany . . . and then returned to London again." Margaret recorded that he returned to Swarthmoor one last time for another convalescence from 1678 until 1680, "having had many sore and long travels, beatings and hard imprisonments. But after some time he rode to York, and so passed on through Nottinghamshire, and several counties, visiting Friends til he came to London." Fell wrote that in Fox's later years he focused his ministry to Quakers in and around London, often living with one of his daughters-in-law who lived in the London environs. There is a wistfulness in Margaret's words about her spouse living his last years far from Swarthmoor.

> And though the Lord had provided a habitation for him, yet he was not willing to stay at it, because it was so remote and far from London, where his service most lay. And my concern for God and his holy eternal truth was then in the north, where God had placed and set me; and likewise for the ordering and governing of my children and family; so that we were willing both of us to live apart for some years upon God's account and his truth's service, and to deny ourselves of that comfort which we might have had in being together, for the sake and service of the Lord and his truth. And if any took occasion or judged hard of us because of that, the Lord will judge them; for we were innocent.[19]

Knowing George Fox well by the time of their wedding, Margaret Fell was aware that marriage to a wandering, charismatic religious leader would not be an easy life. Although she felt regret over his ten-year absence from Swarthmoor after 1680, she had resigned herself to that reality, for she recounted without further account her second widowhood, "he [Fox] staid there [London] and thereabouts til he finished his course and laid down his head in peace."

One of the commoner lessons in the history of Christianity is that persecutions serve to stiffen the faith of believers, and Margaret Fell and her Quaker compeers were no exception. Fell kept a record of the years of Quaker persecutions in the 1670s and 1680s by recounting the local hardships that she, as mistress of Swarthmoor, and her Quaker neighbors endured at the hands of local constables. She wrote that the justices of the county

> were much bent against me, because I kept a Meeting at my house, at Swarthmoor Hall. So they did not fine the house as his [Fox's], he being absent, but fined it as mine, as being the widow of Judge Fell, and fined me £20 for [meeting in] the house and £20 for speaking in the Meeting; and then fined me the second time £40 for speaking; and also fined some other Friends for speaking £20 for the first time, and £40 for the second time; and those that were not able, they fined others for them, and made great spoil amongst Friends, by distraining and selling their goods, sometimes for less than half the value. They took 30 head of cattle from me. Their intentions were to ruin us.[20]

As a result of these persecutions, Fell wrote that she "was moved of the Lord to go to London in the seventieth year of my age" to plead with the king for protection of religious conscience, and to "bear to him my last Testimony, and let him know, how they did abuse us to enrich themselves." She prepared a paper to present to Charles II on their deep suffering. However, Charles lay dying at that moment. In due time she went to the new king, the Roman Catholic James II, with her paper. She wrote that the new king's response to her was, "Go home, go home. So after a few weeks I went home." Harassment continued for awhile from local persecutors, and then gradually died down. Fell concluded, "They [king and Council] gave our persecutors a private caution, for they troubled us no more."[21] In May 1689, with the beginning of the reign of William and Mary, Parliament passed the Act of Toleration. Although this did not begin to eradicate all disadvantages for Quakers, it did reduce the persecutions. Quakers could worship without loss of property or fear of imprisonment. During her visits to London in 1690, Fell wrote a letter to William III, praising him for his "Gentle Government and Clemency and Gracious Acts."[22]

Margaret Fell was seventy-six years old when she visited London in 1690. It was the last time she saw her husband, for he died six months later. In January 1691, shortly after participating in a London Quaker meeting, George Fox felt ill. He succumbed after a brief sickness. Margaret, at

Swarthmoor, was sent the news by her long-time friend William Penn, who wrote, "A prince indeed has fallen today in Israel."[23]

After Fox's death and twenty-two years of marriage, Fell wrote of Fox, "He was the instrument in the hand of the Lord in this present age, which he made use of to send forth into the world to preach the everlasting gospel, which had been hid from many ages and generations."[24] Margaret's devotion to her second spouse was unremitting for the half-century that she knew and loved him. She referred to him often as her "dear husband," and she used her wealth, prestige, and literary ability to promote his leadership *vis-à-vis* other prominent Quaker leaders such as James Nayler, John Wilkinson, and John Story in the early years, and William Penn in the second generation. Her adulation of Fox was first expressed in 1652 when she and her newly-converted household begged him to return to Swarthmoor Hall. After he had left Swarthmoor the first time, she wrote these lines to a man she hardly knew:

> My own dear heart . . . thou knows that we have received thee into our hearts, and shall live with thee eternally, and it is our life and joy to be with thee. . . . My soul thirsts to have thee come over, if it be but for two or three days. . . . And if thou do not come, it will add abundantly to our sorrow . . . and now dear heart do not leave us nor forsake us, for our life and peace is in thee.[25]

Although the fervent nature of her earliest letter to him was toned down in subsequent letters, that original devotion never subsided; rather it took a different form. Fell's tribute to Fox almost forty years later described him as one who was Christlike in his special calling and earthly ministry: "It having pleased Almighty God to take away my dear husband out of this evil troublesome world, who was not a man thereof, being chosen out of it, and had his life and being in another region, and his testimony was against the world." The close parallel of Fox to Christ was intentional. As Christ had not been accepted as God's anointed in his own country, so too Fox had not been received in his own territory of origin. Margaret recalled, "When he declared it in his own country in Leicestershire and in Derbyshire, Nottinghamshire and Warwickshire . . . and his declaration being against the hireling priests and their practices, it raised a great fury and opposition amongst the priests and people against him." In accord with the story of the earthly ministry of Jesus, she added, "yet there was always some that owned him several places, but very few that stood firm to him when persecution came on him."[26]

Margaret Fell remained at Swarthmoor during her second widowhood, only traveling once more to London in 1697 at age eighty-three. One of her final epistles to Friends was written in 1700; it remains one of her best-remembered letters among Quakers. The epistle criticized the growing Quaker interest in outward conformity of dress. She called the gray costume that was becoming popular a "silly poor gospel." This outward appearance that was setting Quakers apart from the world was not an indicator of one's interior religion and purity. It was for her a sign of hubris on the part of Quakers to outwardly set themselves apart as God's special people. She warned her circle that Christ preached salvation for all peoples of the earth:

Let us beware of this, of separating or looking upon ourselves to be more holy, than in deed and in truth we are. Away with these whimsical, narrow imaginations, and let the spirit of God which he hath given us, lead us and guide us; and let us stand fast in that liberty wherewith Christ hath made us free, and not be entangled again into bondage, in observing proscriptions in outward things, which will not profit nor cleanse the inward man. . . . This narrowness and strictness is entering in, that many cannot tell what to do, or not to do. . . . Poor Friends is mangled in their minds, that they know not what to do. . . .

But Christ Jesus saith, that we must take not thought what we shall eat, or what we shall drink or what we shall put on; but bids us consider that lilies how they grow in more royalty than Solomon. But contrary to this, they say we must look at no colours, nor make anything that is changeable colours as the hills are, nor sell them nor wear them. But we must be all in one dress, and one colour. This is a silly poor gospel. It is more fit for us to be covered with God's eternal Spirit, and clothed with his eternal Light, which leads us and guides us into righteousness and to live righteously and justly and holily in this present evil world.

This is not delightful to me, that I have this occasion to write to you, for wherever I saw it appear, I have stood against it several years. And now I dare neglect no longer. For I see, that our blessed precious holy Truth, that hath visited us from the beginning is kept under, and these silly outside imaginary practices is coming up, and practiced with great zeal, which hath often grieved my heart.[27]

Margaret Fell's religion was the fundamental wellspring of her life, and it superseded all other values and relationships in her adult life. After exactly fifty tempestuous years devoted to the Quaker cause and eighty-eight industrious years of life, she departed this world with her youngest daughter, Rachel, in attendance at her deathbed at Swarthmoor.

She commended her soul to God: "Oh! my sweet Lord, into thy holy Bosom do I commit my self freely; not desiring to live in this troublesome, painful world; it is all nothing to me: For my Maker is my Husband. . . . [I am] very weak in Body, but alive to God."[28]

Margaret Fell's personal reminiscences reveal a multifaceted life as mother, wife, and sectarian religious leader through half a century of political and religious change. In Fell's case study the history of non-conformity, politics, and ideas is intertwined with gendered and domestic concerns. It is a gender experience worthy of reexamination.

2 A Family History

Although Margaret Fell traveled widely, she was destined to spend most of her long life in the Lancashire fells of Furness in northwest England. Born Margaret Askew in 1614, she grew to adulthood on a remote manor of the lowland plains in Furness during the reign of Charles I. Her childhood home, Marsh Grange near Dalton-in-Furness, county Lancaster, was a rather spacious stone manor house with a front courtyard, which today is surrounded by a high stone wall with a large iron gate at the entrance. It overlooks the ocean and Duddon Sands to the north and northwest and the rolling Furness fells to the southeast.

FURNESS ECONOMY AND DEMOGRAPHY

Margaret Fell's home territory of Furness was predominantly hilly with some coastal lowland plains. It was far removed from major markets and main lines of British commerce. Furness was mainly agricultural in Fell's lifetime. As a girl, Margaret grew up at Marsh Grange in the pastoral farming region on the coastal lowland. As a bride, she moved to Swarthmoor, an agricultural region in the highland zones but with some mixed farming area in the lowlands, which were near Swarthmoor. In this area of northwest Lancashire, Margaret Fell was accustomed to seeing poverty around her. Because the region was isolated economically, it was considered backward compared to other regions of England. Barry Levy has found in his recent study of the Quaker family that northwest England was considerably poorer, less arable, less commercialized, and less economically competitive than the counties to the southeast in England in the seventeenth century. A study of inventories of fifty-eight yeomen in Cambridge in the Midlands in 1660 ranged from £2 to £1132. An inventory of Margaret Fell's estate at her death in 1702 totaled £500. Levy maintains that, traditionally, southeastern Britons considered northwestern Britons as inhabitants of one of the "dark corners" of Britain, where poverty and religious superstition reigned. Another historian of northwest England, Bruce Blackwood, has pointed out that the ship money valuation of 1636 designated Lancashire as the poorest county of England except for Cumberland county just to the north. In the neighboring county of

Cheshire to the south, Quakers in the 1660s through 1680s comprised only 3 percent gentry while 60 percent were yeomen and husbandmen, 15 percent retailers, 8 percent artisans, and 4.5 percent laborers.[1]

The peasant population of Furness supplemented their agricultural income with spinning and weaving although in Fell's lifetime the textile industry was declining. During Fell's years in Furness, the iron industry was revived, and Fell herself became involved in the iron industry. There was some coal mining done, and this, along with textiles and metal industries, constituted the major commercial activity of an otherwise agrarian district. The population in Lancashire during the Civil War and Interregnum years probably did not grow and may have declined. Along with war, the plague left its destructive demographic mark between the years 1644 and 1657. Further, harvest failures in 1646–50 and 1657–61 undoubtedly contributed to demographic decline in this era. However, the population of Lancashire began expanding during the later years of Margaret Fell's life. The hearth tax of 1664 indicated a population of about 150,609. By 1690 the county population had expanded to 180,909, according to Bruce Blackwood's reckoning.[2]

In this rural and economically poor region of northwest England, Quakerism took hold in 1652 after a six-year gestation in the midlands. It came to have considerable visibility in the northwest of England, with organized monthly and later quarterly meetings, confrontations with Anglican churches, and tithe testimonies in court. Early Quaker ministers of this area began by the mid-1650s to take their messages to other parts of England. By 1681 the Quakers numbered approximately 1 percent of the population of Britain, and they lived throughout the nation. The exodus of Quakers from northwest England in the 1660s and 1670s caused a demographic decline in the northwest in favor of London and the Bristol area, as well as colonies in Pennsylvania and West Jersey in the Delaware Valley. Barry Levy has found in his study of out-migrating northwestern Quaker families, especially from Cheshire and parts of Wales, that there was a 49 percent reduction of the original northwest Quaker community by 1675.[3] Margaret Fell's women's monthly and quarterly meetings reflected this decline in membership, as will subsequently be seen. Margaret Fell's family did not migrate. As a pillar in the Quaker community from 1652 onward, she remained in this isolated region of northwest England, raised her family, buried her first husband, married off her children, and watched them all move away except for her youngest daughter who inherited Swarthmoor Hall.

FELL'S BACKGROUND AND FIRST MARRIAGE

Margaret Fell's family background remains in shadow. Little is known of her origins except that her father, John Askew, was of gentry status. Margaret had one sister whose name is unknown. At their fathers death, each sister inherited the sizeable legacy of £3000. In 1632, at eighteen years of age, Margaret married Thomas Fell of Swarthmoor Hall, a man almost twice her age. Only a few miles to the southwest of Marsh Grange, Swarthmoor Hall stood on the edge of a grassy treeless moor or fell, from which it takes part of its name. The Fell family had recently achieved gentry status, having become landowners in the wake of the Henrician dissolution of the monasteries and chantries in the sixteenth century. George Fell of Hawkswell near Ulverston, father of Judge Thomas Fell, was one of these parvenu gentry. George Fell purchased the Swarthmoor property from the Crown sometime well after 1569. By 1632 his son Thomas owned Swarthmoor, and in that year he brought his eighteen-year-old bride to live at the Hall. The Fell family occupied Swarthmoor Hall continuously from 1632 until 1759. In the seventeenth century this gentry family was part of a minuscule 2 percent of the English population who belonged to the well-to-do landowner class.[4]

At the time of his marriage to Margaret Askew, Judge Thomas Fell was a barrister of Gray's Inn, London. He subsequently became a justice of the peace. As a member of the county elite, he participated prominently in local politics. In 1643 Judge Fell served on the county Committee for Compounding and in 1645 was a recruiter member of Parliament for the town of Lancaster. He was Justice of the Assize for the North Wales and Cheshire circuit by 1651, as well as vice-chancellor of the duchy of Lancaster. He held these positions until his death in 1658. Judge Fell's local prominence was indicated in the 1646 list of the nine "classical presbyteries": he was ranked as first in social status among the laymen of the ninth presbytery in the Furness district of Lancashire.[5]

Judge Fell's position on the Committee for Compounding in 1643 gave him early knowledge of forfeited or sequestered lands in the county, and it appears that he took advantage of this strategic information. A recent study of the Lancashire gentry during the English Civil War offers some illuminating insights on the ruling county elite between 1640 and 1660, which in turn yields some glimpses of Thomas Fell. According to Bruce Blackwood, Thomas Fell belonged, in the 1640s and 1650s, to a group of lower income parvenu gentry who were "war profiteers" among the Lancashire landed families. Blackwood argues that Fell, working in close collaboration with another Lancashire M.P.,

Thomas Birch, used his political connections to augment his properties.[6]

A vignette of Judge Fell found in the calendar of the Committee for Compounding tends to confirm these findings. In 1651 Thomas Fell became interested in the sequestration of Halton Hall, an estate near Lancaster, owned by Thomas Carcus. Carcus and his son submitted a request in May 1647 to pay the fine to lift the sequestration. Later that year Carcus complained that agents of the county committee had not obeyed the order to suspend sequestration and that an agent was occupying his house and damaging it. The petition was renewed by Carcus in August 1650 and again in January 1651 when he begged the county Committee for Compounding not to "let his estate until his cause be heard." By 12 August 1651 the record reads that Mr. Carcus was to "enjoy his estate in security, and Thomas Fell, M.P., to show cause within 28 days against the same; if none be shown, petitioner is to have his security again, and enjoy the estate absolutely." As of 2 December no cause was shown and the estate was discharged.[7] Whatever Fell's personal motivations in this sequestration case against Carcus may have been, we do not know. However, it is possible that if Fell was a "war profiteer" as Blackwood has suggested, he may well have had a coveting eye on the Carcus estate.

On the eve of the Civil War, Thomas Fell was a gentleman of small means with an income of £150 per annum. By 1676, according to the household account book of Sarah Fell (ABSF), the Swarthmoor estate income, largely based on farming, was £450, a substantial increase eighteen years after his death. It appears very likely that Fell did engage in active estate enlargement in the 1640s and 1650s. Based on Blackwood's table of incomes of the Lancashire gentry in 1642 and allowing for inflation, the Fell wealth as gentry farmers thirty years later would fall into the income bracket of the lower to middling gentry. The Thomas Fell family from mid-seventeenth century until the estate was sold in 1749 was, in terms of land, estate value, and birth, or lower to middling gentry status, insofar as this group has received definition.[8]

As the young wife of a judge, Margaret Fell may well have lived an entirely typical gentry woman's life. She attended St. Mary's Anglican Church in Ulverston, a mile from the Hall. She produced seven children – six daughters and one son – at approximately two to three-year intervals, to carry on the family line and patrimony. She managed the estate during her husband's frequent absences from home. Margaret took an active interest in religious matters and often entertained traveling ministers at Swarthmoor. Thus it was not an unusual event when the itinerant preacher, George Fox, arrived at the Hall one day in June 1652. Simply stated, 1652 was a revolutionary year in the life of Margaret Fell, and it altered permanently her gentry way of life.

In that year, as mistress of Swarthmoor Hall, Margaret presided over her household with apparent maternal affection. Despite being typically nuclear, the family was atypically female-centered and female-dominated. Female piety and religious discipline in the Quaker manner were in evidence and, indeed, dominated at Swarthmoor from 1652 onward. Mother, daughters, and several of the household servants became active proselytizers of Quakerism and traveled extensively on its behalf. Moreover, the Fell daughters were also content to live at home and displayed no driving interest to seek mates of their own social class. They married rather late (with the exception of Mary), and they married men of their own faith. Each daughter received maternal approval prior to her marriage. Margaret's son alone married outside the Quaker community.

The Fell women were not unusual in their economic involvement at Swarthmoor. Like other rural English women who held land, they managed the Swarthmoor household and agricultural enterprise after Judge Fell's death. Margaret, as dowager, became financially independent and could do largely as she pleased; the responsibilities of wifely duties were over and childrearing duties were diminished. Her matriarchal role was not unique among women of landed wealth. For example, an aristocratic widow, the Countesse of Warwick, ran her own manor in Essex and enjoyed "the status of monarch in her own local world." Other autocratic women such as Lady Joan Barrington and her daughter-in-law Lady Judith Barrington, of the Essex gentry, lived as matriarchs, presiding over their domestic worlds in a formidable manner. Lady Joan Barrington reigned as the acknowledged head of the clan following the death of her husband, Sir Francis, in 1628. Her eldest son, Sir Thomas, and other family members deferred greatly to her. Her various Puritan clerics who ministered to her as "physicians of the soul" in the capacity of family chaplains and lecturers likewise deferred to her unchallenged headship of the family. Her eldest son's wife Lady Judith Barrington, with whom Lady Joan got on very well, was a strong-willed and self-possessed matriarch in her generation. Lady Judith was an astute businesswoman who managed her husband's financial affairs and used her court connections to help her spouse avoid being saddled with the burdensome office of sheriff of the county.[9] What rendered Margaret Fell highly unusual when compared to other women of her class was that she maintained complete legal rights over her own income and property throughout her second marriage.

Thus, when Margaret Fell's world opens to us in the year 1652, we see that she lived in advantageous social and economic circumstances. As heir to her father's estate, she was mistress of two estates, Swarthmoor Hall and the nearby Marsh Grange, and thereby presided over a considerable number of servants. Her spouse, who was frequently away on business, left her in command at the Hall. There is only fragmentary evidence from which to draw an understanding of the conjugal relationship of Thomas and Margaret. Epistolary material indicates that after Margaret's conversion or "convincement," her Quaker Friends attempted to convert the Judge. Alexander Parker wrote to Margaret Fell on 19 August 1656 expressing his hope that "the Lord [would] open his understanding." William Dewsberry expressed a similar sentiment to Margaret on 15 October 1655, "Oh that he would stand in the counsel of the Lord."[10] James Nayler wrote several letters to Margaret. Letters from Nayler in November 1652 and again in November 1655 referred to the Judge as "dearly beloved" and expressed Nayler's hope that the "seed" would grow. A letter from Thomas Aldam, another first-generation Quaker, written directly to the Judge in 1653, expressed it very directly:

> Beware to Lookinge forth to the wisdome of the flesh to be thy counseller, but minde within how thou stands free in his wisdome. . . . O Bee valiant for the truth uppon earth, and treade uppon the deceite; the heathen are in greate rage about you, through the deceitfull clergie, who hath the people in bondage, and doth bewitch them with the wisdome of the flesh . . . who live in blindness and darkness . . . they are blinde, in that they live 'in Pride and oppression . . . bewitcherize the people, draweinge them out to outward worshipp, where as the true worship is within.[11]

Margaret Fell's personal account of her marriage to Thomas was written in 1690 at the age of seventy-five. In it she described her first husband as a man of

> Wisdom, Moderation and Mercy . . . a Terror to Evil-doers, and an Encourager of such as did well. . . . We liv'd together twenty-six Years in which time we had nine Children [one died early]. He was a tender loving Husband to me, and a tender Father to his Children; and one that sought after God in the best way that was made known to him. . . . He left one Son and seven Daughters, all unpreferr'd; but left a good and competent Estate for them.[12]

Judge Fell lived harmoniously with his Quaker wife in his last years, showing a friendly disposition to Quakers coming to the Hall and to George

Fox in particular. Fox related in his *Journal* that he was examined before Judge Fell, Colonel West and other justices of the peace at the Lancaster quarter sessions in 1652, having been called there to answer the charge of blasphemy made by six priests of the area. Judge Fell protected Fox from the ire of the priests, for in Fox's words,

> Judge ffell and Coll West reproved ye preists seeinge there darknesse: and tolde ym yt then they might carry ye spiritt in there pocketts (as they did ye scriptures) and then all ye preists rusht out in a rage against ye Justices . . . and then Judge ffell spoake to [Justice] Sawrey and Thompson . . . and superseded there warrant and showed ym ye errors of it.[13]

Fox summed up his feeling for Thomas Fell in a statement written some thirty years later:

> Judge Fell was very serviceable in his day and time to stop the edge of the priests, justices and rude multitude, who often fell upon Friends, beat them and persecuted them . . . Judge Fell and Justice West stood up nobly for us and the Truth, and our adversaries were confounded. And afterwards he came to see beyond the priests: and at his latter end seldom went to hear them in that parish.[14]

Most historical accounts of the Fell marriage suggest that despite the age difference they were entirely happy together; he seemingly trusted her judgment and actions. It would be useless to speculate on Judge Fell's private feelings about George Fox, for he left no known record. What can be said is that Margaret showed a deep religious attachment to Fox from 1652 onward, and she became totally immersed in her new religious fellowship.

The actual experience of the Fells' conjugal relations is hidden in shadow, but we can learn something about the relationship by turning to personal letters. One of the rare examples available to us of Margaret's handwriting is preserved in a letter to her husband dated 18 February 1652–3. The letter opens with a religious exposition couched in terms of endearment:

> My dear love and tender desires to the Lord runs forth for thee. . . . Dear heart, mind the Lord above all with whom there is no variableness nor shadow of turning and who will overturn all

powers that stand against him, [they?] shall be as chaff before him. Therefore be faithful unto death and he will give thee the crown of life, and stand firm and close to the Lord and be not afraid of man.

There is little or no sense of formality, distance, or subordination on the part of Margaret Fell in this letter. She continues with news of "dear brother" James Nayler who was suffering in prison, to whom she had sent 20s (shillings), yet "he took but 5." She claimed the local constables were looking for Fox and bidding "£5 to any man who would take George anywhere they can find him within Westmorland." Her letter expressed great enthusiasm for the current south-bound wave of northern Quaker itinerant ministers, the first "Publishers of Truth." In the country there were some local justices such as Judge Thorpe who were "favourable to our friends there and did take notice of the priest's tyranny" at the sessions in Lancaster.[15]

Fell enclosed with her letter a copy of a declaration of Quakers and their principles of faith, and a note from George Fox who had been moved to write to "any in Parliament that is a friend to the truth." Her request followed: "Dear heart, I pray thee do not let it lay at the door but show it to any that is anything loving to the truth. It will stir up the pure in them which is one with that which it came from." Moreover, she implored her spouse to "put in print" the enclosed Quaker materials – "a declaration of these things that we live in for the truth [and] will stand when all other things shall be as stubble, so hoping that thou will be faithful to me and to the Lord, farewell." She devotes two lines to the children: "the children are all in health, praised be the Lord. George [Fell or Fox?] is not with us now but he remembered his dear love to thee." Then she added three further requests: first, the printing of more materials including a note from James Nayler; second, a "query and answers that was put to George [Fox]," and third, that one on her work of "John Lawsons be printed in another book." Upon signing herself "Thy dutiful wife till death," she added a postscript, "I pray thee sweetheart do not slight these things for they are of great concernment . . . so let them be published." The letter expressed wifely affection, yet no grief at being left alone, for her mind and energies were given to a new all-consuming interest. She appeared solicitous of her spouse's concern for his family's health but not subservient or ingratiating in her stance toward her husband.[16]

What an historian of Quakerism would trade for a return letter from Thomas that would enlarge our understanding of his attitude toward her newfound religious enthusiasm. A glimpse of Thomas Fell's thoughts *en famille* would be a treasure, but alas, we must be content with only a very partial picture. According to the slim evidence at hand, there was apparently some marital discord after 1652. Sometime after Margaret began writing and sending out her epistles concerning Quakerism, the Judge opposed his wife's wishes to publish her statements. In October 1657 she sent a letter to Gerard Roberts in London requesting that he send her book abroad to the Netherlands before her husband could stop its publication. She wrote, "Let it come forth speedily and be sent abroad, before my husband come up to London, lest he [get] light of it and prevent the service of it." Some of her epistles in this period were sent to William Caton and John Stubbs in the Netherlands. The particular book of October 1657 was not published at that time, which suggests Judge Fell may have stopped its publication.[17]

Although Thomas Fell cannot speak for himself, two letters, one from his wife and the other from his daughter Margaret in 1656, do shed some light on family attitudes as they relate to father and eldest daughter on the subject of matrimony. These two letters stand in contrast to patriarchal attitudes that prevailed in the upper gentry Verney family of Claydon House, as well as in the yeoman family of Ralph Josselin of Earles Colne of the same period. In the Verney and Josselin families the father or eldest brother played an important role in the marriage arrangements of the daughters (and sisters). In 1656 Margaret, Jr., then about twenty-two years old, received a proposal of marriage from Colonel William West, a fellow justice of the peace and friend of Judge Fell. Like Judge Fell, Colonel West had a reputation for kindness to Quakers, but he was not one himself. Margaret Fell wrote West in 1656 to coax him to "follow God's law and so win peace, even through suffering." Another letter written by Margaret, Jr., addressed both her father and Colonel West on the subject of the marriage proposal. She refused any offer of marriage with one not "come into the Unity of the Spirit." Margaret, Jr., added, "If we should sell the Truth of God . . . for a fading inheritance . . . we should lose the peace which the whole world cannot purchase for us." George Fox later docketed these letters: "The trial of Margaret Fell and her children concerning her children's marrying with the world." If that indeed was the case, then in this instance Judge Fell was overruled by mother and daughter. West's offer was refused.[18] In Judge Fell's family the entering wedge of Quakerism seemingly modified the paternal power.

This incident in itself does not yield a clear view of family hierarchy, and in particular of father-daughter relations. It does, however, suggest an attitude that when seen in the larger perspective tells us something of the family power structure. Miriam Slater has argued in her study of the seventeenth-century Verney family that the "givens of social organization" in upper gentry family life were, to a significant degree, patriarchy, primogeniture, and a tendency toward minimal family affect, causing a hierarchy of dependence and importance in which daughters were on the lower end of the scale.[19] In the Fell spousal relationship an attitude of wifely importance and power seemingly was present that countered age difference and social norms. We may merely speculate whether Margaret, Jr., was valued as an individual with feelings of her own or whether she was to some degree coerced by her mother not to marry "with the world." In any event she was not overruled by her father. Five years later (1662) she married John Rous, who was a Quaker. At least in the case of the eldest daughter, Margaret, Sr., as mother held a strong opinion as to the religious persuasion that her daughter's husband should have. Just how much actual control the mistress of the Hall exerted in this area is unknown. However, all of her daughters did marry Quakers, while her son, who was far removed from her at an early age, married a non-Quaker who subsequently became Margaret Fell's antagonist. Taken as a whole, the evidence suggests that the social "givens" of patriarchy and primogeniture that applied to the Verney family, and to the Josselin family farther down the social scale, did not hold true in the Fell family.[20]

The image of marriage portrayed in the Christian conduct books of the period and the official marriage ceremony was one of male domination and female submissiveness in the home. The wife was seen as the weaker vessel not only physically and intellectually but spiritually and morally as well. Common law placed control of the property in marriage in the hands of the husband. This theory of marriage was undoubtedly modified in actual practice in many instances. Nevertheless, as the evidence of literature, law, letters, and diaries of the period indicate, for most people marriage was an unequal institution, with greater power given to the husband.[21]

Although the record of the marital relationship of the Fells is a brief one, two striking features emerge. Thomas Fell, who never converted to Quakerism despite his close encounters with it, appears as a man who experienced ambivalent and occasionally negative feelings concerning his wife's involvement with her nonconforming co-religionists. Margaret's letter imploring him to publish Quaker tracts, along with the constant comings and goings of Quakers at his home, suggest that the Judge may have been beleaguered by his wife's newly acquired religious zeal. Secondly, it appears that Margaret's religious attitudes affected the traditional hierarchy of family relationships. Nothing is known of their relationship in the early years of marriage except for her account of it late in life. However, from 1652 onward Margaret's actions turned the patriarchal order topsy-turvy. She became insistent in her beliefs, with little or no pretense of spousal deference.

CHILDREN

The parish records of St. Mary's Church in Ulverston, Lancashire, where the Fell children were baptized, are incomplete for the period in which the children were born. Isabel Ross has estimated that the births occurred in the following order and years: Margaret, Jr., 1633 (?); Bridget 1635 (?); Isabel 1637 (?); George 1638 (?); Sarah 1642; Mary 1647; Susannah 1650 (?); Rachel 1653. Ross asserts that she was a loving and devoted mother and grandmother. However, Margaret Fell's relationship to her seven daughters was different from her relationship to her only son. The family correspondence, the Swarthmoor Hall accounts, and the early local Quaker meeting records show this to be true except in the case of her daughter-in-law and son after his marriage in 1660.[22]

The seven daughters were intensely loyal to their mother and her religious ideals. There never appears to have been any breach of a personal or religious nature at any time among them, except for a granddaughter of Margaret Fell, Margaret Rous Manwaring, who left the Quaker meeting at the time of her marriage.[23] Growing up together in Quaker unity at Swarthmoor, the Fell children did not experience any ostracism or disruption of family ties as was common for other offspring who chose Quakerism against family wishes. The Fell family became an extended family circle that included those of like religious faith. One instance of the undisrupted mutuality among household members occurred one year after their father's death, when the Fell children underwent a religious fast together. "Bridget Fell fasted twelve days, Isabel . . . fasted seven and [was] to fast nine, little Margaret . . . fasted five, and a little maid that is a servant in the house . . . fasted twenty."[24]

Margaret Fell began her extensive travels on behalf of Quakerism after her husband's death. One month after the Restoration of Charles II in May 1660, Margaret made her first visit to London with her servant Mary Askew. Her eldest daughter Margaret had been in London several months and her daughter Isabel had visited her sister there, seeking medical aid for a chronic knee problem. Margaret, Sr., and Margaret, Jr., resided at Pall Mall in London, apparently not returning to Swarthmoor until September 1661. Her business in London included appealing to Charles II and the Duke of York on behalf of imprisoned Quakers and writing papers to Parliament defending Quaker principles.[25]

During her sixteen-month sojourn in London, Margaret's six younger daughters were at home in Lancashire. Bridget, who was then about twenty-five years old, was left in charge of the farm. The youngest, Rachel, was seven years old. Margaret's letters to her children during this long absence also indicates that her religious zeal took precedence over family concerns. Margaret wrote home in September, "Take care of the three little ones [Mary, Susannah, and Rachel], but especially Rachel." She added that she "would like to come home but cannot until George Fox comes."[26] In a letter dated August 1660 she wrote "to Dear and eternally beloved Lambs and Babes of God," telling them of her efforts to convince Charles II of their peaceable principles and to lobby for George Fox's release from prison at Lancashire Castle.[27] A letter of 24 July 1660 indicated that Bridget was unwell and discontented, for Margaret's advice to her daughter was to "drink her Janesse drink [an herbal medication] and I would have her take a quantity of anise three mornings together and keep warm to do [drink] warm broth . . . and to be patient and be subject to ye will of the Lord." Bridget's letter of 22 July 1660 to her mother passed hers enroute. Bridget reported "all the children are well, and Rachel is very well and contented; have felt weak since coming home, but made Gandess' drink and feel much recovered." Bridget also informed her mother that she had been to Lancaster the day before to visit George Fox in prison and that he was well and "desired to be remembered to Friends."[28]

In August Margaret reported on the trial of the regicides and persons closely connected with Charles I's death, including, "[Major General John] Lambert, [Sir Henry] Vaine and [Sir Arthur] Haslering," which she concluded, "may occasion a longer stay there." Margaret, Jr., added a postscript to her mother's letter, "My dear sisters, my true unfeigned love is dearly to you all in the bowels of endless love. The Lord keep us in His holy fear in obedience to the spirit of truth that He alone may be exalted." In early November 1660 Bridget made a more mundane request of her elder sister in London: "I have written to thee several times concerning the buying of me a petticoat. I desire thee, if thou have not done it, to do it for me, for I am in great want for such things."[29]

George Fell, who had been living in London since 1653, also appears in comments in the correspondence. Margaret, Jr., indirectly revealed the distance that had grown between them in his seven years' absence from home when she noted in August 1660: Brother George "is well and comes here sometimes." On another occasion Margaret, Sr., requested that Bridget send more money to London, for her son George "had taken £20 which she did not know of until after the lease was sealed." Margaret continued in the same letter to inform her daughters that she hoped "to get brother into the country as he has been more idle than ever, but the Lord will prevent his wickedness." The younger sisters left at home were very curious about their one brother whom they no longer remembered well. George Fell married Hannah Potter in London in December 1660, and Bridget wrote in January making a special request: "I would have thee let us know the manner of my brother's marriage, what the woman is [like?] every way: compose as much as thou can in a little room."[30]

The unattended children left at Swarthmoor felt abandoned in a time of increasing local persecution, despite Margaret's frequent assurances that she remained in London doing the Lord's work. Bridget wrote her mother in late August 1660 an expression of filial devotion mingled with a tinge of dejection:

> My dear and endless love is dear unto thee, who hath not only been a mother and nurse to the natural birth but hath also travailed for me in the spirit, that the life of righteousness may be formed in me which work the enemy in my own particular hath greatly withstood, but of late I have found his power hath been weakened, and my peace hath been preserved more than formerly, for which my soul praiseth the Lord.[31]

Margaret wrote in December 1660 that she had received her daughter's letter

> wherein ye mentioned that there is a report in the country concerning my [long] stay, which is little to me. I am so well used to them [her non-Quaker neighbors]. I know how to bear them but the Lord of heaven and earth knows that if His hand had not holden me here, I had been with you ere now; but to that arm in which I stand am preserved must I be subject.[32]

Margaret placed much responsibility on Bridget, who as an adult was expected to mother the younger children and to keep the family finances in order. Margaret asked her to send to London all the deeds concerning their land rents with Colonel Thomas Birch, as well as funds for their living expenses in the capital. Margaret, Jr., wrote to Bridget that mother "desires you to be very exact in every particular concerning the rents." In the same letter her mother confessed, "The last week we omitted writing to you which we never did before since I came here, but we writ twice the week before. G.F. and we are all well here in the Lord."[33] Bridget's fear and depression had not lifted by January of 1661 for she began a letter to her sister in London, "We have ever greatly longed to hear from my mother or thee, but hath not received one line this long time, neither do we know whether you be in bonds or at liberty." She reported on the local persecutions of Quakers:

> Our friends in all parts here away is most part of them in prison, and men servants from all places is gone except John Taylor and a lad at Marsh [Grange], which we do expect shortly to be taken away but that doth very little trouble us, if we could but hear from you. The soldiers told me there came letters from my mother which they said was suspicious, but I know not who kept them from us. . . . Our dearest love to my mother, farewell.[34]

Although Margaret claimed to write her children weekly from London, Bridget's letter of 3 February 1660–1 mentioned not receiving a letter from her since December 1660. The interception of correspondence by the county authorities who were intent on persecuting local Quakers appears to have been effective. Bridget's sense of isolation and loneliness was acute by February. Having been in charge of the estate and family since the previous June, she wrote:

> No letters could pass from us quietly nor none was willing to carry us any, so that we had no way but to rest in patience, being willing to submit to whatsoever the Lord did suffer to be acted against us, and to be content if we should a been separated from you in body for many days, which was much more to us than the loss of outward liberty and enjoyments . . . if we should be always from you outwardly, there was a pretty hard time with us, yet praised be the Lord who did strengthen us and gave us boldness in the time of need.[35]

Bridget reported the effects of the persecution on the family farm income, which further disclosed her feelings of forlornness:

> Truly it is a very hard time for money with us; corn of no sort is anything saleable, and being that we have none to send to the markets it stays all on our hands, and we have pretty much use for money here, seizements and wages to workmen and servants must needs be paid, and them which bought corn last year hath several of them unpaid as yet. . . . if the things continue that they keep [Quaker] men in custody that should till the ground, the time of seeding being now, it cannot be expected that we should be able to maintain them in prison.

Giving another small but telling window into her tension and apparent depression, Bridget continued,

> Truly the burden of outward things hath been very heavy here. . . . if our comfort had been in outward things we had utterly been undone. And indeed I cannot but admire the goodness of our God many times, to see how He hath fitted us . . . for His service, for if this trial had come to some of us a little while ago it could scarce a been undergone, but now can truly say that we are made partakers in measure of that which is able to bear all things.[36]

Their mother responded on 27 February 1660–1, that "in the everlasting fountain of dear love do I dearly remember you and never let it enter into your hearts that I should stay here upon any account or concerning any outward things, but upon the Lord's account and the Truth." Then Margaret added, as if to assuage her own feelings of guilt for leaving her daughters to bear such hardships alone: "The Lord of heaven and earth knows . . . that if I were clear in His presence and free before Him, no outward things should have stayed me from you, therefore have no such belief of me." Margaret, Jr., added a postscript reassuring her sisters that they partook of the younger sisters' "sufferings" and that they in London were "sensible of your great weights and burdens . . . chiefly in regard of my mother's long absence. . . . And as for the time of our coming I am not able to give you any account."[37] It is evident that Margaret expected her second daughter to be self-sufficient and resourceful. Nonetheless, in view of the conventions and attitudes concerning unmarried women in this period, it was undoubtedly unusual for six unmarried sisters to be running a family farm alone. Their work was made more difficult by constant local persecution. In contrast, in the crisis years of the Civil War, Sir Ralph Verney lived in exile and he placed his steward in charge of family finances and estate oversight. His estate was, admittedly, much larger than the Fells'. This left his younger children dependent upon the steward's goodwill and generosity for their living expenses. Neither the brothers nor the sisters in the Verney family were considered sufficiently trustworthy or competent to take charge of the family estate. It is probable that the role played by the Fell sisters was unrepresentative of their class.[38]

Margaret Fell's first (of ten) visits to London occurred at a time of high political tension, and she became deeply involved in the local political-religious scene in London. In addition to visiting the king, she had an audience with the queen mother, Henrietta Maria, and presented her with Quaker books. She also visited Anne, the queen of Bohemia, daughter of James I, who was staying in London at that time. After Fox's release from Lancaster Prison in October, he went to London, where Margaret also remained, continuing her work to obtain freedom for imprisoned Quakers. In January 1660–1 Margaret began preparing for her journey north by purchasing a small coach and two horses. She was, however, delayed in her return by the Fifth Monarchy uprising in the same month in London. Although it involved only a small number of persons, all of whom were quickly arrested, it did cause a furor in the city. A violent reaction against Quakers and Baptists ensued, and as a result of the royal proclamation of 10 January prohibiting meetings of "Anabaptists, Quakers and Fifth Monarchy men, or some such like appellation," about four thousand were quickly imprisoned. This helps explain the break in the flow of correspondence from mother to daughters between December and February 1660–1. Neither Fell nor Fox was arrested in this period although they apparently attended outlawed meetings. In a proclamation of May 1661 Charles released Quakers allegedly involved in the January rebellion. Then in August Charles issued another proclamation freeing Quakers arrested for conventical worship. It was after this second proclamation that Margaret Fell felt "freedom in spirit" and returned to Swarthmoor in September 1661.[39]

Were there other factors that may have influenced Fell's extended stay in London? It is evident that her religious principles were uppermost in her mind at this time in her life. However, she was also caught up in the political excitement of London. In addition, as a widow, she experienced a whole range of new alternatives open to her and for the first time in her life felt free to make her own choices.[40] Her sons's proximity was undoubtedly another reason for lingering in London. Finally, local rumors were abroad concerning the Fox-Fell friendship, the reports of which she dismissed as meaning little to her. Nonetheless, Fox's presence in London after October 1660 may have influenced her prolonged stay. As a widow with newfound freedom and head of a family with sufficient wealth to provide for her needs, she may have found it difficult to tear herself away from the lure of the city and Fox's presence there. Finally, it was her first experience mixing with those in high places, and it may well have given her a heady sense of purpose in her newly won public and rather prestigious political-religious role. Whatever her relationship to her only son may have been, it did not curtail or thwart her style of life. George Fell was either indifferent to his mother's new Quaker lifestyle at the time or unable to assume a dominant male role and thereby circumscribe her activities in 1660–61. Fell had assumed the patriarchal role of the family. She was in an invulnerable position, every family member being dependent on her instructions, patronage, and goodwill.

MARRIAGES OF CHILDREN

Between the years 1660 and 1691 all of the Fell offspring married. With the exception of the fifth daughter, Mary, and the Fell heir, George, all of these marriages were rather late, even according to the average late age of the contemporary English pattern. The ages of marriage ranged from 22 to 41, and the number of offspring of each of the Fell sisters and brother numbered from none to ten.[41]

Little is known of the courtships and actual weddings of the daughters. There is an absence of evidence to indicate that either parent sought to arrange these marriages with the possible exception of the eldest daughter, Margaret. Evidence of third-party involvement, as recorded by Samuel Pepys, who helped arrange the marriage of the daughter of his patron, Lord Sandwich, in 1665, is entirely absent in the records of the Fell family. The very late marriage ages of Sarah Fell at age 39 and Susannah Fell at age 41 suggest that no arrangements were made by their widowed mother to marry off these two daughters. Of the five other sisters, only Mary, who was wed at age twenty-one, married early. There is no evidence to indicate that a dowry was given or land exchanged for any Fell daughter. However, the daughters had their own legacies and their own expense accounts deriving from their father's estate. Thomas Fell left each of his daughters an equal share of the residual of his personal and real property. When Thomas Fell's only son George was relatively young, he married Hannah Potter, a woman of his own choice, in London. She was a widow and the daughter of the London barrister Edward Cooke. George Fell's family did not know her. This match was presumably opposed by his mother, for Hannah Potter was not a Quaker.[42]

Although the Fell sisters, other than Mary, did not marry early, they did not endure the demeaning hardships frequently associated with spinsterhood.[43] Both Susannah and Sarah, who married very late, did not appear disadvantaged. Sarah, as will be seen in the next chapter, ran the family farm productively until her marriage to William Meade of London in 1681, at which time she left Swarthmoor permanently. Her mother and sisters relied heavily on her accounting and farm management skills. This expertise made Sarah a valuable member of the Fell household, and she was sorely missed by her mother and Rachel after her marriage. Thus, in her unmarried years, Sarah avoided the experience of being the useless spinster sister in a patriarchal family. Susannah, who never had children of her own, visited her married sisters and helped as a nurse during the births and infancies of her nieces and nephews. She traveled sometimes with her mother. For ten years she lived with her sister and brother-in-law, Sarah and William Meade, near London, through whom she met widower William Ingram. She returned to Swarthmoor for at least one visit in 1687 to visit Rachel during her confinement. Susannah married Ingram in 1691 and was widowed in 1706, at which time she returned to the Meade home, where she spent her last years. Family letters reveal a warm sisterly relationship with no hint of tension expressed by Sarah, William Meade, or Susannah. Had Meade been an impoverished gentleman, as opposed to a wealthy one, this might have been different. Susannah wrote a letter to her younger sister Rachel from Sarah's home in February 1685, greeting Rachel with her "love to thee and my duty to our dear mother, for your remembrance is very pleasant to me, and I can truly say its neither length of time nor distance of miles make void or alienate my love to you." Sarah's letters to mother and sister Rachel at Swarthmoor after her marriage speak warmly of Susannah, often expressing a concern for her sister's health. In 1689 Sarah wrote her "dear and honoured mother" that "sister Susannah has her health better of late than I have known for some time, which I am very glad of, and she is very cheerful and hearty."[44]

Margaret Fell not only encouraged her daughters' self-sufficiency, she protected them from the disadvantages of spinsterhood. Fell did not play the role of a mother anxious to arrange advantageous marriages for her daughters. Her only requirement for her daughters' "preferments," as she labeled it, was that they marry Quaker men. Additionally, there is no evidence that the sisters were in competition for their shares of the family wealth. There was enough family income from the farm enterprise to allow all to live comfortably, although rather frugally, on the farm. Sarah Fell's account book indicates that each sister had her own expense account, which suggests a definite degree of autonomy for a seventeenth-century single woman. For example, while Susannah was living with Sarah Meade in the 1680s she wrote Rachel at Swarthmoor asking her to collect on her behalf three local loans totaling £100, a considerable sum.[45]

The later family letters between mother, stepfather, daughters, and sons-in-law from 1680 to 1702 give a continuing picture of family harmony. During the busy decades of the 1650s, 1660s and 1670s, when Swarthmoor was the pivotal center of nascent Quakerism, most of the Fell daughters married and left the Hall permanently to raise their families in different parts of England. However, they did return home for visits, some of which were extensive. Some of Margaret's grandchildren were born at the Hall or lived there for lengthy periods. The oldest granddaughter, Bethia, was born at Swarthmoor in 1666. Further, Margaret's daughters displayed great regard not only for their mother, but also for their stepfather. In October 1667 Margaret Rous wrote to her widowed mother, who would soon marry George Fox in 1668, that she was caring for her "deare ffather" George Fox, who was ill, in her home. Over the years John Rous frequently wrote to his mother-in-law dutiful letters informing her of family, politics, and business. Overall, the family correspondence conveys the picture of a closely bonded family.[46]

When the eldest daughter, Margaret, married John Rous of Barbados in 1662, they settled in London. Distance did not strain or sever family ties between the Fells and the Rouses. In February 1664 when Margaret Fell was imprisoned in Lancaster, her seventeen-year-old daughter Mary went to stay in London with her older sister Margaret. While there, Mary worked with others to secure her mother's release. Margaret, Jr., Mary Fell, and John Rous all had audiences with the king to this end, but they were unsuccessful in their efforts as they could not promise the king that they would desist from holding Quaker meetings at Swarthmoor. They encountered well-placed persons in court circles who were ready to take over their estate if it were forfeited to the Crown, an ironic replay in the reverse of their father's former position.[47]

Margaret Fell's second daughter, Bridget, who ran the farm and survived the 1660–61 persecution, married John Draper from county Durham in 1662 in Swarthmoor Hall. She apparently died in childbirth early in 1663. Her mother crossed England in the winter of 1662–63 to Bridget's home in Headlam, Gainford, in order to attend her in her confinement. From that date on Bridget's name disappears from all family correspondence; however, several grandchildren bore the name Bridget.[48]

Isabel Fell was, next to her mother, probably the most extensive traveler of the family. In 1664 she married William Yeamans, a merchant of Bristol. He died ten years later, and she did not remarry until 1689. In the interim she visited Swarthmoor for extended periods and left her two children, Rachel and William, in the custody of her mother and aunts. Isabel devoted much energy to visiting meetings in the north during the 1670s. Her travels took her also to London, Barbados, and, in 1677, to Holland with George Fox, George Keith, and William Penn.[49]

Sarah Fell, the fourth daughter, is best known for keeping the Swarthmoor farm accounts and for her role as clerk of the Swarthmoor women's monthly and quarterly meetings. Although Sarah was an astute businesswoman, as seen in her meticulous keeping of the Swarthmoor accounts, the family letters reveal another side of her that was warm and maternal as well. In 1681, after her marriage to William Meade, she resided near London. She wrote home at regular intervals right up to Margaret's death in 1702. Sarah's letters are among the most interesting of the Fell family correspondence. Her parsimony, evident in the farm accounts, was balanced by generous affection for her only child, Nathaniel. This unusual woman also displayed the Quaker trait of throwing off deeply ingrained social customs. For example, according to the correspondence, she broke with prevailing class tradition by nursing her infant son. When Nathaniel was born, Margaret wrote Sarah motherly advice, and husband William wrote and thanked his mother-in-law: "about suckleing my child, [Sarah] hopes her milk will increase; the child feeds pretty well, his sore mouth being over." When Nathaniel had teething problems, Sarah wrote that he was "pretty weak and low . . . and I am fearful of him about it, though he is pretty cheerful, we give him strengthening things, but he drinks spring water yet, which we have thought has done well with him."[50] When the boy was seven, William Meade hired the Quaker scholar and botanist Thomas Lawson to tutor him and his cousin Richard Lower.

Sarah sent gifts to her mother and sister Rachel at Swarthmoor in December 1683. The list is quoted at length because it reveals much about the level of affection in the Fell family.

> Dear Sister Abram, I have endeavoured to fit my dear Mother with black cloth for a gown, which is very good and fine . . . (five yards and a half) and what material . . . [is] needful to send down; viz., silk both sewing and stitching, gallowne ribbon and laces, and I was very glad to know what she wanted, for it has been in my mind a pretty while to send her and you something. . . .
> 3 pairs of doeskin gloves, such as are worn in winter for Mother, Sister Lower and thyself; the thickest pair for Mother if they fit her, but I leave that to you to agree on as you please;
> 1 pair same sort of gloves for Brother Abraham;
> 4 ells of holland [cloth] for Sister Lower and thyself. . . .
> 2 pots of balsam, one for my Mother, the other for Sister Yeamans;
> 3 pocket almanacs for Sister Yeamans, Sister Lower and thyself;
> 1 muslin nightrail for Sister Yeamans, which she sent for;
> 100 needles, of which half for Sister Yeamans which she sent for, the other half hundred for Sister Lower and thyself.[51]

When Margaret was seventy-six years old, she made her ninth visit to London, arriving in April and remaining until June 1690. In March 1689

Sarah expressed her concern about her ancient mother's endurance prior to her long journey:

> As to what thou writes about thy coming to London, we writ plainly our thoughts of it, and we are glad that it so much agreed with thy most serious and weighty considerations of the journey. Not but that we should have rejoiced to have enjoyed thy company with us here, if it might have been with safety for thee, but thy years are considered and the journey so long, and especially the way on horseback before you come to the coach; this made it uneasy to think on, fearing how it would be. But blessed be the Lord, who gives a sense of that which is our places and best for us, whose dictates and ordering hand I desire we may all know and follow, and then all will be well.[52]

Margaret undertook this journey contrary to the advice of Sarah, and in 1697, at the age of 83, returned to London for a tenth and final visit.

Margaret Fell's fifth daughter, Mary, married Thomas Lower at Swarthmoor in August 1668, two months after Margaret's release from Lancaster prison. Thomas Lower became a fellow traveler and favorite amanuensis of George Fox. Lower sometimes shared Fox's imprisonment and was the recorder of Fox's *Journal*. Fox left Lower several of his prized articles in his will. Margaret Fell also was fond of Thomas and enjoyed his letters to her in her old age. Rachel wrote her sister Mary from Swarthmoor in 1694, "my Mother bid me write thee [she] would be glad to hare from Bro[ther] Lower, being he is so intelligible."[53]

While in London at the home of John and Margaret Rous in December 1664, Mary Fell became ill, probably of the plague. Her mother, in Lancaster Prison, was informed of Mary's critical condition by John Rous. She wrote a letter to the Rouses that tells us something of Fell's attitude toward death and family adversity.

> My Deare Mary, her spiritt neare and Deare, and present with mee, whether in the Body or out, with the Lord and to the Lord of heaven and Earth, she is given freely, and his heavenly and holy will, I freely submitt too. . . . Solace thy Soule in that and in the life and power of Almighty God, wch thou Injoys, rest satisfied, and be Content And as I have said often to thee, give up to be Cros'd, and yt is the way to please the Lord, and so follw him in his owne way and will, whose way is the best. . . . lett not sorrow fill yor hearts, for we have all cause to

rejoyce, in ye Lord evermore, and I most of all, that bought her Deare and brought her up for ye Lord.[54]

Her letter continued on a more mundane level, referring to business matters concerning George Fell and Swarthmoor, as if she had passed the thought of her daughter's possible death from her mind. This seeming cavalier attitude toward death is somewhat unnerving to the modern reader. Some historians have explained it in terms of low familial affect that was typical up to and throughout most of the seventeenth century. Within families, it has been argued, "there was a general psychological atmosphere of distance, manipulation and deference; . . . high mortality rates made deep relationships very imprudent."[55] The teleological thesis of growing emotional bonding in family life in Western society from the early modern to the modern era is inadequate as an explanatory model for the Fell family. It is true that the Fells expressed their family sorrow in religious terms, for it gave meaning to the relationship between suffering and salvation to those concerned. Certainly the belief in a life after death was an ameliorating factor in parental grief over the loss of children. But the stoic attitude expressed here by Margaret as well as the total silence in the family records about Bridget's death one year earlier is not due to lack of emotional bonding within the family. Quaker sensibilities reacted against all outward forms, as seen in the absence of gravestones in their burial grounds, for example. Likewise, conventional Quaker piety disparaged overly demonstrative behavior in grief, and Fell is a representative example of one observing that religious taboo. To express grief meant to be weak in the faith that promised eternal life. As a seventeenth-century sectarian woman, Margaret cloaked herself in Quaker spirituality to buffer herself from personal adversity. Puritan and Quaker piety alike called for a stoic attitude toward death.[56]

Rachel Fell, the youngest daughter, was her mother's life-long companion, for she resided permanently at the Hall. Rachel cared for her aging mother in her last years at Swarthmoor, and she was the only daughter present at her mother's death. Margaret initially objected to Rachel's marriage to Daniel Abraham but consented to it in 1683. Daniel Abraham, whose father was a merchant in Manchester, moved to Swarthmoor Hall upon his marriage. In 1691 he purchased the Hall from Charles Fell, son of the late George Fell, in order to end a long family dispute. He was gradually impoverished due to heavy tithe persecution. Daniel lost Hawkswell, the Fell ancestral home. His son John Abraham fell into debt and finally lost Swarthmoor Hall itself in 1759.[57]

A letter, probably sent to Margaret in London during her last visit there in

1697–98 by young John Abraham, one of her favorite grandchildren, reads in part:

> Dear Grandmot thou art oftener in my mind then I can menshon and my love and Respect for thee is great I hope to see thee next Springe at Swarth wh I shall bee Exceed glad of: this being moest at present but my Dearest Duty to thee and my Respects to all my Vnkell, Antts and Cosen, From Duttiful Granch, John Abn.[58]

Undoubtedly in his old age John Abraham deeply regretted the loss of his grandmother's home, as he had held her in great love and esteem in his childhood.

GEORGE FELL

George Fell, the fourth child, followed his father's example in both religion and profession. As the only male child, George Fell was treated according to seventeenth-century gentry norms for the grooming of an inheriting son. Named after his grandfather, who had purchased Hawkswell from the sale of monastic lands in Furness, George was given the education proper to a country gentleman-to-be. His father arranged for a resident clerical tutor and brought in another youth, one William Caton, slightly older than George, to be his companion and fellow student. William Caton retrospectively described his boyhood experience with George in Judge Fell's home:

> When I was about fourteen years of age my father took me to Judge Fell's there to learn with a kinsman who was preceptor to the aforesaid judge's son; and thereby I came to have an opportunity to be conversant with them that were great in the world. I was in due time promoted to be a companion, night and day, to the judge's son, and did eat as he did eat, and lodged as he lodged, and went after the same pleasure who he went unto, as to fishing, hunting, shooting, etc. . . . After we had learnt some time together in the judge's family we were removed to a school in the country.[59]

William Caton became a Quaker along with much of the rest of the Fell household on Fox's first visit there in 1652. He became Margaret Fell's secretary in the 1650s, transcribing many letters and documents for her.

Their relationship became a very warm one, for his letters to Margaret after he left the Hall as a first Publisher of Truth were adoring. He regarded her as his surrogate mother, and in the subsequent mother-son dispute, he attempted to be a mediator on her behalf. He showed total loyalty to her rather than to his boyhood companion, George. For instance, in 1664–65, while Margaret was in Lancaster Prison, William Caton wrote a letter to George Fell. He addressed it first to Margaret, however, with the instruction, "let it then be sealed and sent him."[60]

In February 1652–53 when he was about fourteen, George Fell matriculated at Gray's Inn, London. He lived in London until the mid-1660s when, partly due to financial necessity, he returned to Swarthmoor to play the role of the country squire. His years in London apparently changed George from one who was mildly influenced by Quakerism as a child to one who, much to the concern of his mother and erstwhile Quaker friends, could not espouse Quaker principles. Margaret's concern for her son's changed ways is evident in a letter of 1657:

> My dear love, take heed of wildness, lightness and vanity, and take heed of pride getting hold of thee. . . .
> My dear love, all the ways of the wicked will come to naught and perish though never so delightful for the present, yet woe and misery will be the end of all sin and wickedness; therefore, my dear love, turn from evil and sin and take heed of rashness and forwardness and headiness. Keep these down and strive for patience, and thou will see the blessing of God will be upon thee. My dear one, I cannot forget thee; my cries to my Heavenly Father are for thee that thou may be kept and that the measure of Him in thee may be preserved . . . .
> So my dear love, the Lord God of power be with thee and keep thee in His fear. Read this often and as thou readest thou will be with me.[61]

Among the many letters to the Fells in the 1650s and early 1660s, there are frequent greetings to George Fell from traveling Quakers, thus reflecting a certain deferential treatment of the heir. Among those who attempted to visit young Fell in London was Margaret Fell's friend Thomas Rawlinson, who, in 1657, reported, "I went this day to George Fell. . . . I was with him three or four hours in his chamber. We spoke of many things. The spark is not quite out."[62]

George's father died in 1658, leaving him, at the age of twenty or twenty-one, possibly heir in accordance with the customary common law rules of inheritance. However, the customary rule of primogeniture was left unstated in his father's will. His mother was accorded a devise, or gift of real property, namely, use of the Hall and surrounding buildings and 50 acres until her death or remarriage. As a devisee under her husband's will, she received a life interest in his property as a widow. Therefore, the right of dower or jointure seemingly did not apply in her case.[63]

The future of the Swarthmoor estate was placed in jeopardy during Margaret Fell's four-year incarceration in Lancaster Prison. At that time George Fell apparently took action to keep the Hall in the family, petitioning the king, as the Judge's only son and proper heir, to take possession of Swarthmoor. Prior to this petition for Swarthmoor, George Fell in 1660 had petitioned the king for pardon as his father had been a member of parliament but had not acted against Charles I during the Civil War. Charles II granted his petition by 1664.[64] Meanwhile, he returned to Swarthmoor about 1664 and was apparently on friendly terms with his mother and sisters. He and his wife and daughter made Marsh Grange their residence, while the Fell sisters continued to live at the Hall. Margaret gained her freedom in June 1668, whereupon she returned to Swarthmoor.

The Fells were in accord concerning the Hall's occupancy until 1668 or 1669. Quaker historian Alfred Braithwaite has suggested that two things probably happened upon George Fell's return that caused the rupture between mother and son. First, George may have discovered, as he mingled with the local gentry, that his mother was considered a religious rebel and pariah in the neighborhood, causing his family to become an embarrassment to him. Second, George Fell had been reared in the deferential social system of his time that relied on patronage to get on in life. He probably was outraged by the Fell-Fox marriage in 1669, considering it a deep humiliation that his mother had married a social inferior. George Fell at this juncture appears to have proceeded to carry out his father's last will, which stipulated that the Hall was only to belong to Margaret as long as she remained a widow.[65] Even though George Fell had Marsh Grange and Hawkswell, he commenced his move against his mother's possession of Swarthmoor.

Family correspondence conveys the bitterness of the battle for Swarthmoor that ensued. In April 1670, Margaret Fell was recommitted to prison; as the correspondence states, family members were convinced that this was the work of George Fell. He wanted to dispossess his mother of her home and made threats to this effect to his sister and brother-in-law, Margaret and John Rous. They tried to persuade him not to follow this path but to no avail. In a letter to his mother-in-law, dated December 1669, John Rous exclaimed, "My brother Fell was with us and we cannot prevail anything with him about thy part of Swarthmore." Thomas Lower called his brother-in-law "barbarous" in his "unnatural actions" toward his mother.[66] George Fox, Margaret's husband of only a few months, had always claimed he wanted no part of his wife's wealth. He did, however, have the following advice for her, written from Enfield, 23 November 1669: "Here hath been a great noise about thy son George Fell," adding that this could injure the reputation of "Truth." He cautioned her,

> thou may make as little noise of it as may be. And thou may speak to thy broth[-in-law] Richardson [an attorney] about these things (in the seed and life) which are below. And reason quietly with him and them. As concerning the house keep over it and give both it and him to the Lord's ordering. And so if thou canst preserve a part to thyself, the interest thou hast already whereby thee may not be banished out of the country but him . . . for if he should wholly put thee out of the house, it might hurt himself and be the destroying of himself turning the Lord out of doors.[67]

The dispute did not abate with the untimely death of George Fell in October 1670. His will of that same year bequeathed the lands of Swarthmoor to his infant son Charles, and the use of his estate to his wife as long as she remained a widow.[68] The will made no mention of his mother or sisters except Isabel, to whom he bequeathed Hawkswell and two other parcels of land. He requested to be buried "in ye parish church of Ulverstone, as near to my father as w[i]th conveniency it may."[69]

In April 1671 the king granted Margaret Fell a pardon and awarded the "Estate of the said Margaret Fell, alias Fox, unto Susan Fell and Rachel Fell, the daughters of the said Margaret."[70] According to Sarah Fell's household account book, following George's death, litigation commenced between Margaret and Hannah Fell. John Rous wrote to his sister-in-law, Sarah, concerning the family feud: "In regard to my sister [Hannah] Fell's wilfulness and foolishness, who, I always feared would do as bad, if not worse than her husband, she hath a crooked generation to be her advisers if she take counsel of her kindred, who no doubt will do the worst they can."[71] Among Hannah Fell's annoying acts against her in-laws was locking the gate that opened on the footpath from the Hall to the town of Ulverston one mile away, ostensibly to prevent further Quaker meetings at the Hall. William Meade wrote his mother-in-law that she should demand the gate be reopened, "as thy right, which will be greatly to her [Hannah Fell's] shame and reproach if she refuse it." But he also advised her not to commence litigation against Hannah on this issue. Although the sisters of George Fell at Swarthmoor continued to have some contact with the widowed Hannah, the antagonism experienced by both sides did not disappear for at least a generation. In her will of 1702, Margaret Fell bequeathed one-half less to the Fell grandchildren than she did to her Quaker grandchildren. In 1691, Charles Fell received compensatory money of £3,900 from his uncle Daniel Abraham to put an end to the family dispute over the Hall.[72]

Knowledge of this conflict is limited by the fact that we have letters only from those in sympathy with Margaret Fell. In reassessing the case, it appears that George Fell probably was embarrassed by his mother's radical sectarianism and her marriage to Fox. However, it is not surprising that the only son, traditionally expected to inherit, should have fought for his right. His mother, having remarried, had no real legal ground to oppose him. It appears that Margaret Fell's desire to retain the Hall was motivated largely by her religious interests, Swarthmoor being the center of a whole network of Quaker activity in 1670. She also did not like her daughter-in-law who was non-Quaker, noncooperative, and nondeferential. Margaret Fell's family, as well as her wider Quaker circle, treated Margaret with great solicitude; hence Hannah Fell's surly attitude was behavior to which Margaret was unaccustomed.[73]

George and Hannah Fell, acting out of a policy of self-interest, competed against the interests of mother and daughters for the limited assets available in the Swarthmoor estate. Their sparring for a greater share of the family resources was a reflection in microcosm of the larger patronage society of the seventeenth century. It may well have represented common behavior between heirs and the siblings of landed families when family assets were at stake, as seen in the case of the Verney family.[74] In light of this, George Fell's behavior ceases to appear either irregular or scandalous. George Fell considered himself the rightful heir of the estate. He was displaced by a shrewd mother who wanted to retain Swarthmoor Hall for personal and religious reasons, and who outmaneuvered him for the spoils, partly by craft and partly by the coincidence of his early death.

MARGARET FELL'S SECOND MARRIAGE

On 27 October 1669, George Fox married Margaret Fell, who at 55 was ten years his senior. Fox's own account of his marriage is expressed in his *Journal*, "And there [at Bristol] Margarett ffell: and her daughters: and son in laws mett me: where wee was marryed." Having consulted with her daughters "whether they was all satisfyed" that Margaret had fulfilled the obligations of her first husband's will in regard to their legacies, Fox continued, "And so when it had been layde before severall meetinges (both of ye men and women . . .) and all was satisfied there was a large meetinge appointed of purpose: where there was severall large testimonyes." At that time Fox claimed that he would not concern himself with his wife's wealth.[75]

One week after the wedding, Fox left Bristol, passing into Wiltshire and then on into Berkshire to continue his roving preaching life, while Margaret "past homewards toward ye North." A curious man, hearing of his recent wedding, asked Fox why he had married. Fox responded,

> I tolde him as a testimony yt all might come uppe Into ye marriage as was in ye beginninge: and as a testimony yt all might comme uppe out of ye wildernesse to ye marriage of ye lamb.
>
> And he said hee thought marriage was onely: for ye procreation of children and I tolde him I never thought of any such thinge but only in obediens to ye power of ye Lord: and I Judged such things below mee: though I saw such thinges and established Marriages but I lookt on it as below me and though I saw such a thinge in ye seede: yett I had noe commande to such a thinge till a halfe yeere before though people had longe talkt of it: and there was some Jumble in some mindes about it but ye Lords power came over all.[76]

The couple spent very little time together during the twenty-two years of their married life, except for two extended periods Fox spent at Swarthmoor, one in 1676–77 (twenty-one months), and one in 1678–80 (eighteen months).

A few extant letters exchanged between George and Margaret, mostly written in endearing terms to one another, represent the bulk of the evidence available to enlighten us on their relationship. It is evident from preserved comments in other letters and documents that the daughters and sons-in-law all revered Fox and wrote very warmly of him, usually referring to him as their "deare father." For instance, Sarah Fell saw to it that his sojourn at the Hall from June 1676 to March 1678 was marked by a well-stocked larder; more white bread was purchased during his stay, runletts of wine were frequently purchased, and other delicacies such as oranges were sent from Lancaster. Medicine in the form of juniper berries was bought "for father," and Margaret purchased a white horse for George before he recommenced his travels in March 1678.[77]

There were moments of tension between Fox and Fell, however, especially during the trying period of Fox's Worcester imprisonment from December 1673 to February 1675. Some letters from Fox to Fell during this time indicate that she was annoyed with her husband for allowing himself to be arrested by the constables when he could have eluded them. During his imprisonment the record indicates that she left him alone in Worcester gaol and went home, not returning for several months to care for him, as was customary among the Quakers. Fox may have been somewhat piqued by this behavior, for he wrote his wife from prison that she should tell people where she was, for most people thought she was with him at Worcester.[78]

Students of early Quakerism have speculated on the spousal relationship of Fell and Fox. It has been suggested that the marriage was intended to offset rumors of an "illicit relationship" that could have injured the reputation of the struggling new movement. Another explanation has been that it was "a symbolic union of men and women Friends to give greater force and impetus to the forthcoming establishment of women's meetings." Such an assessment helped offset any questions on why Fell and Fox spent so little time together during their twenty-two years of marriage. Another interpretation centered on Fox's ideas of his marriage as having a "mystical fitness at a time when the church, through the Quaker movement, was come up out of the wilderness, and the gospel order had been set up again." That is, it was part of Fox's restorationist theology, the restitution of human harmony and perfection as before the Fall. Fox explained it "as a testimony that all might come up into the marriage as it was in the beginning; and as a testimony that all might come up out of the wilderness to the marriage of the Lamb."[79]

This view of marriage denoted a mystical union or spiritual rationale for the marriage of these two Quaker leaders. What of the physical union of this couple? The family correspondence is entirely silent on the matter relating to their cohabitation – or sharing of bed and board. Quaker biographer Isabel Ross claims that their relationship was the result of a long friendship "of close spiritual and mental cooperation in the work of the Church, which both believed was the true primitive Christianity revived" and thus was "pre-eminently spiritual." Their contemporaries had talked of their relationship for a long while, and rumors of an illicit liaison caused what Fox called a "jumble in some minds." Ross does not specifically touch the subject of "due benevolence" in this conjugal arrangement, while others have speculated that the marriage was never consummated. Quaker historian Henry J. Cadbury found a curious letter of 16 July 1670 in the *Calendar of State Papers Ireland* addressed to William Penn, who was in Ireland in that year. The letter, written by a Quaker woman minister in London, was intercepted by government agents. It included the following allusion to the Fell-Fox conjugal relationship:

she [?] haeth bene here to mete Margret Rouse About her Mouther's Besenes that is in order to Geat her relese and . . . ouderstanden Margrett Foxes Condeshon and that she being weth child and so nere ass she is her time beingen out All Most they are veary endorstret [interested] to procure her liberty which I hope thay well done.[80]

This letter was either unknown or ignored by Ross, Crosfield, and Braithwaite. Henry Cadbury maintained that Margaret Rous was at that time pregnant and delivered a son, Nathaniel, in September 1670, the implication being that this was a misprint, and the pregnancy referred to Margaret the younger. However, Cadbury also pointed out that Margaret's marriage to George occurred eight months prior to the writing of this letter. After her marriage in October 1669, Margaret was reincarcerated in Lancaster gaol the following April and not released until April 1671. She was then fifty-five years old and was apparently in ill health during this confinement.[81]

This perplexing comment on Margaret Fell's possible late-in-life pregnancy supports a hostile account printed in 1707 by Francis Bugg, alluding to a mystical conception by this couple. Much ridicule of Fell's and Fox's alleged belief in such a mystical conception circulated in the north country even after her death. Bugg claimed that the pair were deluded into thinking it was a miracle event. The account printed by Bugg read in part:

After many years' cohabitation, when both George Fox and Margaret Fell were grown old, whether the spirit of delusion to whom they had given themselves up possessed them in a vain conceit that they were Abraham and Sarah I know not. But it is very certain that they both persuaded themselves that Margaret Fell, alias Fox, was with child, and the Lord would raise up Holy Seed of them. All preparations were made for old Maximilla's lying in, baby-clouts were prepared, the midwife was called and gave attendance for about a month together, but there came nothing forth, all proved wind."[82]

None of the family letters at this juncture mentions this incident, and considering Margaret's advanced age and her imprisonment at Lancaster during this time, it would seem sensible to conclude that the above reference in the letter to Penn was indeed a misprint. However, considering the persisting rumor that circulated years later among those in the north country hostile to Margaret and her Quaker circle and who as neighbors might have known about a midwife being called, the possibility that something occurred – perhaps a menopausal miscarriage – should not be totally ruled out. If the letter is read at face value, the incident also suggests that prior marital sexual activity may have occurred for Fell to have made such an assumption. The Restorationist imagery, frequently used by Fox, which signified a pure unity of humanity before the Fall, would not be hostile to the notion of physical consummation to complete the mystical union.

Evidence does not permit more than speculation on this aspect of the Fell-Fox conjugal relationship. What can be concluded here is that as a widow, Margaret Fell had a wider range of options in selecting a marriage partner. She exercised this greater freedom by choosing a man of her own liking and, despite her son's apparent anger, of lower social origin. Remarriages in seventeenth-century England were to a significant degree based on considerations of social connections and property advantages. Nevertheless, widows were known to choose second partners based on their personal preferences, which offered greater personal satisfaction.[83] There was obviously a personal affinity between Margaret Fell and George Fox since their first meeting, as indicated by her earliest letters to him in 1652. That Fell may have held a deeply sublimated romantic attachment for the charismatic preacher and vice versa is certainly not impossible. There is simply not enough evidence to prove this from family letters. What the correspondence does reveal was their mutual aim to usher in the Kingdom of God on earth as they understood it. Shared religious ideals and personal affinity seem the most likely reasons for the bond in this marriage of twenty-two years.

SERVANTS AT SWARTHMOOR HALL

When George Fox convinced Margaret Fell to follow the Quaker path in 1652, she was influential enough in the euphoria of her conversion to carry along with her the children and servants of the household, despite Thomas Fell's refusal to become a Quaker. Fell continued to maintain close ties to household servants who were converted to Quakerism. Several servants, such as Leonard Fell, Richard and Ann Clayton, and William Caton, subsequently left Swarthmoor as Quaker itinerant ministers, writing endearingly to Margaret on many occasions, saluting her as "nursing mother" to the struggling missionaries scattered throughout England and beyond the seas.[84] Richard Clayton wrote to his Quaker mother-superior: "Deare and precious thou art, unto all ye flocke of God . . . my unity is with thee and with the flock of God with thee." The devotion of a youthful servant girl, Mary Pease, rings clearly through three letters to her erstwhile mistress, as she punctuated her travels with reports to Margaret, expressing great warmth in that mistress-servant relationship. In June 1659 Mary wrote, "oh my deare my deare Mother, Thy deare Love to me is not to be forgotten. . . . Love to my deare Sister Bridget and to Sarah and all the rest of the children to Mary Ashew [a household servant] and all the rest in family wth thee." In another letter she asked her "deare Mother" advice concerning an incident that occurred "long before I came to thy house," and signing the letter "thy Daughter and I hope shall be found obedient to that wch shall be required of mee." A final glimpse of Mary Pease comes to us from December 1663 when she wrote from London to Margaret at Swarthmoor: "I cannot forgitt but must acknowledge yt thou hast been as a nursing mother unto mee and I hope thou wilt bee rewarded for it by him who is ye rewarder if ye faithfull of which thou art." Her wish was that, "I may still grow as a branch in ye true vine yt [that] ye fruit I yeald may bee to his glory."[85]

In contrast, those servants of Swarthmoor that chose to "marry out" of the Quaker community moved outside the close circle of Quakers who gained the loving concern of Margaret Fell and her daughters. For example, Abigail Trindel (alias Curtis) wrote a letter to Rachel Abraham after Margaret Fell's death in 1706. Due to its revealing nature, it is worth quoting in full:

Dear & Loving Mistress [Rachel Abraham]
    Thou might Justly think me Ungratefull to those many favours which I have Received from thee & my Good Master, [John Abraham], because I have not paid my Acknowledgments to you Both as those great undeserved kindnesses did require But I trust your wonted Goodness will put another construction upon my Actions, Since my heart will ever bear a most dutiful Remembrance of your many favours, For the true Reason of my not paying my Respects to you both in a few lines oftner than I have was principally my Consciousness of having disobliged you by my Marrying out off the meeting, being confirmed therein by never having ye favour of hearing from you tho' I wrote twice, And tho' I have had the misfortune to mary so, yet I have got a verry honest Man who never debarrs me off going to ye meetings of ffriends as oft as I please; So that I want for nothing but ye Conversation of a Husband of my own Principles which yet I must Own is a great Loss. I hope thou wilst be pleased never the less to continue thy favourable Opinion towards me and to condescent to Accept this Small toke of a dozen & a half of Lemmons. . . . [It is] my humble desire of the favour of a Line from thy hand (if not too great a trouble) that I might hear ye are all well.[86]

Not all the servants found Swarthmoor a haven. A young servant at the Hall, Edward Herd, discovered that Sarah and her mother were strict taskmasters. He was warned by the Swarthmoor Women's Monthly Meeting in February 1674, "to stay att Swartzmoore and to behaue himselfe as besemes truth and a servant that truth does not suffer by him." It appears he continued at the Hall as a servant throughout 1673–74, for Sarah supplied him with clothing, a "frocke" that was made up for him by Matthew Fell, tailor, for 2d (pence).[87]

Fell's special concern for Quaker servants is seen in her care of Mary Benson, a Quaker orphan of ten to whom we will return in a later chapter.[88] Mary had been raised by Quakers and had, for several years, been passed from one Quaker household to another to protect her from being tainted by the "world." Margaret Fell offered to take Mary Benson for a year as a servant. Fell displayed a parental concern for the servants in her home who became or remained Quakers. In fact it is highly likely that all the servants in her household were children of Quaker families. Her religious ideology directly affected her sense of loyalty and her mothering role to her servants; believing that in the Kingdom of God all were alike, the mistress of the Hall attempted to treat her inferiors with kindly indulgence. Those who moved outside of the Quaker flock ceased to be members of her Quaker family.

* * *

Nonconformist religious behavior in early modern England and Europe had common characteristics, providing useful parallels to conceptualize and understand Fell's religiously motivated behavior and that of her co-religionists. In investigating the emerging Protestant ideology in mid-sixteenth-century France, Donald Kelley has interpreted the social aspects of sectarian religious behavior of the French Protestant movement. He has argued that the concept of the "flock" or religious community of like-minded persons in early French Protestantism was able, through group identity, to transcend the generational, educational, and class differences of their times. The religious "flock" provided, in spite of political and religious turmoil, the soil or social foundation in which French Protestant ideology took root. By charismatic leadership, pastoring, and preaching, the early French Huguenots bonded the flock into a mutual community of antihierarchical, antiestablishment outlook. Thus a nonconformist congregation was gradually forged into a group "with its own rules and ideals, living in a world apart." Contemporaries perceived this as an internal threat to the religious-political status quo.[89] Kelley has found that in sixteenth-century France, such a religious sectarian phenomenon was disruptive to the family and local community. Likewise, in the following century, Quaker sectarian loyalties disrupted familial bonds and tended to replace the natural kin loyalties with a new family of the faithful. This family with a new consciousness and new values based on like religious goals, has its theoretical foundation in the Quaker statement, "it's better to forsake wife and children and all that a man hath . . . for Christ and the truth sake."[90]

Quakers used the word "flock" to express that special spiritual bonding strong enough to replace natural family ties. The Quaker missionary, Ann Curwen of Lancaster, a contemporary of Fell's who traveled to the American colonies, expressed the religious conviction and understanding of home and family in this transformed sense of consciousness. In an epistle to her own Swarthmoor meeting from Huntingdonshire in January 1677, Curwen stated her loyalties:

> My dear Friends [Swarthmore Meeting], with whom I would be glad that I might set down with you once more, that we might sing together of the Mercies of the Lord . . . but I do not see my service [missionary travels] over that I may come to my Place of Abode in the outward. . . . Oh my Dear Friends! breath with me to the Lord for the Prosperity of the great Work, which the Lord hath made me Instrumental in; though but weak and little. . . . I dearly salute you, . . . and remain as one of God's Flock, and that Family where we were begotten to God, for which my Soul shall Bless the Lord.[91]

On the practical level, the new religious fellowship offered Quaker women ministers such as Alice Curwen greater freedom from the family hierarchy. They found new opportunities to express themselves and gain influence through their ministry. If married, they could avoid continual pregnancies and practice the ascetic life within their newfound religious community. Fell was atypical in her religious community insofar as she was already head of the family hierarchy. Aside from that, the Quaker fellowship appealed to Fell for reasons of self-expression, enhanced personal influence, and pious living that rejected the Puritan doubts, fears, sense of sin, self-introspection, and morbidity of more traditional Protestant female piety.[92]

The existing evidence of Fell's relationship with her first husband, Thomas Fell, suggests that some tensions existed regarding her religious beliefs. Had the Judge lived longer, it is entirely possible that he would have been heavily pressured to convert to Quakerism in order to preserve family harmony, or his marriage with Margaret would have become increasingly strained once she gained notoriety as an authoritative writer, preacher, and prison reformer for Quakerism. The other side of this speculative coin is the question of whether Margaret could have carried her religious work as far as she did had he lived. It is apparent that her only son had her loyalty until he opposed her religious principles and social liaisons with the "dispised Quakers." From that point on, she looked upon her son with what appears to be little or no maternal affection and fought, without firm legal grounds, to win the Hall in order to continue her base for Quaker missionary activity. For Margaret her "eternally beloved lambs and babes of God" were her Quaker daughters and grandchildren and the local "flock" of believers. Her natural daughters became as devoted to Quakerism as she; thus their relationship remained exceptionally close. As the story of Fell's life and influence unfolds, I hope further evidence will support my theory that their mutual loyalties were transformed and, to a significant degree, became rooted in a new family fellowship where "spiritual communion and group identification"[93] took precedence over natural family ties and age differences.

Part II
The Domestic and Economic World of the Fells

3   The Swarthmoor Farm

Swarthmoor Hall, a late Elizabethan or early Jacobean house, is situated among beautiful ancient cedar trees allegedly planted by Judge Fell at the birth of each of his children. The Hall is located one mile southwest of the old market town of Ulverston in the Furness district, north of the sands of Morecambe Bay. The Furness district, and Ulverston, lie northwest of the city of Lancaster, and due west of North Yorkshire. Surrounding Swarthmoor are treeless rolling hills or moors, a transitional agricultural zone, which gave rise in the seventeenth century to mixed farming.

At the death of Thomas Fell in 1658, his estate went to his widow along with some provision for each of his seven daughters. Judge Fell's will stipulated that Margaret was to inherit fifty acres surrounding the Hall, with its barns and buildings, as long as she remained a widow.[1]

THE HOUSEHOLD ACCOUNT BOOK OF SARAH FELL (ABSF)

After Margaret Fell's conversion to Quakerism, and especially after her spouse's death, the pattern of life in the Fell family centered around their Quaker fellowship and its social, economic, and religious commitments. The evidence of the Swarthmoor farm accounts, kept by Margaret Fell's middle daughter Sarah, is of great value as a first-hand source in attempting to reconstruct the household economy and living pattern of the Fells.[2] These accounts were undoubtedly kept for several decades but now exist only for the years 1673–78. The ABSF yields information of an economic, social, and, indirectly, a religious nature regarding the Fell family between 1673 and 1678. Although the picture given by this ledger is incomplete, it does tell us something about the inner organic structure of this Quaker landed family and its network. It reveals something of the Fell women's day-to-day business transactions, their farming and managerial skills, their charitable behavior, their hospitality, their social interactions and aspirations, and their religiosity. Moreover, it sheds some light on approximately two hundred women and numerous men whose names recur in the accounts, revealing something of their productivity and life circumstances in an underpopulated area of northwest England.[3] This extraordinary ledger provokes several questions. In the decade of the 1670s what resources did the Fell women have and how did they use them? As an active Quaker woman, how did Margaret Fell's domestic and religious experience dovetail? What persons constituted the social network operating around the Fells, and how deeply did Quakerism penetrate Margaret's social network? Evidence of the female and Quaker component of the Fell network is culled also from the Swarthmoor Women's Monthly Meeting Minutes (SWMMM), for a substantial number of women appear in both sources. Using the female and Quaker aspects of the farm accounts and minutes of the Swarthmoor women's meetings, this chapter will provide a clearer understanding of Quaker women's roles and attitudes, as they relate to the Fell family economy during this period of intermittent and sometimes severe Quaker persecution.

THE FELLS' NEIGHBORS AND KIN

The account book of the Swarthmoor farm yields information that strongly suggests that the Fells pursued an undeviating Quaker lifestyle, maintaining a close Quaker network – social, religious, and economic. However, the ABSF also reveals that despite the separate lifestyle of this Quaker family and the local persecutions of Quakers, some non-Quaker social and business connections did occur between the Fells and their neighbors although on a limited basis. For instance, the Fells had some involvement with neighborhood squires such as the Doddings of Conishead and the Preston families, one of Holker Hall and one of "the manor" – both related. They were even on friendly terms with one branch of the Kirkby family, despite the fact that Colonel Richard Kirkby of Kirkby Hall and Daniel Fleming of Rydal Hall were notorious as persecutors of Quakers. However, "old J[oh]n Kirkby of Connistone," an elderly bachelor and nephew of Colonel Kirkby, lent Sarah Fell money on at least three occasions. An adjacent neighbor of the Fells, Miles Dodding of Conishead Priory, owned a house of thirteen hearths, the same number of hearths as Swarthmoor Hall. These two houses were the largest in the area. Miles Dodding and Thomas Fell both served in Parliament during the 1640s and, although known not to share the same political and religious views, the families maintained a friendly social exterior. During the 1670s these neighborly connections were still intact, for Sarah Fell purchased coal and peat moss from Dodding, while he paid her for some bottles she purchased for him. In July 1674 Sarah Fell loaned Dodding £30 for ten days, while in 1675 Sarah borrowed and repaid £40 from his mother, Sarah Dodding. This reciprocal trust was in accord with traditional neighborly gestures among social equals. Robert Sawry of Broughton Hall, eight miles north of Swarthmoor, was an outspoken Baptist nonconformist whom Sarah and Rachel Fell visited at least once, giving Sawry's servants 1s. During the period covered by Sarah's account book, the Fell women made a visit to the Roman Catholic Anderton family of Bardsea Hall and paid 6d to either the owner or the servants.

Although the Fells maintained some semblance of neighborliness with the other nearby gentry, the overall picture given in the accounts is one of the Fell family's isolation from their own social strata. For instance, the Prestons of Holker Hall near Cartmel opposed Fox and the local Quakers. Preston's wife Katherine visited Fox when he was in Lancaster Castle about 1660 and, in Fox's words, "used many abusive words to mee and tolde mee my tongue shoulde bee cutt out and I should bee hanged." In 1663, while Fox was staying at Swarthmoor, he visited Colonel Richard Kirkby at Kirkby Hall to confront his adversary face to face. Colonel Kirkby, a local constable, was a known opponent of Quakers. It is also known that a local justice of the peace, Curwen Rawlinson, relative of Fell's friend Thomas Rawlinson, opposed Quaker activity at Swarthmoor. Neither the Prestons of the Manor nor the Andertons of Bardsea Hall shared Fell's religious views. The traditional neighborly network of reciprocal social obligations among equals appears to have been minimal.[4]

We have seen that within the family the Fells retained a tight nuclear loyalty among the married and unmarried daughters, sons-in-law, and their children. The married daughters returned home for visits and the grandchildren were left in the custody of their aunts and grandmother. The wider kinship network, although in evidence, was of limited importance. Paternal kin are never mentioned in either account book or letters; maternal kin are mentioned only infrequently in the accounts. Judging from Sarah Fell's account book, Margaret Fell's brother-in-law, Uncle Matthew Richardson of the nearby town of Dalton, was closer to the Fell women then his wife, their blood aunt, first name unknown. Margaret Fell's sister appears to have been incapacitated by illness during the 1670s. Uncle Matthew Richardson sold iron ore to his niece for Force Forge, and Sarah took care of her uncle's funeral arrangements in 1677. As an attorney, he was occasionally consulted by the Fell women, and George Fox suggested that Richardson advise Fell on her rights to Swarthmoor and Marsh Grange in 1669–70 in the dispute with her son. Margaret Fell also drew her brother-in-law into her long-term business dispute with her erstwhile Quaker friend, Thomas Rawlinson, a subject to which we will turn later.[5]

GENTRY LIFESTYLE AT SWARTHMOOR HALL

The most apparent characteristic of the Fell women's gentry lifestyle is seen in their style of dress and their financial ability to make frequent travels. Fell was known to dress according to her social status, and letters written to her give effusive descriptions of her appearance and manner, reflecting in part the style of the times. Although the compliments seemingly refer to Fell's spiritual beauty as opposed to physical appearance, they do contain a gendered component. However, considering the fact that Quakers rejected "superfluities," the various descriptions of her appearance are curious. Examples of her personal impact are captured in the expressions of Richard Waller and Richard Roper, two Quakers in Waterford, Ireland, who wrote in 1657: "O thy beauty and thy comlyness passes and exceeds many." Also, Quaker John Audland addressed her, "thou art clothed with beauty, with glory and wisdom from on high. Oh thou fairest among women, blessed art thou for evermore." William Caton was perhaps most effusive in his letters to his mentor and adopted mother. To him she was "a precious jewel in the hand of the Lord, and in the eyes of His lambs," who possessed "glory, excellency, wisdom, knowledge, understanding, and manifest virtues." He remarked further, "these make thee amiable in my eyes, and in the eyes of the whole Israel of God that beholds thee." Thomas Salthouse called Margaret "a vessell of honour" and "glorious one: Thy glory is within, and thou art made gloryous in Righteousnesse." Salthouse wrote from Exeter gaol in 1655 that she was a "dear and tender-harted nurseing Mother to all the babes in Christ who are begotten of the Immortal seed . . . in thee is my life."[6] It is apparent that the descriptions of Margaret by her co-religionists were cast in biblical and formulaic style.

These effusive remarks to Margaret are curious because the early Quakers seemingly repudiated outward forms of speech deference only to replace them with biblical effusions when the need arose. In light of the early emphasis on an economy of words when speaking, and avoidance of the superficial forms of address, these numerous outpourings regarding Fell give a strong indication that she was an imposing figure who awed people in their personal encounters with her. When John Camm and Francis Howgill visited Oliver Cromwell in 1654 and delivered a letter to him on her behalf, Cromwell "receive[d] it thankfully" and described her as "a good woman; he had heard much of [her]." A hostile description survives in the State Papers of John Thurloe. The non-Quaker Timothy Langley, writing in 1658, describes Margaret's outward bearing and appearance thus: "Margaret Fell, who livs about Lancheshire, who, they say, is judge

Fell's wife; and that she is one that is past the cloud, and hath liberty to were [wear] satins and silver and gold lase, and is a great galant." It is apparent that Margaret was concerned about her appearance, for Fox sent her some lace from Ireland in 1669. Also, daughter Sarah, in a letter of December 1683, described a dress to be made for her mother which was in keeping with current fashion: "We advise you to make my mother's cloth gown without a shirt, which is very civil, and usually so worn, [in London] both by young and old, in stiffened suits." Fell was fond of dressing in black, for Sarah's account book described a black silk gown to be tailored for her mother. Once in 1673 she sent her second husband a gift of £3 when he was on his travels, but he did not spend it on himself. Rather, Fox wrote Margaret that he "did speake to a frend to send the[e] as much spanesh black cloth as will make the[e] a goune and it did cost vs a prety dell [deal] of mony."[7]

Margaret Fell never agreed with the Quaker plain dress code that became an issue in the 1690s. Late in life she wrote an epistle against legalistic outward rules, claiming that it was the inward work of God in the heart that counted. Fell and her daughters dressed according to their social station, claiming "gospel freedom." Their accounts substantiate this with continuous tailor bills for mother and daughters. The Fell women were fond of colors, as is evident in bills from the dyer describing "dying sea-green and sky colored stockings for Rachel and Sarah." There were frequent purchases of "scotch cloth," or colorful tartans, many yards of multi-colored ribbons, black silk points, black silk laces, black and ash-colored ferrets, and sky-colored worsted. Sarah recorded a typical purchase in October 1676: "2d for 3yrds of white ribbin 6d for 11 yrds of greene ribbin 3s 8d for a P[air] of garters 6d forr a lethr purse and a lase 3d." Although some Quakers, such as William Penn, were criticized for their outward dress, there is no indication that Fell and her daughters ever forsook dressing according to their station in life or that they were openly criticized by their co-religionists for doing so.[8]

The Fell women traveled probably from the beginning of their Quaker affiliations, if not before. In 1652 Margaret's newly convinced Quaker daughter Bridget visited York, probably with one of her sisters. An observer in York commented that it was a "rough, profane place" for Judge Fell's daughters to visit; "better that they be at home with their mother." Margaret Fell made ten trips to London between 1660 and 1697–8, and when she traveled the length and breadth of England on her prison mission in the 1660s, her daughter Rachel accompanied her. Some recent historical studies have revealed that the English were surprisingly mobile, and that country folk moved in a much wider radius than their village or parish in their economic and social activities. However, according to some comments in Quaker correspondence from 1652 onward concerning the traveling Fell women, the mobility of a married woman and subsequently a widow of gentle birth and her unmarried daughters may have been considered somewhat out of the ordinary.[9]

SARAH FELL'S ACCOUNTING METHODS

The transactions recorded by the keeper of the accounts between 1673 and 1678 were not a casual work. Sarah Fell took great care to create a well-ordered domestic record, which is clearly reflected in her accurate computation and legible hand. She recorded all income and expenditure with precision, including a brief explanation for each item. The document therefore gives us a valuable microscopic view of what life was like in the primary entity of seventeenth-century English society, the household unit – albeit a variant form thereof.[10]

In the 1670s while Margaret Fell and George Fox may have been sitting in the manor house conversing, thinking, and writing on religious matters (when Fox was periodically in residence), Sarah, as chief bailiff or steward of the farm, was providing them with their wines, tobacco, oranges, and other luxuries and necessities. As bailiff, she lived seriously her religious philosophy, that of being a good steward of the Lord's bounty. Her accounts show the assiduous counting of every penny and farthing with considerable savings accruing through her careful management. Undoubtedly, Sarah Fell was also motivated by the practical fact that in a gentry family in which preservation of the patrimony and economic solvency were entirely in mother's and daughter's hands, hard work and scrupulous conservation of resources was necessary in order not to sink into the poverty that surrounded them. There were other affluent families in the Furness district – such as the Doddings, Prestons, Kirkbys, and Flemings mentioned above – but Furness was remote and economically underdeveloped. Poverty was a way of life for the majority of the families whose names come to light in Sarah's accounts.

In the 1650s Judge Fell had employed Thomas Salthouse of Dragley Beck, Ulverston, as his bailiff at Swarthmoor. Salthouse, one of the Fell household whom George Fox convinced in 1652, subsequently became one of the early traveling preachers, but he continued to maintain a close personal bond with Margaret Fell through letters long after leaving the Hall. Chief bailiffs of landed estates were often men, but women were known to keep the accounts of their family estates. Margaret Fell undoubtedly taught her daughters the requisite farm management skills needed by any mistress of a large household in order to keep the farm productive and solvent. Sarah Fell stands out as a seventeenth-century woman of superior managerial skills and literary ability. Margaret was often absent from home during the period the ledger covers, but the farm enterprise under Sarah's management showed no stagnation or reduction in productivity during her mother's frequent religious pilgrimages. Sarah probably took charge of the farm while her mother was imprisoned in Lancaster Castle between 1664 and 1668. Bridget had married and left home in 1662; a year later Isabel married and moved to Bristol. By 1664 Sarah was about twenty-two years old; she remained bailiff of Swarthmoor until 1681 when she married William Meade, a London Quaker.[11]

Sarah Fell kept her cashbook in the ordinary double entry format of credits and disbursements. For example, the first complete entry column of the accounts, dated 18 through 30 October, shows, on the left-hand "credits" page, a previous balance of £59.10.01, followed by a list of receipts for that two-week period with a total of £81.19.03 ¾ on 30 October. On the right-hand "disbursements" page for the same period, dated 16 October through 1 November, the money disbursed in goods, services, and loans totals £81.19.03 ¾. Sarah Fell balanced the account every two weeks, with occasional variations to this schedule. She followed this procedure throughout the book, making the account book a source that speaks with considerable authority. However, its value is somewhat limited because the ledger offers us only three complete years of accounts to use as the basis for any analysis. The remaining three years contain large time gaps where the record is missing. The book opens on 25 September 1673. The year 1674 is complete, but 1675 has two gaps amounting to almost six months. The years 1676 and 1677 are complete, but 1678, the last year on record, lacks seven months as indicated in Table 1 of the Appendix. As the accounts are incomplete, discrimination in their use is necessary. With this caveat in mind, the accounts have been used both quantitatively and descriptively. I have chosen, from the three complete years, the year 1676 as a reasonably representative sample of the accounts. During the year 1676 there were approximately 210 persons who carried on a social or business relationship with the Fells. Of this number, at least fifty-two were women, and fifty-four persons were known Quakers. This 25 percent Quaker total undoubtedly underestimates the numerical strength of the Quaker network surrounding the Fell family in 1676.

When the account book opens (with old-style dating) in March 1676, the following persons were residing at Swarthmoor Hall: Margaret and

George Fox; three unmarried daughters, Sarah, Susannah, and Rachel; and Sister (Isabel) Yeamans, a recent widow with her two children, Will and Rachel Yeamans. Little Rachel Yeamans died on 20 June 1676; one simple entry for death expenses marks her apparently unexpected death. At the other manor house, Marsh Grange, which was part of the estate, sister Mary Lower was in residence with her husband Thomas and at least one living child, Humphrey, who was born in 1676. It is possible, but doubtful, that Margaret Fell's estranged daughter-in-law, Hannah Fell, and her two children were also at Marsh Grange. The household servants probably numbered twelve in residence at either the Hall or the Grange in 1676, six males and six females.[12] Among the farm laborers hired on a day-to-day basis and fed but not boarded by the Fells were approximately twenty-six men and eighteen women, as seen in Table 2 in the Appendix. The comings and goings of thirteen artisans or craftsmen doing specialized jobs such as carpentry, tailoring, chimney sweeping, smithwork, and joining appear in Sarah's cashbook in that year. There was a midwife called for Mary Lower as well. The total number of male and female servants and day laborers employed in both households was close to fifty-five persons with 43 percent of the total being female.[13]

One atypical feature of 1676 was the presence of George Fox at Swarthmoor for the entire year. It was his longest sojourn at the Hall in his forty-year relationship with Margaret Fell. In analyzing the disbursements for 1676, a category for Fox's expenses in clothing and other items for that year indicates that his living expenses were not excessive. In terms of the Fell family economy, he was a peripheral figure who did not participate in the workings of the farm and whose presence did not alter the farm enterprise aside from his basic food and clothing costs. There was one observance recorded in 1676 that probably occurred because of his presence at the Hall; namely, the visit of William Penn, for whose visit Sarah recorded the purchase of a "fatt sheepe." The accounts also record a postage payment of 1.2d (pence) for a letter written to Penn among the other postage payments for Fox throughout the year. The only other observable changes apparently due to Fox's presence at Swarthmoor were more frequent purchases of white wheat bread; possibly a somewhat greater variety of food, such as oranges sent from Lancaster; medicinal juniper berries for "Father"; and runlets of wine, purchases made infrequently or not at all in the other two complete-year inventories. Fox's frugal lifestyle at the Hall conformed to his philosophy of avoiding "superfluities," not only in speech but also in the fashions of the worldly.[14]

THE SWARTHMOOR FARM LABOR FORCE

Swarthmoor was a part of the local cash economy, deriving its income largely from cattle raising and mixed farming. The Fell farm was not self-sufficient, for there were weekly and biweekly excursions to Ulverson market. On Thursdays the Fells purchased provisions at the Ulverston market such as meat, fish, salt, white bread, eggs, butter, wine, and yeast.

Entries in the accounts show that perishable products produced on the Fell farm such as cheese and butter were peddled at market or from house to house by women of the neighborhood. Alice Clark's study of seventeenth-century working women uses the term "regratress" to describe this class of women. This word, which had its origin in the Middle Ages, refers to a class of peddlers often consisting of widows and wives of wage-earning laborers, who, lacking their own plot of land to produce foodstuffs to sell, sold the goods of larger farmers, hoping to make a small profit. Clark points out that this peddling business was a nationwide, predominantly female phenomenon in the seventeenth century. Although regratresses were frequently looked upon by hostile business people as "hawkers" of large profits for the produce they sold, Clark argues that it was a bare survival enterprise. The ABSF supports Clark's argument, for the women that appear as regratresses in the accounts were largely unskilled and poor. Their families depended upon them to supplement the laboring wage of the husband, or as widows with families they were forced to eke out a meager existence by whatever means available. With the fluctuations of the economy, demand sometimes exceeded supply in seventeenth-century market towns. As a result, Clark indicates, the cheese, fish, and poultry peddlers and the sellers of butter, salt, eggs, and other small perishable items were charged with creating shortages. People believed the peddlers were responsible for purchasing large quantities to sell, thereby raising the price.[15]

The Fells, as larger farmers, supplied neighborhood petty retailers with produce, often on credit with easy payment terms. Approximately fifty-two women had business dealings with the Fells in 1676, and a good number of these women were Quakers. Exact ages of these women are unknown, but the age range was apparently wide. As a group, the women who sold for the Fells appear to have been a working network of individuals whose dealings with one another operated on the economic as well as social and religious levels. For example, Ann Dixon of Cartmel, whose name appears in the minutes of the Swarthmoor Women's Monthly Meeting and the account book, paid the Fells in installments, indicating that she marketed the produce before paying off her creditors. Dixon purchased, on credit, oats from Marsh Grange, as well as cabbages from Sarah Fell at the Hall. She was also in the employ of the Fells until 1674 when she was given a 1s gift before her departure. Sarah Benson and Bridgett Cowell purchased cloth from the Fells, giving us a glimpse of the home textile production at the Hall. Furthermore, these two women come to light again in the SWMMM between March and May 1674, both having been reproved for "disorderly walkeinge wch gives occasion to ye world to speake evill of ye truth." They were at odds with the Women's Meeting for their peddling activities, which included the "sellinge of lase, which is needlesse, and ffreinds cannot owne them in it, nor ye coveteous spiritt, that sells it for advantage."[16] This comment is especially interesting considering that Fell was described by a non-Quaker as a woman who wore lace, and the account book indicates frequent purchases of lace.

Other petty retailers enter and exit Sarah's household ledger. Jenny Colton was a member of the Horslydown monthly meeting and was frequently supplied with produce that came from the Fell farm. She made payments for chickens in July 1674, for beans and butter from Marsh Grange in August 1674, and for "new wheate of Mothers" in January 1674, April of 1675, and June 1676. Sarah Fell advanced her credit without interest at regular intervals. For example, in March 1673 she gave Jenny Colton a loan "upon her manner [manure]" of 1s. In 1674 she loaned her 6d to buy butter, and on one occasion in 1674 she was advanced credit by the "women's stock" or treasury of the women's meeting.[17]

Ann Geldart, whose name recurs most frequently in the Swarthmoor ledger, was the wife of George Geldart of Ulverson. She too acted as a regratress, and her family received aid at various times. Ann sold cheese for the Fells, making several payments between 1673 and 1674. Sarah lent her 3s in December 1674, which she repaid without interest in January. A versatile person, Ann sometimes peddled honey for the Fells and occasionally was a wage laborer "setting and dressing peates." She also sold chickens, butter, and eggs; delivered items for the Fells; pulled hemp and worked hay; darned stockings; and brought manure to the Hall. Geldart borrowed money on several occasions, often making partial repayments over extended periods, and always without interest. At the same time she continued to receive payment from the Fells for her other work. Also in May 1674 Sarah "pd 6d. to Ann Geldart for Jenny ffel [Fell] reeinge [sifting grain] here 3 days." It appears Sarah Fell trusted Ann to carry the money to Jenny, indicating a bond of honesty, characteristic of the Quaker community.[18]

Ellen Pollard, a servant of the Fells, acted as a cheesemonger, selling, over a three-year period, cheese amounting to £6.19.06. Although she was employed at both the Hall and Marsh Grange for at least one year, her household service included learning the regratress technique. Ellen paid for at least part of her own clothing, for the account book shows her repaying Sarah 3s for cloth dyed in 1673. It appears that Sarah supplied female servants with the necessities – "shifts" and shoes – but anything extra was not provided. Ellen's annual wage was £2, and although her year ended May 1673, payment was not given until the following January, suggesting possibly a tight money supply or disciplinary action. Not only was her wage paid nine months after she left the Hall, but 8s was deducted to pay "Margt Kirkby for ye time Elin was sicke." However, Ellen was given a parting gift of 1s, a Fell tradition to their departing servants.[19] The Swarthmoor bailiff attempted to be fair but was obliged to be parsimonious as well in order not to overspend the limited income of the farm.

Women receiving wages or aid from the Fells were expected to recompense this credit in responsible ways. According to Sarah Fell's reckoning, Ann Standish failed to be responsible. In 1675, while she was in their employ as a household servant, she lost a silver spoon and broke a pot. At the year-end tallying of her wage of £1.17.16, 7s6d was deducted to cover the cost of the missing spoon and broken pot. Nevertheless, Rachel gave her the usual going-away goodwill bonus of 1s.[20]

Peggy Dodgeson frequented the Fell farm looking for work. She was as strong as a man for she was one of a few female employees who did men's work. Nonetheless, her wages usually conformed to the female laboring wage, about 1d (penny) per day. She toiled "filling manr [manure], scailering manner, washinge, harrowing, and dressinge meadowes" for seventeen days, being hired also on a day-to-day basis "rubbinge, as well as dayworke of peates, or working at hay, or pulling hempe and line." Peggy Dodgeson also washed "closes" and did other jobs for six days, for which she earned 8d. Her religious persuasion is unknown. Her name does not appear in the SWMM records as either member or pensioner, and whatever her religious persuasion, her loyalty to the Fells won her credit. She too was given small interest-free loans.[21] These women whose lives were ordered by the daily need to work in order to augment their husband's income or to eke out a living as single women looked to the Fells to employ them, sell to them, extend loans to them, and, as a last resort, give them charity.

ECONOMIC ACTIVITIES OF THE FELLS

The Fells derived the principal portion of their income in 1676, as in the other five years, from agriculture although there were some other possible sources of income. One was rental income, which, according to the account book, was minimal.[22] A second source of income was Sarah Fell's business ventures, including family ownership of an iron forge and a shipping venture via Bristol. These appear not to have been long-term, steady sources of income. The shipping venture in farm produce, it appears, ceased by 1676, while only partial income and expenses for Force Forge appeared in the accounts.[23] Swarthmoor, as a 156-acre farm, did not appear to operate on the manorial rent system, as did Sir Daniel Fleming's lands situated nearby to the north, for example. The only safe observation one can make is that the Swarthmoor accounts present a picture of a mixed farming enterprise, which derived its maximum income from livestock and arable production. The Fells probably were not land renters as the gentry commonly were. They did not acquire new lands during the 1670s, nor did they make wealthy local marriages to augment their property holdings, typical of seventeenth-century landed families and affluent yeoman families.[24]

The agricultural income of the Fells during the year 1676 derived heavily from livestock raising. They sold at least thirty-one sheep, nine cows, six steers, three horses, three colts, and three oxen. They also raised chickens and pigs and were well known in their area as cattle breeders – a steady income was earned from "cow bulling" over the period covered by the accounts. The farm produced several grain crops, namely, wheat, barley, oats, and corn. Margaret Fell's wheat brought in a steady monthly income, indicating that barn storage was necessary either on her own farm or in rented barns of local farmers. She sold "old wheat" into March when the "new wheat" sales began. According to Sarah's record, a total of eleven bushels, thirty-three pecks, and nine hoops were sold over that year.[25] Other produce sold from the Hall or Grange included peas, cabbage, beans, and potatoes. The house also had a garden, an orchard, and a beehive from which produce was sold at market. Low-lying, undrained marshes or peat bogs existed on or near the estate, for there was a plentiful supply of peat moss that laborers or servants cut and dried for fuel. The Swarthmoor farm was situated in the transition zone between the northern moor pasturelands, ideal for cattle and sheep, and the mixed farming lowlands where arable agriculture took place. The most pronounced characteristics of this mixed farming enterprise were its livestock breeding and its grain and crop production, requiring the fields to be annually manured. Dairying was important because substantial amounts of cheese and some butter produced on the farm were sold at market. The Fell's yearly income conformed to the agricultural cycle, showing greater expenses at seed time and harvest time each year.[26]

WOMEN'S ENTERPRISE AT SWARTHMOOR

These accounts shed some light on the enduring relationship between home and work in a seventeenth-century woman's world. At Swarthmoor, women's tasks included baking, spinning, and carding. The spinning of farm rope and sacks was undertaken, presumably to carry iron ore, grain, and produce. The making of homespun cloth for blankets and clothing and the knitting of work socks and garments indicate what Alice Clark terms both a "domestic industry," where goods were produced for family use, as well as "family industry," where goods produced by the household members were sold or bartered. Such work not only contributed to the local economy, it also provided the entire Fell household with necessary items.[27]

It appears that Sarah was not especially innovative in her farming methods. She did not cultivate increasingly popular fodder crops, like turnips and clover, which increased manure production and in turn increased corn or grain production. Instead Sarah paid her poorer neighbors to cart in manure. The Fell women did not specialize in sheep-corn farming as practiced by larger specialized corn farmers in the seventeenth century. They did not grow dye crops; instead they paid rather dearly for imported dyes such as indigo. However, the Fells were not completely isolated from outside agricultural-economic advice. They grew flax and hemp on the farm, probably having been influenced by earlier seventeenth-century economic pamphlet advice, which inveighed against paying money for foreign imports, especially for Dutch goods during the war years. Likewise, their orchard and spice garden was something encouraged by the farming pamphlet literature of the later seventeenth century.[28] Population in the latter half of the seventeenth century was on the rise, contributing to a general trend for farms to increase their quantities of produce, grain, and livestock. Productivity on the Swarthmoor farm appears to have fluctuated in the 1670s, but data is insufficient to determine any trend.

INCOME, DISBURSEMENTS AND SAVINGS IN 1676

The total Fell farm income for 1676 was £406.09.05 (see Table 4, Appendix). A useful contrast to the Fell's income is the income of Ralph Josselin, a farmer in the south of England. Josselin was a clergyman and substantial yeoman landholder of Earls Colne, Essex, and his calculated income was at £160 per annum in the 1670s. For both the Fells and the Josselins there was a large untallied per annum income in terms of food grown and consumed on the farm, as well as homespun clothes made on the premises. The hidden expenses, which include the cost of feeding the family and labor force at Swarthmoor, are therefore only partially listed in the accounts, leaving significant cost gaps in our measurement. Macfarlane estimated that the maintenance cost of adult members of the Josselin family was £10 p.a.[29] If we use £10 as the basic cost figure per adult p.a. for the Fells in 1676, then the estimated cost of maintaining the household of eleven immediate family members (adults and children), and twelve servants, amounted to £230 p.a. In Table 4, of the Appendix, a breakdown of the expenses of the Fells for 1676 into nine categories of disbursements gives us a picture of how they spent their income for that year.

Table 4 points to the fact that the total annual expenditure of £230.10.01, was far less than the Fell p.a. income of £406.09.05. The discrepancy seen here can be accounted for by at least four possible reasons. Firstly, much of the domestic industry is not recorded in the expenditures. Secondly, there was a credit of £94 tallied to George Fox's account in December, which was incorporated into the Fell's annual income. Thirdly, several transactions in the accounts were of a miscellaneous nature – small gifts and items that each sister purchased on her own account – and were not figured into the categories in Table 4. Fourthly, Sarah Fell's investments in Force Forge yielded some income from iron ore during the year. Between 1658 and 1681, first George Fell, then his mother Margaret and four of her daughters, owned Force Forge, of which more will be said in a later chapter. Sarah's account book records iron sales in the 1670s, but the record appears to be an incomplete measure of forge income, as Sarah seemingly recorded only southbound sales. Sarah acted as intermediary for sales in her areas, keeping some of the profit and reinvesting a portion of the receipts.[30] Many items produced at the forge were used at the Hall, including windmill spindles, axes, chisels, hammers, square iron forgrates, a grindstone axeltree, a waterdish, curtain rods, window stanchers, nails, and a frying pan. Fifthly, Sarah acted as banker for her Quaker neighbors and friends, dealing with people as far away as London. Her financial operations in loans, receipts, and payments, maintained mostly in a local radius, may well have required a significant amount of capital from which to draw at a moment's notice. While she entered into several business transactions with Thomas Curwen of Lancaster and London, the bulk of her loans and debts was contracted with the local Quaker community. More will be said of Sarah's banking activities in the next chapter.

The total income, disbursements, and savings for 1676 support at least one reasonable conclusion: The Swarthmoor farm productivity was sufficient to allow the Fell family, as lower-middling gentry farmers in terms of wealth and land, to be reasonably well-off especially in their district and to be influential in the surrounding neighborhood. Margaret Fell held enough land in the 1670s to employ a sizable labor force and to maintain the family within their gentry status. Her daughter overseer was content with that.

THE FELLS' SOCIAL WORLD

Although the Fells were farmers of lower-middling gentry status, in the decade of the 1670s the Fell family style of life was not consistent with that of the gentry. Because of their Quaker theology, a contrasting pattern of rural social relationships developed. Prior to Judge Fell's death, there is no reason to believe the Fells were not included in the social network of the local gentry. Thomas Fell was on friendly social terms with other county and north-country notables and officers, such as Justice Anthony Pearson of Ramshaw Hall, Durham, a man of some worldly stature. In 1652 or 1653, after presiding over the trial of James Nayler, Pearson met Fox and then Margaret Fell at Swarthmoor Hall. Margaret Fell's impact on him was considerable, and Pearson became a Quaker in the 1650s.[31] His friendship with Thomas Fell was a close one, for Fell appointed Pearson executor in his will. Fell named another Quaker, Gervase Benson of Heay-Gray, York, and justice of the peace and squire, as his second executor. Both of these prominent north-country Quakers were considered by Thomas Fell to be gentlemen and his "very true friends." Other gentlemen friends included Thomas Knype of the Manor and William West. Colonel West, like Thomas Fell, was also on friendly terms with the local Quakers but was not one himself. Fell's last will conveyed a salute of deference to Lord Bradshaw, "my very honorable and noble friend." He bequeathed Bradshaw £10 to buy a ring, "as all the acknowledgement I can make, and thankfulness, for his ancient and continued favors and kindness . . . since our first acquaintance." Lord Bradshaw was most likely John Bradshaw, a regicide, and second son of Henry Bradshaw, gentleman of Stockport, Cheshire. It is possible that John Bradshaw and Thomas Fell became friends while attending Gray's Inn in London, for Bradshaw was called to the bar at Gray's Inn in 1627, four years prior to Fell.

One feature of the lifestyle of the rich and powerful was owning or renting a town house in London. Thomas Fell probably did not own property in the city, but as an M.P. he signed in 1648 a seven-year lease for £5 p.a. for chambers in a fashionable area close to Gray's Inn.[32]

After Margaret Fell's conversion, and especially after Judge Fell's death in 1658, she excluded herself, with few exceptions, from the world of the gentle born. Her Quakerism initiated a radical reshuffling of her social and emotional bonds. As Richard Vann has pointed out, families were often fragmented in early Quakerism, especially landed families where the eldest son remained true to the Anglican church. We know that bitter estrangement between the Fells, mother and son, occurred because of religion and money matters. Following her son's death in 1670, Fell became totally estranged from her daughter-in-law. Margaret Fell's will made permanent these family divisions. She singled out her Quaker heirs in bequeathing the estate to Daniel and Rachel Fell Abraham and in giving small legacies to her grandchildren, with discrimination based on religious affiliation.[33]

Margaret Fell maintained close ties to household servants who were convinced Quakers from 1652 onward, such as Thomas Salthouse, Leonard Fell, Richard Clayton, William Caton, and Mary Pease. Her wider social circle included neighbors, laborers, friends, and strangers who were predominantly Quakers. Her associations, although highly selective, reached out a great distance, wherever either penniless or wealthy Quakers were traveling or in prison. She maintained a lifelong close friendship with co-religionists William and Guilelma Penn, with whom she exchanged correspondence and gifts.

Thus Fell's altered neighborly and wider social network did not conform entirely to Keith Wrightson's picture of traditional reciprocal neighborly relations of the agrarian "moral community." Wrightson has defined the positive and negative limits of neighborliness of the moral community, that is, the standards of behavior that were acceptable as goodwill actions that characterized local neighborly relations. Conforming to Wrightson's definition, Margaret did not seemingly forego these minimum standards of behavior expected of a good neighbor. She was not, as far as the evidence shows, a gossiping person. Neither was she contentious, profane, given to drinking, or abusive toward others. Likewise, she was not litigious toward her neighbors, except in her argument with a former friend, Quaker Thomas Rawlinson, which we will examine in another chapter. Further, in accord with Wrightson's definition of positive neighborly relations, the Fells loaned money both with and without interest. However, their loans were confined mostly to Quakers, as will be seen in the following chapter. Although Margaret may have been interconnected with the other gentry of her locale in a few minor business activities, she did not seem to share in their sociability. For example, the Flemings of Rydal hall were given to card playing for entertainment. There is no mention of cardplaying or gameplaying in the Fell household. Neither is there any mention of attending such amusements at neighborhood homes. If the northern gentry resorted to evening gameplaying, it would have been out of character for the Fell women to indulge in such amusements. Margaret's relationships to social inferiors such as local laborers whom she hired on the farm, may have conformed more closely to the norm of patron-client relationships. Here too, however, her economic network apparently included a significant number of Quakers.[34]

Like the pious Ralph Josselin, Margaret maintained close ties with the nuclear family, her daughters and grandchildren, and to some household servants as well. She was unlike the affluent yeoman priest Ralph Josselin in that she had neither patrons nor the traditional "spiritual kinship" of godparents surrounding her and her children. Her circle of kinship included only those of like Quaker belief, for ties outside the nuclear and religious family were largely noneffective. While the will of Thomas Fell revealed token bequests to esteemed friends near and distant, Margaret Fell's will did not include such bequests toward friends and neighbors. Although her credit amounted to £249 in her estate, her bequests were kept entirely within her Quaker family. Indeed she was very clear in her preferences among her kin, for she singled out her three non-Quaker grandchildren by giving them less than the others.

Margaret Fell's Quaker family and larger network marked her social boundaries and were of primary importance to her within the rural agrarian community of Furness. Thus, while Wrightson found that parish subcommunities of persons existed within larger towns and cities, Fell's Quaker subcommunity existed within a wider rural neighborhood, and it was oriented toward the matriarch of Swarthmoor Hall. Her Quaker support system replaced other external ties normally operating in a seventeenth-century person's world, as defined by Alan Macfarlane, Keith Wrightson, and others.[35]

Frederick B. Tolles asserts in his study, *Quakers and the Atlantic Culture*, that Quakerism was fundamentally a way of life, not a system of thought or doctrine.[36] This rested on the central Quaker belief in the Inner Light and testimonies that claimed it could only be realized by personal experience. Margaret Fell's religious beliefs deeply affected her mothering and domestic experience as well as her gentry manner of living. The character of her religious behavior as mother can be seen in the lifelong loyalty of her seven daughters to her and to Quakerism.[37] This carried over also to the servants in her household, both young men and women, who lived under her matriarchy at Swarthmoor. We have seen that the Fell women were competently running their own farm in the 1670s as well as traveling and speaking as "public" Friends.[38] These activities took up all the energies of the Fell women. In so doing, they were thwarting traditional female stereotypes and the deferential patron-client relations of gentry life.

Fell and her daughters did hold onto their gentry status, but in a significantly modified form. Their landed lifestyle was permanently changed from the conventional concerns of gentry women, namely childbearing, social entertainment, preservation of the patrimony for the heir, and arranging advantageous marriages for the other sons and daughters. While widowhood caused Fell and her daughters to engage in the outward businesses with great diligence, their conversions caused a reshuffling of their social network. The Fell women probably became increasingly removed from the traditional neighborly network, described by Wrightson, as nonconformist persecutions continued into the 1680s, and as the Fells publicly and privately persisted in their distinctive Quaker lifestyle. We turn now to examine more evidence of this distinctive social experience of Fell and her Quaker network of neighbors.

4 "Poore and in Necessity": Margaret Fell's Charitable Activities[1]

A contemporary of Margaret Fell, Thomas Lawson, gave the following warning to Parliament at the time of Charles II's Restoration. "He that stoppeth his Ears at the Cry of the Poor, he also shall cry and not be heard." Lawson, a Quaker school teacher, notable botanist, and friend of George Fox, addressed his special appeal for poor relief to the Parliament of 1660. He urged those in authority "let not a Settlement for the Poor be forgotten." In so doing, Thomas Lawson expressed a typical Quaker view of social obligation to the impecunious. The Quakers, from their earliest organization in 1652, consistently addressed the problem of poor relief within their community in practical terms.[2]

This chapter analyzes a pattern of poor relief that unfolds in Margaret Fell's family and in her local Quaker community in the 1670s and 1680s, confirming this Quaker social awareness but with a distinctive twist. Swarthmoor Hall became a focal point in a network of neighborhood Quaker female-directed charity. Sarah Fell's carefully kept ledger tells us something important about the Fell women's religious ideology and their charitable activities. The account book of the Swarthmoor farm together with a second source, the minute book of the local Swarthmoor Quaker Women's Monthly Meeting (SWMM) covering the period from 1671 until Margaret Fell's death, reveal a substantial number of the same women. By interweaving evidence from these two sources, this chapter will interpret the particularly female Quaker aspects of the Fell circle philanthropy in a period of increasing economic hardship, due in part to intermittent and sometimes heavy persecution. It will be suggested that the pattern of charity practiced by Margaret Fell, her daughters, and the SWMM on behalf of local Friends in penury expressed a bias in favor of poor women in an underdeveloped area of northwest England.

The SWMM was the first continuous Quaker women's meeting formed outside London, and its first meeting took place at Fell's home in October 1671. The first order of business was the taking of a collection for the poor. Sometime between the years 1675 and 1680, the SWMM sent out an epistle on discipline and order to the other newly formed women's meetings. It advised various actions in the ordering of women's tasks; among them the obligation to their own poor:

> And also all friends, in their womens monthly . . . Meetings, that they take special care for the poore, and for those that stands in need: that there be no want, nor suffering, for outward things, amongst the people of God, for the earth is the lords, and the fullness of it, and his people is his portion and the lot of his Inheritance, and he gives freely, and liberally, unto all, ...
>
> And so let Care be taken for the poore, and widdows, that hath young Children, that they be relieved, and helped, till they be able and fitt, to be put out to apprentices or servants.
>
> And that all the sick, and weak, and Infirme, or Aged, and widdows, and fatherless, that they be looked after, and helped, and relieved, in every particular meeting, either with clothes, or maintainance, or what they stand in need off. So that in all things the Lord may be glorified, and honoured, so that there be no want, nor suffering in the house of God, who loves a Chearfull giver.[3]

It is first necessary to see the Swarthmoor Quaker relief system in the wider context of the problem of poverty and poor relief in Tudor and Stuart England, and in contrast to the background of general Quaker attitudes toward charity. Historians have long noted the problem of poverty in Tudor and Stuart England which, in the eyes of contemporaries, posed a serous threat to the social order. It has been estimated that the poor in sixteenth- and seventeenth-century England numbered nearly one-third to one-half of the population, with averages increased to 50 percent in the years of bad harvest. Tudor and early Stuart paternalism attempted to relieve this immense social problem by parliamentary statute, Privy council intervention, and the implementation of poor relief through local justices of the peace and parish aid. Government and regional charity was based on a philosophy that combined compensatory relief to the settled or deserving poor but punitive action for the migrant or masterless poor. The gradual growth of a national system of poor relief in this period represented an end to the older medieval system of voluntary charity. Although Parliament passed several poor laws before 1597, two acts passed in 1597 and then modified and made more precise in 1601 combined several aspects of earlier poor law legislation. These acts reflected the two-sided nature of the government philosophy of poor relief. One statute of 1597 entitled, "An Acte for the Reliefe of the Poore" (39 Elizabeth, c. 3) called for parish overseers to give relief to the local poor and employ those who could work, especially children of the poor. The overseers were to help the poor find domiciles and obligatory poor-rates were established. A second statue of 1601, "An acte for punyshment of Rogues, Vagabondes and Sturdy Beggars" (39 Elizabeth, c. 4), sought to establish, in all cities and counties, houses of correction for masterless, unemployed, wandering persons who were considered potentially subversive in the eyes of contemporaries. Vagabonds who were convicted for vagrancy were to be whipped and returned to their place of origin.[4]

It is generally agreed that the plight of the poor worsened in the sixteenth to seventeenth centuries because of a complex set of economic and demographic factors. Demographic pressure from rising population began in the first half of the sixteenth century and lasted until the 1650s when the population began to decline somewhat. Likewise prices, reflecting the pressure of population growth, also rose then leveled off later in the same period. Between 1580 and 1620 inflation caused real income to decrease by half. This in turn affected the fluctuating levels of poverty. According to Paul Slack's analysis, in the earlier period the numbers of poor children were greater, while in the later seventeenth century there was a larger number of the elderly who were indigent.[5]

At the local level, various solutions were devised to solve the economic problem of the poor. Towns sometimes assumed the responsibility for poor relief by purchasing and reselling grain at moderate prices on market day. Local parish churches, part of the state machinery of relief, financed direct poor aid. Also, the more affluent families of a parish gave direct charity and provisions to the local poor. The rich were known to hire more servants than a household needed, and they sold grain to the poor at less than the market price. The wealthy grew crops that required a large labor force for their harvest and frequently acknowledged the poor in their wills. It has been suggested that local outdoor relief declined in the latter seventeenth century as the statewide parish relief system grew more efficient and benevolent. However, local informal poor relief never vanished, nor could it be completely limited by public policy.[6]

Among the poor of Stuart England, women often predominated. This was especially true among the deserving poor, such as young widows with children and elderly widows. As we shall see, the Quaker female pensioners of the SWMM were a case in point. That numerous women, especially those who were single, existed on the ragged economic margins of life in seventeenth-century English society is a subject that calls for further study. Knowledge of poor women's lives in the period is sketchy at best.[7]

For a considerable portion of the population of the rural northwest county of Lancashire, poverty was a condition of life. In the preindustrial period, Lancashire was traditionally considered backward in economic wealth. In the civil war years trade decreased, especially in some towns in the southern section of the county. For example, the year 1649 was marked by poor harvests as well as by the ravages of the second Civil War. In the towns of Wigan and Ashton, approximately 2,000 poor people had no relief for more than three months as no justice of the peace had been elected to oversee poor relief. In contemporary language this was because "most men's estates being much drained by the wars and now [are] almost quite exhausted by the present scarcity. . . . There is no bonds to keep in the infected, hunger-starved Poore, whose breaking out jeopardeth all the neighborhood." Another contemporary described the poverty in Lancashire during this same year in vivid terms: "[It would] melt any good heart to see the numerous swarms of begging poor, and the many families that pine away at home not having the faces to beg." In the same year the suffering of the poor in the wake of the war and bad harvests was experienced in the neighboring county of Cumberland. A report from a Cumberland county committee of justices of the peace registered that some 30,000 families had "neither seed nor bread, corn nor monies to buy either." This picture of dire poverty in rural northwest England is not out of line with the estimates of W. K. Jordan, that philanthropy in Lancashire county dipped to a low ebb during the Civil War and Interregnum.[8]

Beginning about 1652 the Quakers distinguished themselves by their own internal attack on the problem of poverty. They developed a system of poor relief for their membership that gradually became established on a nationwide basis and was administered through the network of Quaker meetings. The Quaker system operated as a locally based, essentially autonomous sectarian mechanism that existed outside the state-organized system of poor relief. It is an example of successful institutional charity at the local level, which has largely been overlooked by those who have studied the wider context of state and parish poor relief in the Tudor-Stuart period.[9]

The first generation Quakers' concern for the poor was manifested in their rhetoric and in their cohesive fellowship that went beyond a purely religious bonding. George Fox looked upon charity as having "apostolic precedent," and he frequently preached about the problem, warning the nation that, "oppression and . . . fulness are in your streets; but the poor are not regarded . . . woe be to them that drink wine in bowls and the poor is ready to perish." This socially conscious outlook was expressed by other Quaker leaders as well. James Nayler scolded England's magistrates in 1654, "the wants of all the poor in the nation cry out against you, . . . you have exceeded all that ever went before you. . . . Covetous and cruel oppressors, . . . you grind the face of the poor."[10] Margaret Fell expressed a similar outlook in her attack on the tithe-taking Anglican prelates in 1668:

> The bishops . . . rob [the king] both by their power and there greate benefits and liveings that they keepe . . . to mainetaine there prid and haughtiness wch the kinge may justely . . . take from them, and might thereby ease the heauie burdens and apprehsions of the people, that he is forced to lye upon them by tacssations . . . if the kinge had that wch you usurperers [the bishops] . . . have, the poore people of the nation would find ease thereby.[11]

In 1654–58, Margaret Fell organized the Kendal Fund in Lancashire for Quaker prisoners and families in financial distress. In her letters requesting contributions, she articulated the Quaker sense of cohesion and charitable concern: "You may know the Mystery of the Fellowship, and be in that Mystery, in Unity with the Brethren; that so you may come to be one with them in their Sufferings, in their Tryals, Travels, Troubles, Buffetings, Whippings, Stockings, Prisonings."[12]

As early as 1659, Richard Hubberthorne could assert that within the Quaker Fellowship there "was no beggar among Friends, and they needed no maintenance from any people or profession in the nation." Quakerism offered emotional companionship that was based in part on an economic policy of sharing, thus satisfying deep human needs in a period of religious persecution, political and economic transition, and turmoil. Moreover, the Quaker economic "cocoon of protection" shielded members from poverty while at the same time discouraging them from the acquisitive spirit. Early Quakers favored the ideal of a simple lifestyle and "modest sufficiency," accompanied by internalized philanthropy, based on the principle of "help to self-help," when needed.[13] The Quaker attitude toward internal philanthropy was summed up in George Fox's view on the collection of the "stock" or treasury for poor relief in the Quaker fellowship or meeting.

> Concerninge Collection:
> Let ffreinds in every particular Meetinge bee orderd to Receive all Collections, And to bringe them to the Monthly Meetinges' . . . lett an Account be kept of what is Received, and to whom it is Disbursed. . . . And in every Monthly and Quarterly Meetinge, to Inquire what poore there are, and who are fitt to goe, to bee Apprentices or Servants. And ffreinds then and there, may order them to bee

Apprentices, to such trades as they in the wisdome of God shall see fitt.

Likewise, widows and orphans were given aid by the Quaker meeting:

> And alsoe all widdows . . . lett them be taken notice of and informed and encouraged in their outward businesse, that there be not any hinderance to ym [them] . . . and soe carefully looked after that they may be Nourished and Cherished . . . and if they have many Children, to putt out Apprentices or Servantes, that may be a burthen to ym to bringe upp – then lett ffriends take Care.[14]

The Quaker philosophy of charity had a disciplinary side as well. This may have been influenced to some degree by the widespread hostility toward masterless persons and the general condition of penury in seventeenth-century England.[15] Quakers had an aversion to bankruptcy with the result that persons within the fellowship who overextended themselves in business debts could be disowned by the meeting. The Quaker attitude toward charity was not that of a simple dole or unexamined program of relief to persons exhibiting poor business management. George Fox warned his own people involved in trade and commerce in 1661:

> And all, of what trade or calling soever, keep out of debts; owe to no man any thing but love. Go not beyond your estates, lest ye bring yourselves to trouble, and cumber, and a snare. . . . For a man that would be great, and goes beyond his estate, lifts himself up, runs into debt, and lives highly of other men's means; he is a waster of other men's [means], and a destroyer."[16]

This attitude toward poor relief was not unique to Quakerism. What was very unusual about early Quaker thought and practice was, first, its sectarian practical application was implemented on a nationwide scale, and second, the bulk of the administration of poor relief was left in the hands of the women's meetings in the areas where they existed. Due to internal resistance from some quarters, women's meetings were not organized everywhere as parallel meetings with the men's monthly meetings. It is most probable that in those cases the men's monthly meetings, with the informal help of the women, cared for their own poor. Thus, Margaret Fell and the SWMM, as the earliest Quaker women's charitable organization outside London, had few female precedents to follow except the example of the London women Quakers.

Charitable Activities

With this background on Quaker philanthropy, let us return to Margaret Fell's works of charity. Sarah Fell's account book of Swarthmoor Hall from 1673 to 1678 gives a picture of the priorities of expenditures for both her family and the SWMM reflecting this Quaker philanthropic thought and practice. As shown in Table 4 of the Appendix, the total Fell farm income for 1676 indicates that the amount spent on real charity to neighbors was 3 percent and for loans to neighbors and business associates in a wider radius was 16 percent. Together the real charity of Fell and daughters combined with their loans constituted £77.24.06, or 19 percent of their annual income of 1676. The reason for including the Fell's loans as part of their charitable activities will become clear as this chapter unfolds.[17]

As we have already seen, Sarah Fell's account book reveals that she acted as a banker for people in her area. She was well located to do so because Swarthmoor Hall was only one mile from the market town of Ulverston where produce and goods were purchased weekly. Artisans and craftsmen who were specialized and required credit extension to survive in the local cash economy could contract short-term loans or give and receive credit in order to buy and sell goods. Judging from the record of tailors, shoemakers, coopers, saddlers, smiths, peddlers, and carpenters who appear in the pages of the Swarthmoor account book, Sarah accommodated people who came to the Ulverston market and perhaps those who wished to go to the nearby markets at Kendal and Dalton.[18]

Sarah's banking activity was, at least in part, a profitable enterprise for the Fell women. Sarah was a businesswoman of no mean ability as seen in her co-ownership, with her mother and sisters, of Force Forge, an iron bloomery a few miles to the north of the Hall. Additionally, Sarah, together with some Furness neighbors, ran a corn and grain shipping enterprise via the port of Bristol. Further, her banking expertise was based on experience gained through watching the collection and disbursement of the Kendal Fund.[19] To a degree, Sarah Fell's banking enterprise was typical of the traditional neighborliness of wealthier families toward the less wealthy in seventeenth-century England. Private lending was, in fact, a way of life in this period. Among families of enormous wealth, it occurred on a grand scale. Although evidence of larger interest-paying loans is somewhat scanty of the year 1676, Sarah's banking activities over the whole period of the accounts indicate that along with the larger loans at interest, she patiently recorded small short-term, non-profitable loans to the poor, many of them to women. A mere farm steward might well have found such detailed bookkeeping for the impecunious too burdensome.[20]

The early Quaker rejection of excessive profit or debt in business dealings was expressed by George Fox who frequently reiterated, "Owe nothing to any man but love." On the subject of usury, Fox wrote, "no one . . . run into debt, usury and exaction, for many people have been wronged thereby . . . be serviceable in the creation, serving one another in love and not in oppression and taxation."[21] However, it seems that neither Fox nor other early Friends defined precisely what Quaker business policy should be concerning charging interest on loans. In the marketplace, Quakers were to hold to a fixed fair price, refrain from bargaining, use exact measures, and let their "yea be yea" and "nay be nay." It appears that Sarah Fell operated within the Quaker warrant on business ethics, not exacting interest from the poor but borrowing and lending money on larger loans at an interest rate current for the period.

In 1676 Sarah loaned money to several of her laborers and servants at Swarthmoor. Some examples include Edward Braithwaite, a Quaker servant at the Hall, to whom she loaned a half penny in June of that year, and in October she loaned him £3 "for his Cousin J[oh]n Braithw[ai]t[e] till ye 11 of ye next month." Sarah employed a woman, Alice Atkinson, who did spinning for the Fells and on occasion borrowed small sums of money, such as in February 1676 when she borrowed a half penny. As of the following month she had not repaid it, for an entry reads, "pd. to Alis Atkinson for spinning . . . 3s.4d (pence) . . . s^d Alis . . . owes me . . . 1/2d." Sarah loaned 6s in June, 5s in September, and 3s more in November of 1676 to Jane Colton, a Quaker widow of the SWMM. These small amounts were repaid without interest. Sarah also dyed and made a petticoat for Jane's little girl at her own expense. On occasion, Jane baked for the Fells, darned stockings, sewed shirts, and sold them "manner" (manure). Sarah Fell both borrowed and loaned for interest among her Quaker neighbors and business associates. To William Salthouse, a shoemaker of nearby Dragley Beck, she paid interest in April on a £10 loan from William's late mother at a yearly rate of 12s, and in May she repaid the principle. Sarah paid Bryan Osliffe interest of 11s.6d on a £10 loan at the end of one year for her brother-in-law, Thomas Lower, then living at Marsh Grange with his family. Likewise Lower owed interest of 12s to Robert Salthouse, brother of William, on another one-year loan of £10. Even though the Fell and Salthouse families were Quakers and close personal friends, they charged one another interest on larger loans such as these, at a rate of about 6 percent. In 1676 Sarah and the steward of Marsh Grange, Joseph Sharpe, undertook several business transactions. She gave him four loans over the year, which may have been related to either their mutual investment in the grain shipping business or coal purchasing for Force Forge. She received money "in earnest" (interest) from Sharpe for two colts she sold him in the previous year for £4.3s.4d. Those persons who took out larger loans could afford to repay with interest and were charged accordingly. The very tiny allotments, meted out for crisis needs of the poorest neighbors, were not charged interest. Sarah seemingly regarded this as charity.[22]

In Sarah's cashbook and the SWMM minutes, numerous incidents come to light that involve individual women. Of the 198 women having some social or business relationship with the Fells, whose names occur one or more times in the account book over six years, sixty-three, or one-third of the total, also appear by name in the SWMM minutes. Of the remaining 135 women whose names are not mentioned in the SWMM minutes, some may have been more silent and poorer members who remained unrecorded. Also some may have been from other women's monthly meetings in the district and not of the SWMM. Finally, some may have been non-Quakers. However, this last possibility seems unlikely because the minutes of the Swarthmoor Women's Meeting and of the Lancaster Quarterly Women's Meeting consistently express a highly exclusive attitude toward non-Quakers.

In Chapter 3 we learned of some female servants and peddlers or regratresses who had contact with the Fells. Along with this group there was another group of women whose names recur with regularity in the accounts and minutes. These were the pensioners, whose bare subsistence standard of living forced them to look to the Fell women for aid either in work, kind, or financial doles in order to survive. Such women, elusive figures of the past, take on a flesh and blood reality through Sarah Fell's careful notations of their comings and goings. For instance, in the SWMM there was a poor woman, Jane Hudson, whose relatives owed her money, which they were apparently unwilling to repay. The SWMM, through Sarah who was the clerk of the meeting, wrote to one of these relatives, Robert Shaw, of Kendal, requesting that he pay Jane Hudson. This intervention was successful, for an entry in the minutes of March 1676 reads: "rec'd of Anthony Shaw of Kendall to give to his A[u]nt Jane Hudson p[ar]t of some money his ffather Rob: Shaw owes her 5s." A second payment of 5s was made two weeks later via the banker at Swarthmoor Hall.[23]

Mary Taylor was another local Quaker who both worked for the Fells and participated in the women's support system. She serviced the poor women of the SWMM by carrying funds from Sarah, who kept the women's "stock," or treasury, to those in need, such as the widow Ellen Braithwaite of the nearby town, Cartmel. Mary received 10s from the meeting's stock in June 1676 "towards buying Ellen a cow." Subsequently a note records that Ellen's uncle died and bequeathed her some money; she thereupon returned the money to the SWMM. On at least one occasion, Mary Taylor knit stockings for the Fells, receiving 1s.6d for her labor. When she became ill in November 1677, Margaret Fell spent 10d in travel expenses to visit her in Cartmel. Mary, who was married to James Taylor, hosted the women's meetings in her home on numerous occasions, indicating her financial ability to absorb the cost of a large group of women requiring sustenance for themselves and their horses. Mary Taylor's signature appears on many minutes of the SWMM. Among her assignments, she inquired into "marriage clearnesse," enabling the SWMM to certify Quaker women and men who wished to marry.[24] Dorothy Beck, wife of William Beck, a glover of Lonethwaite near Hawkshead in Furness, also delivered aid for the meeting, as well as being a life-long member thereof. Dorothy carried the money of the SWMM to impecunious women like Mabel Gunson, whose husband had gone to Ireland in 1676, presumably as a travelling Quaker minister, leaving her at home to care for two small children.[25] Dorothy's signature of attendance at the women's meetings appears in steady frequency next to that of Margaret Fell for almost thirty years.

Two steady female pensioners of the women's stock, and recipients of Fell aid, were Ann Birkett and Jane Woodell. These two women's lives present a picture of permanent or "deep" poverty with great dependence on the good will of the Fells and the SWMM. Ann Birkett of Cartmel may have lived too far from Swarthmoor Hall to have been employed by the Fells. However, she received 5s of the women's stock in April 1674, 2s in December 1674, and again in February 1674–5 she was given 5s as one who was "poore and in Necessity." In April 1676 Ann received 2.6d, in July 1676 2s, and in April 1677 2s again. In May 1677 the stock paid "some house Rent in arreare" for her, and 2s.6d in February 1677 before the account book closes. Presumably, Ann Birkett continued as a pensioner throughout the last ten years of her life, dying in 1686. Her dire poverty apparently wore her patience thin, for she was accused of having a "cross and peevish spirit" when some women of the meeting called upon her to assess her need. Their advice to her was to be "more loving in receiving her charity." However, on a later occasion, when Ann attempted to borrow 6s from the stock, she was reported in a "cross spirit again," being "angry toward friends." The women of the SWMM decided not to supply her in such circumstances and called upon Ann's particular meeting in Cartmel to admonish her and ask her to attend the next SWMM, in the meantime laying aside her "past evil spirit."[26]

Jane Woodell, who was a pensioner of the SWMM until her death in 1678, apparently exhibited a less petulant spirit. Her name appears in Sarah Fell's accounts as "old Jane Woodell," to whom Sarah lent 3s early in 1673 and 4s in December 1673. In February 1673–4 she paid Sarah back the 7s. Jane took several other loans from the Fells, which she also repaid.

The Fells gave to general Quaker collections for her assistance and at one time the stock paid for mending her shoes. The minutes of the SWMM indicate that she was in need of a petticoat. Sarah Fell purchased material for the petticoat, which was given to the elderly Jane. However, the minute stipulated that if she were to die before she wore it out, it was to be returned to the SWMM. Following her death in 1678 a note was entered into the SWMM minute book: "theres A petticoate; and an under Wastecoate in Issabele Brittaines custody, that old Jane Woodell had not wore out when she died, wch the Meeting had given her; and the Meeting orders, that they bee sent to ye said Jane ffisher to serve her for the present."[27]

The women of the Swarthmoor meeting did not look kindly upon ingratitude. When a recipient was insufficiently grateful, they not only took umbrage but feared a hostile public's disapproval as well. A case in point is Jane Strickland, a widow of the Cartmel meeting to whom, in October of 1678 and at her own request, the women sent some corn. It was delivered to a non-Quaker neighbor for the sake of convenience, and Jane was instructed to pick it up at the neighbor's home. However, a minute of the meeting informs us that Jane was negligent in picking up the grain. Instead she "lett it lie there severall weekes, [in] wch . . . shee did not [act] well, not carefully in soe doeing." The meeting responded:

> Wee desire some that are her neighbours, yt [that] are now here; [at the meeting] to Reprove her, for being soe Negligent in not sending for it, nor seeing after it, in soe longe A time – which might open the Mouthes of the worlds people to speake evill [of us]. Therefore wee desire that shee bee more carefull for the future, and not sleight ffriends charity to her, and care over her, but to value their Love, and goodwill to her.

They called on Jane to be of "A thankfull minde, and [to be] in the same Love that it is Administered to her."[28]

Charity was given, but sparingly, as the example of Margaret Geldart demonstrates. On occasion she sold manure to the Fells for a few pence, suggesting that she was a cottager who did not own enough land or the resources to cultivate extensively herself. When the women's meeting received reports that Margaret was sick and needy, they sent two women to assess her situation. The minute of 7 May 1678 reads as follows:

> [M. Geldart] is out of health, and hath been for sometime, and seemes to bee in some want; but we are also informed that shee hath some household goods of her owne or something that shee may make money of, to supply her necessity with, but out of A covetous minde, seems

unwilling, to supply herselfe with her owne: Neverthelesse, wee doe desire Jane Cowell and Sarah Fell both of Ulversone, to goe to her, and Enquire into her condicon and see what ability, she seemes to have of her owne, and to acquainte her, that its friends minde and Judgment, That shee ought first to supply her selfe, with her owne, while shee hath it, and then when that is gone, if ffriends see need, and that shee bee worthy; wee cann reach forth A hand of charity to her.[29]

Some appeared in need of assistance for temporary periods when dearth or hardship caused by persecutions required them to turn to the meeting for aid. Isabel Holme and her spouse suffered financially because her Quaker husband was "a Suferer for the Testymony of Jesus against tyths." They were impoverished in 1681 when Isabel was "neare her time of lieing In." Dorothy Beck gave them five shillings "toward their necessety they beeing members of the Sam body with us." Isabel soon thereafter delivered twins, and the SWMM "being Sencabell of her . . . Condishon and allso of her pore family of Children" willingly contributed aid "Sutabell to yer [their] nesety." In June 1681 the meeting sent Mary Walker to visit Isabel with some relief. The meeting stipulated that Mary bring back an "account of her Sencerety to Truth of her zeall and feruancy ther unto." In July another minute of the SWMM reveals that Isabel received her aid with "much gratfullness Acknowledgin[g] ffriends care and love." One year later, in July 1682, Isabel's husband, in prison in Carlisle for "truths testymony," was very ill. He sent for his wife to care for him. This was very difficult for Isabel due to her "lowe condishion . . . and young child yt Shee can not leave behind." Once again she turned to the SWMM for aid for her journey to Carlisle. The meeting again sent her money. In June 1689 Isabel Holme reappears in the meeting minutes. When the impoverished Margaret Geldart was in "want for 2 shifts," the women's meeting ordered that 4s4d be given to Isabel Holme "whoe hath cloth to sell convenient for her [Margaret]." It is apparent that the women willingly aided those loyal to the Quaker cause and that they kept a close record of all relief given. Further, their network of relief combined both ouright doles in money and kind, as well as petty retail patronage.[30]

The collective approach to poor relief among the Quaker membership reveals what Michael Mullett has called a Quaker welfare state. This sectarian welfare state that existed within the larger society was, by necessity, "distinctively rational, cautious, discerning and indeed legalistic [in its] attitude to poor relief." Although the SWMM appeared very rigorous in its screening of poverty cases, it had to be parsimonious due to limited resources and plentiful local poor. Moreover, a fair number of pensioners constituted the permanent poor, those whose poverty was "deeper" than the seasonal or life-cycle phases of poverty experienced by some. Consequently, careful attention to each charity case was necessary before taking on a new, potentially long-term pensioner. Quaker poor relief also demonstrates that seventeenth-century perceptions of poverty were "a relative an culturally determined phenomenon," differing markedly from our modern definition of poverty. In rural Lancashire the Quaker perception of distress worthy of relief essentially meant complete destitution.[31]

Charity took another form that contributed to the emotional and religious cohesiveness of the Quaker community. Early Quakers expressed an abiding concern for young orphans. Mary Benson is a case in point. Mary, orphaned in 1679 at age eight or nine, was taken under the wing of the SWMM. Between 1679 and 1686, Mary's interests were protected by the women who inquired into a legacy left to her, of which no one knew the whereabouts. When found, the Swarthmoor women held it for safekeeping. The women decided that Mary should not go to live with her great uncle, Anthony Strickland, who offered to keep her, a decision explained in a minute of February 1679:

> Wee doe not know whether it may bee safe or noe to send her to him, being noe ffriend, least hee should traine up ye childe in vanity and folly; therefore wee are more free, to continue our supply for her mentaineance awhile longer, untill wee are satisfied that the child may have her Liberty to goe to Meetings and amongsts ffreinds, according as her Parents would have desired.[32]

However, subsequently the SWMM did send Mary to her uncle for a quarter of a year. She was received "loveingly and kindly" with a promise to let her attend Quaker meetings. However, after three months had passed, Anthony Strickland was willing to let her go although the reason for this is not specified in the minutes. Elizabeth Birkett offered to take the child into her home, at which time Mary was in need of a pair of shoes, which the SWMM stock supplied. Birkett reported to the meeting that Mary was "a willing orderly Child," and the women of the Cartmel meeting liked her so much, "they [were] not willing to part with her yet from Amongst them" as of July 1680. Their wish was to "instruct her and correct her, as they see occasion; soe as shee may bee kept out of Wildeness: And that shee may keep her Reading and knitting, or rather Improve it." Mary was next a guest at James Taylor's house in Cartmel and appears to have been passed around to various Friends' homes, always being kept under close supervision and supplied with clothing as needed. She was employed at £2

96   Part II – The Domestic and Economic World of the Fells

a year and then was sent to Manchester to better her learning and to "tend to her ffuture advantage." The final glimpse of Mary Benson is a negative one. It is entirely possible that Mary was a difficult child to handle, for the SWMM appears to have grown weary of their charge. Nonetheless, the meeting found her a home and employment.

> Margett [Fell] ffox is willing to keeppe her this year and to give her 12s waydges if shee bee of a capasity to earne it; but shee hath ye caractor of a very Staborne froward child so yt it is an Exercize to meadell with her.[33]

The situation of Mary Benson is the only example available of the long-term oversight of an orphan by the SWMM in the period investigated. As Quaker philosophy toward widows and their children was one of "nurturing and cherishing," this example may well have been representative of Quaker policy concerning orphans in general. For instance, the Women's Six-Week Meeting of London was known to discuss and create policy on the care of orphans of the poor. In April 1680 a minute of that meeting considered the matter of "ye charge the women friends are at in nursing and bringing up friends Children that died poor, and are left to Friends care to look after." The Six-Week Meeting agreed that "w[ha]t charge shall be contracted by the women ffriends in providing for the said children, be def[r]ayed out of the publick Cash." The example is not inconsistent with Fell's and the SWMM's inclination to help disadvantaged Quaker women.[34]

In contrast to the SWMM, the Swarthmoor Men's Monthly Meeting (SMMM) in the same period concerned itself more with problems confronting a newly emerging sectarian church. The men's meeting spent much of its energy and "stock" on publishing books, settling internal disputes, disciplining "disorderly walkers," and those who paid tithes under duress. Between 1668 and 1674, however, there were occasional comments in the men's meeting minutes regarding specific charity cases. Beginning in 1673 Jennett Woodall received quarterly doles of approximately 14s from the combined contributions of the Swarthmoor, Hawkshead, and Cartmel men's meetings. Likewise, the impecunious Gunson family, who lived in the Hawkshead jurisdiction, received aid from the combined collections of the Swarthmoor and Hawkshead men's meetings. Moreover, there were sporadic appeals for funds in order to "suply a pour ffrend" of Sedberg and to supply a meeting in Cumberland to aid some Quakers "banished out of the Ile of man" in July 1669. The SMMM also arranged to give aid to those most severely penalized for refusal to pay tithes. The household goods of some Quakers were distrained in September 1671, and the SMMM minuted a "suply accordinge to their necsseity and faithfull nesse in their Sufferringe."[35] Aside from such occasional examples from the SMMM minutes, the bulk of the oversight of poor relief was the bailiwick of the women's meeting.

PRIVATE PHILANTHROPY

Along with the charity of the SWMM, the Fell philanthropy during the 1670s was given in small amounts over an extended period of time. It is useful to compare Fell's charity to a few of her contemporaries and nearby neighbors. Sir Daniel Fleming of Rydal Hall lived not far to the north in Westmorland and opposed the Quakers. William Penn was a Quaker and close friend of the Fell family. The accounts of Fleming and Penn, although exhibiting greater financial resources than Fell's, display a more sporadic charity. This observation must be tentative for the sample is very small. Daniel Fleming's charity doles were more generous than the Fells, usually 1s each given to needy recipients when he went to Kendal as justice of the peace for the county Quarter sessions. However, the evidence does suggest that Penn and Fleming failed to donate to the local poor in as persistent a manner and to as great a degree.[36]

An example of non-Quaker female philanthropy in the 1670s is that of Anne Clifford, Countesse of Pembroke. Anne Clifford was a wealthy Anglican of Appleby Castle, near Carlisle. In the same decade she gave generously to the local poor in Cumberland. Although her accounts cover only a two-month period in 1673, they indicate that she distributed aid to the needy frequently and generously without distinguishing between the sexes (Table 3, Appendix).[37] In October 1673 the Countess paid a deaf woman "of my Almshouse here at Appleby for 18 yrds of bonelace – 12s." In her account was a gift "to Amy Walker the lame woman of my almshouse here at Appleby 2s.6d," a substantial sum by Fell standards. On the anniversary of the death of her mother each April 2nd, Lady Anne distributed among the poor £1.16.04 at the "Pillar I caused to be erected at yt place in memory of my blessed Mothers' and my last parting." However, she also gave Mr. Edmond Sandford, "ye deafe gentleman" 20s, and even more generous sum, perhaps a consequence of his higher social status.

Although Protestant and Roman Catholic giving patterns have not been proven to be distinctively different, we do know that philanthropy was emphasized in Puritan teaching. Alan Macfarlane's assessment of Ralph Josselin's theory of charity offers an example. Josselin was concerned about the poor and felt it incumbent upon him to "doe god service and help the poore." This puritan priest of Earls Colne, Essex looked upon almsgiving as a spiritual investment, once writing, "he that gives to the poore lends to the Lord." Through his charitable activities, Josselin enhanced his reputation as parish priest in Earls Colne; he was known to fast and give the money thus saved to the poor. In 1652 he budgeted his charity and library costs in an effort to tithe his income from all "rents or profits," although he did not include income from his ecclesiastical living and teaching. Apparently he fell short of his aim, for according to Macfarlane, later in life he renewed his efforts. Josselin recorded:

> I sett aprt the tenth of all my incomes in money as minister, the 10th of my rents in money, and the 10th of my profitt by any bargains, the 20th part of the money I take for all corn I sell. To pay my tenths and to serve in gods worship and free charitable bounty to gods peace as neare as I can, allowing out 20s yearly for books.

Macfarlane concluded that it was very unlikely that Josselin ever achieved his tithing goal, which would have amounted to £14 per annum. According to Macfarlane's calculations, Josselin gave away about £10 per annum, spread over his ministry at Earls Colne.[38]

This background helps to assess the Fells' giving patterns from 1673 to 1678. In this period, fourteen received charity, amounting to 2s.7 ½ d in donations (42 percent). Nineteen women received a total of 7s.10 ½ d (58 percent). Outright gifts of money were seldom more than a penny, and men received less charity, frequently receiving only 2 farthings or ½ d (half penny) per person per dole. Women recipients received at least 1d per person per dole and sometimes more. For example, in July 1676 Elizabeth Williamson, an "old woman" whose religious affiliation is unknown, received charity from Margaret Fell in the amount of 3d. Only one woman received ½ d, while eight men received ½ d in donations.[39]

The Fells contributed frequently to Quaker corporate collections. The Swarthmoor account book lists thirty-two Fell contributions to the stock for a total of £6.5.0. Other collections listed in the accounts include a fire brief in Cheshire, a collection for John Towers, who had the misfortune of losing four horses; for two poor men of Dalton; and for Jane Woodell, as well as a contribution to the stock at Lancaster, totaling another £6.19.0.

Of the sixty-three women's names that appear both in Sarah Fell's accounts and the minutes of the SWMM, at least seventeen were needy. Thus, judging from evidence in these sources, about 27 percent of the women of Fell's Quaker network lived in poverty, either permanently or temporarily, and were aided by both the SWMM and the Fells. Although charitable behavior was a way of life for the Swarthmoor Quaker network, it is uncertain whether charity and easy work-pay terms were limited largely to Quaker women or were also given to the other deserving poor in the neighborhood. The minutes of the SWMM as well as the LWQM convey a consistently strong sense of separateness from the world. The account book presents a picture of a social, religious, and economic network around the Fell family that was at least 25 percent Quaker. The fact that this proportion, and probably much more, of the total Fell network of business and social contacts was Quaker, allows the reasonable assumption that Fell and her circle of Quakers were, from their earliest years, a closely bonded and separatist group, not only in their religious life but in their social-economic life as well. The Quaker women around Swarthmoor appear to have spent the bulk of their resources on their own members, with some sporadic charity given to the poor in general.[40]

What other conclusions can be drawn concerning Margaret Fell and the philanthropy of the Swarthmoor women's meeting? Considering the size of the sample under investigation, any answer must be cautious. This notwithstanding, some conclusions are suggested by these female Quaker patterns of charity in rural Furness. First, it has been suggested that informal charitable activities of neighbors were, by and large, replaced in the late seventeenth century by the state machine of philanthropy especially undertaken through parish administrators. This sample suggests that a modification of the interpretation is in order. It is evident that Quaker charity was dispensed outside of the state system of relief. Fell and her Quaker women friends developed their alternative relief system that simultaneously rejected and reflected the state machinery. It was a rational, on-going program of sectarian poor relief, administered through the local Quaker meeting in much the same way as the state system operated but on a smaller scale. It differed from the state machinery in that it was administered mostly by women, and often for women, and operated on a smaller scale. In terms of work, pensions, doles, and noninterest loans, the SWMM implemented an efficient structure of internalized mutual aid that lay outside of and extended beyond the Stuart poor-law system as it has recently been defined.[41]

Second, the Quaker charity of Fell and her circle did not lose its personal character despite being part of a nationwide Quaker network of support. In contrast, the state system became increasingly bureaucratic, with neighborly philanthropy giving way to formalized and compulsory poor relief. Further, judging from this sample, Quaker relief (except in cases of bankruptcy) appears not to have been either punitive or humiliating in nature. The poor were not singled out and badged or publicly punished, thus being forced to endure a public stigma as was the case with the poor dependent on the parish relief system. The Swarthmoor women's charity emphasized "social obligation," and those poor who appeared grateful to their Quaker benefactors were cared for and accepted as active members in good standing in the local fellowship.[42]

The Swarthmoor women's meeting relied significantly on the Fells' generosity. As comfortably well-off independent farmers, the Fells employed several of the pensioners as farm laborers and servants and gave no-interest loans on extended repayment terms. The poor laborers and pensioners, women and men, who appear in Sarah Fell's ledger and the minutes of the SWMM, existed in a relationship of semi or even permanent dependence on the Fells and the women's meeting, much as the English poor in general were dependent on the parish relief system. Margaret Fell's charity, and that of the SWMM was restricted philanthropy on an organized personal and cohort basis, offering a special "cocoon of protection." It especially protected impecunious rural women.

Finally, in Margaret Fell's wider experience of Quaker leadership, social deference was accorded her not only by those in a relationship of semidependence but by rank-and-file Quakers as well. This notwithstanding, Fell laid heavy stress on spiritual equality in her religious writings. This resonated in her published works, such as her best-known tract on the equal spiritual authority of men and women in the Quaker ministry, *Women's Speaking Justified* (1666). Her belief in gender parity seemingly influenced her economic life as well, at least as it applied to Quakers. Fell's commitment to the female Quaker network around Swarthmoor was in part due to her own economic and political interests, but it also went beyond traditional *noblesse oblige*. Her religious principle of gender equality shaped both her philanthropic thinking and activity. Concern for poor women was not unique to Margaret Fell. However, it appears that Fell consistently attempted to offset the economic disadvantages rural laboring women experienced in her area of preindustrial England. Fell's Quakerism gave her a gender bias that was expressed in her private philanthropy. In the next chapters it will be suggested that her belief in gender parity was consistently expressed in other aspects of her fifty-year public and private ministry.

5 Feuding Friends

A widow of independent means who inherited property and who could legally control her own interests for herself and for her children was somewhat of an anomaly in seventeenth-century England. Such a woman stood in counterpoint to English patriarchal theory that a man must head a household and a woman must be subservient to the head.[1] Such was the anomalous position of Margaret Fell after 1658. As matriarch of a lower-middling gentry estate, it was vitally important for Margaret to have an acute business sense, to practice thrift, and to be alert to any opportunities to enlarge the annual estate revenues. Margaret, like other widows in her position, knew only too well the bleak alternative: insolvency, impoverishment, and possible loss of the patrimony. Further, Margaret's position as an authoritative public figure in emerging Quakerism rested to a significant degree on her local social and financial standing in the Furness district.

Margaret Fell and her daughter and business partner, Sarah, experienced the threat of financial loss in a business venture they undertook between 1658 and 1681. While Judge Fell was alive, the Fell family had come into possession of an iron bloomery, known as Force Forge, located a few miles to the north of the Hall in the Lake District. Sometime between 1658 and 1681, the year the Fells handed the forge over to Thomas Rawlinson, Force Forge ceased to be a profitable enterprise. Thomas Rawlinson, a close family friend in the 1650s, became Margaret Fell's iron steward at Force, probably in 1658. According to a recently discovered document written by Rawlinson, this business liaison caused a deep rift in their friendship. Rawlinson and Fell became estranged over a business quarrel that began after he stopped working at the forge in 1663 and was not resolved until 1681. Each accused the other of dishonest business practices. The dispute came to involve the SWMM and the Lancaster Quarterly Meeting, as well as a larger circle of non-Quaker relatives and friends who attempted to act as arbitrators. Thomas Rawlinson became increasingly embittered by what he considered to be highly unfair treatment at the hands of Margaret Fell and the local Quaker meetings she seemingly controlled. Rawlinson wrote a series of lengthy letters about this dispute and threatened to send them abroad to meetings everywhere. It is this threat of exposure that makes the Rawlinson manuscript pertinent to our discussion of the role of Fell in the rise of Quakerism. Although this source is biased, giving only Rawlinson's side of the dispute, it does reveal how pivotal and authoritative a figure Fell was and how damaging a bad image of this public minister could be for the reputation of the nascent Quaker church. Moreover, the document shows that she was an intimidating figure who held a hegemonic power over Quakers in the Swarthmoor area.[2]

Thomas Rawlinson (? – 1689) was the son of Captain William Rawlinson of Graythwaite and Rusland Hall near Hawkshead, Lancashire. A minor gentleman of Furness, William Rawlinson, a parliamentarian during the Civil War, had been granted an indemnity at the Restoration. His son Thomas became a Quaker in 1652, joining the Cartmel Friends Meeting, whereupon he was banished from his father's house. His Quakerism brought him into contact and subsequent friendship with Margaret Fell and her family in the 1650s. In 1663 he married Dorothy Hutton of Rampside. Some of the letters in Rawlinson's document were written from his father-in-law's home. Rawlinson traveled with the first generation Quaker itinerant ministers, endured an imprisonment at Launceston in Cornwall in 1656, and shared another imprisonment with James Nayler of Exeter gaol in 1657. He was also Fox's travelling companion during the year 1657 and was described by W. C. Braithwaite as a very dedicated Quaker, one of the "men of substance and position in the Quaker community."[3]

When the dispute began, Rawlinson had known the Fells for some time and was on very friendly terms with the family. For example, in March 1657 Rawlinson wrote a letter to Fell from London, telling her he had visited her son George at his living quarters, staying three or four hours, to speak on religious matters. In October 1659 Margaret Fell, Jr., traveled to Wapping seeking care for a chronic knee problem. Her letter indicated that Thomas Rawlinson had accompanied her there, for he sent word via Margaret, Jr., that he had sent a letter to Margaret, Sr., in the last post.[4]

During Margaret Fell's absence of more than a year from Swarthmoor in 1660–61, Thomas Rawlinson was very helpful to the Fell daughters who remained at Swarthmoor. Bridget mentioned his name three times in letters to her mother. Rawlinson conducted business transactions for Bridget, for on one occasion in 1660, she wrote that "T[homas] R[awlinson] [had gone] to Adam Sandges before the day of payment to know . . . if he would spare £40 . . . a little time, being that Tho. was to go away and we had it not all ready. So he was willing and I . . . gave him bonds for the £40 till Michaelmas." On another occasion during the persecutions of the early 1660s, when many letters of dissenters were intercepted, Bridget Fell wrote that the Hall had been searched, and Thomas Rawlinson, coming while the searchers were there, was arrested and taken away. However, because his relative, Robert Rawlinson of Cork, a justice of the peace, had originally sent out the warrants for this search of Quaker homes, Thomas was not incarcerated at Lancaster castle as were the others. Bridget wrote, "The sheriff show[ed] more favour to him than to other Friends in respect that he is his kinsman."[5]

Rawlinson felt deep disappointment and anger when, a few years later, Margaret and her daughters began to accuse him of dishonesty. He expressed his sentiments to the Fells in the Prologue of his book:

> [You are] murtherers and betrayers of me; in my simplicety and innocency of my hart towards you all, I have had such confidence in [you] . . . who seemed so faire alwaise to me . . . [you] have betrayed my trust p[ro]veing such miserable Frinds and comforters to me, such unfaithfull people yt are like a brocken tooth and ye foote out of jointe . . . so let him yt readeth [this book] understand . . . if it had been but from you yt are wthout God in ye world wch had delt thus wth me, I could Bett[er] have borne its marke, yt are enemies to Truth, but it hath been from you of ye same househould of faith wth me by p[ro]fession yt hath thus p[er]secuted me and rewarded me evill for good, who have eaten at ye same table wth me and now are my greatest enemies.[6]

When the Fells purchased the iron bloomery in 1658, they hired Thomas Rawlinson to be their iron steward. When Judge Fell died in October 1658, his son George Fell, then of London, took possession of Force Forge. During the period of George Fell's absentee ownership, the forge was run by Margaret Fell and her daughters. In June 1666 George Fell signed the forge over to four of his sisters: Sarah, Mary, Susannah, and Rachel. Rawlinson, as steward between 1658 and 1663, dealt with the Fell women only as his employers. The disagreement between him and the Fell women occurred after his 1663 departure from the Forge.[7] Rawlinson's account blames any shortfall in profit on a fall in iron prices, and that the Fell's forced him to spend his own money to cover operational costs. He claimed that the Fells not only refused to reimburse him properly, but that they altered his account book entries to make him look blameworthy. The Fells accused Rawlinson of stealing profits from the forge and entering unnecessary expenses for running Force Forge in the account book.

The case was initially arbitrated by the local Cartmel meeting; gradually the Swarthmoor Men's Monthly Meeting and the Lancashire Quarterly Meeting became involved as well. The Quaker process of dispute settling followed the format outlined first by George Fox and subsequently by Robert Barclay. Quakers were to bring their personal disagreements to the local meeting where "in the spirit of Christ, a just solution might be found."[8] In July 1668 the SMMM, convening at Cartmel, recorded that Rawlinson was guilty of wronging Margaret Fell and her children. At that juncture, James Taylor, William Wilson, and Reginald Holmes were chosen to confront Rawlinson concerning the problem, and the SMMM ordered him to appear before the next meeting to give an account. The dispute continued in several successive monthly meetings before being passed to the quarterly meeting at Lancaster for resolution. According to the LQM minutes, Rawlinson was notified to attend the meeting in Lancaster in December 1668, but did not appear. Rawlinson explained that his absence was due to a bad horse who refused to cross the channel of Morecambe Bay. Thomas Salthouse, an important member of the SWMM and close friend of the Fells, believed this was a "hide and seek" strategy, used by Rawlinson for his own advantage. By February 1669 the monthly meeting was exasperated with Rawlinson's behavior and read a paper against him.[9]

Yet in 1680 Force Forge was handed over to Thomas Rawlinson, and in 1681 Thomas Lower, who through his wife Mary was a part owner, released his share to Rawlinson. The transfer of property included the forge, an adjoining house and furniture, tools, and machinery. However, after 1681 the sisters continued to have an interest in the forge, in which some of their money was still invested. Ross states that Fox also invested some £144 in 1683. In July 1684 Sarah wrote to her sister Rachel at the Hall:

> Take thy care in getting in the forge money, very kindly, which would be hard to get, if not looked after carefully by some in the country, and since you have got 30/- of that *ill smith*, do not use any severe course for the remainder, but if you cannot get it, by fair means, it must be lost, and I shall abate it, out of that which is due to me of it.[10]

In addition to the new-found Rawlinson MS, the *corpus delicti* of the dispute was Rawlinson's account book of income and disbursements of Force Forge. Accusations and counteraccusations flew back and forth between Fell and Rawlinson based on evidence given in the accounts. The forge accounts help to fill in many pieces of the puzzle of the Rawlinson-Fell case.[11] The account book opens with Rawlinson's impressions of the forge when he became the Fell family's steward.

> When I came to itt; it was so much roten down and in a decay that no man could stand dry headed in no part of the forge when it raint, and

belows and lether stopt, patcht with clouts, and al dames and goeing work and wheles much brocken down in many places and roten, and hutch and all wheels and cases and al implements and appurtences in bad order.

Over a three-year period the forge was repaired at the cost of about £40 per year.[12]

According to Awty's calculations of Force production, approximately 600 quarters or 360 tons of ore were taken by pack horse to the forge over the period covered by the accounts. This is significant because the Rawlinson MS disputes these amounts with Fell. Over the period of Rawlinson's stewardship, the forge produced about 90 tons of iron from the ore. The annual average rate of production for Lakeland bloomery forges in the seventeenth century was about 20 tons per annum. In 1660 Force Forge produced its highest annual tonnage of 23.75 tons of iron. Although iron products were shipped by pack horse all over Furness and south to Lancaster, the majority of the forge's iron was sold in the Cumberland and Westmorland areas. The number of customers peaked in 1660–61. Rawlinson traveled for iron contracts in order to increase sales, and in 1661 a shop was opened in Ulverston. The shop helped to increase iron sales in Furness and Cartmel, but this was accompanied by a subsequent fall in sales to Cumberland and Westmorland. Moreover, in 1661–62 prices of iron, and probably of coal as well, fell and continued to fluctuate through 1663 when Rawlinson left the forge. According to Awty's findings, the forge was operating at an overall loss over the five-year period 1658–63, with the value of iron in stock plus accounts outstanding at £248 in 1663.[13]

In the seventeenth century there were several iron forges operating in the lake district of Lancashire. Awty has suggested that the fall in iron prices experienced by the Fell entrepreneurs was due to local competition and the rise in supply of iron production. The production rate dropped after 1662, apparently not due to any shortage of coal but to the elevated price of ore. Other forges in the area were competing directly with Force, the result being a slight decrease in customers and the consequent opening of a local retailing shop to offset the trend.[14] This suggests that Thomas Rawlinson was not personally responsible for any shortfall in profit in Force Forge iron production.

The Fell women considered Rawlinson's personal expenses to be quite high, and it appears that they refused to accept most of them. For example, in 1659 Rawlinson submitted a personal expense account which included his clothes, food for his horse, and other necessities amounting to £13.6.10.

The two largest single items on his list of expenses included £2.16.0 for a pair of sheets and down for making a feather bed, and £1.18.09 for "buck leather skihnes [skins] to be a doublett and a par of bretches and for dresing them and for 3 calfe skins [for] . . . lyneings and pockits and for all other nesciaries and for makeing them." Although this was not the central issue of dispute, it undoubtedly contributed to the Fell womens' annoyance.[15]

In 1664 Margaret Fell and George Fox were prisoners in Lancaster Castle. Rawlinson, accompanied by his father-in-law, Thomas Hutton, visited the prison on 26 September to show Fell his accounts. Hutton recorded his view of the meeting at which both parties seemingly arrived at some agreement that subsequently proved to be only a temporary truce. Fox wished to remain aloof from Fell's business ventures, for only after the discussion was over, did he enter the room: "George Fox came into the rome from out of an upper rome and spoake theise wordes to them, or wordes to the like effect, vizt. Frendes, seinge you are agreed let all jealicies and prejudices bee put out of your myndes and come into the Unitie."[16] Ross's account claims that Margaret commenced Quaker proceedings against Rawlinson in June 1668, after her release from Lancaster castle. However, the above quotation and a letter to Fell from Rawlinson in May 1667 indicates that the dispute had commenced much earlier. In fact, it probably continued without much abatement from 1663 onward and she actively pursued her case while she was in prison.[17]

A LETTER SHOWDOWN

Thomas Rawlinson explained his side of the story in letters of 1663 and 67, often using terms that were biting. He described Margaret as the "glory of Swarthmoor" who was a "painted sepulcher yt . . . stand[s] all awhiten[ed] on the outside." She was, in his opinion, a woman "puft in mind and so high and pride[ful]." She had been too "longe from home so many years suiteing hunteing, va[u]ntinge, and gadeing abroad." He looked upon her as a person whose religion was vain and who "hath longe secreitly watcht for an opptunety of advantage . . . and for an occasion aga[ins]t me."[18]

Rawlinson maintained that he was innocent of all charges of thieving at the forge. In writing his book, he wished, so he claimed, to show the wider Quaker community the injustice of his treatment by Margaret Fell and the SMMM in the ensuing arbitration. To Fell and the local meeting Rawlinson wrote: "[you] have put out such unjust, unequall order [?] agat me . . . whose waise and doings herein are come . . . to be openly made manifest . . . how yee have all delt wth me wch must no longer be wincked att." He said he was sending his book among Quakers because his many letters and pleadings to Fell and the SMMM had gone unanswered, so "this shameful business of M:ff: . . . will go forth and be read . . . onely among all Frinds ith [in the] Truth to be read in all nations, cities and Countries where Truth is known." Rawlinson saw this as his only recourse. Claiming not to be intimidated by Fell or the meeting, he asserted, "I do not much heed ye said M:ff:s flaterings nor powers."[19] However, he alluded to her status and pointed out his proper deferential behavior toward her and her daughters in the past:

> You are all so p[ro]ud and haughty and full of opression scorneing . . . [and are] high minded . . . . Ye liueing god is witnes this day how faithfully I have alwaise humbled myselfe to yw and gone through much [?] . . . in the countrie and abroad for yw . . . wth my life in my hand amonge robers and theues in my troubles and sorows as aboute yr busnises and through great stormes, drought and snow . . . night and day; and through great and deep waters, many [times] riseing up early and travele[d] late wch many could not . . . to doe y[ou]r seruice.

He concluded his litany of loyal behavior to her over the years with the following:

> I hear so longe after I have left yee . . . yee [are] enquiring abroad amongst ye people of ye world from hous[e] to house yt have had dealeings wth me to see if yee can get any accusion of evill in me by ym . . . as touching my trust . . . amonge ym behinde my back.

Rawlinson asked her why she had not expressed her dissatisfaction with him face to face when he was in her employ. He felt that Fell's latter day efforts against him revealed her to be a "false" person and one who had lost all "piety, virtu[e], godlines, truth – honesty." He added, "I speake bouldly, faithfully and frely not fearing ye wrath of man or woman wt yee can do unto me though [you] be great . . . mighty, fat and stronge and full of this worlds goods."[20]

There are other scraps of evidence that suggest that Margaret Fell had an authoritarian reputation within the movement from the beginning. A few examples will serve to point this out. The early Quaker leadership disapproved of missionaries marrying, having families, and traveling abroad together. The young itinerant mnister, Thomas Holme, married Elizabeth Leavens sometime in 1655 and Elizabeth soon became pregnant. Margaret Fell disapproved of these missionaries marrying, having a child, and traveling together, thereby depending on Quaker hospitality for themselves and their infant. She wrote Thomas Holme of her disfavor and insisted that they remain apart in their ministries. Holme responded, "with tears upon my cheeks, . . . for the charge . . . thou complains of, it is yet to come, and if our going together be the ground of what is against us, the ground shall be removed . . . for we had both of us determined long before thy letter came to keep asunder." The Holmes remained separated through much of their active ministries and their married lives. They had three children whom they left for others to raise on point of conscience. Fell's remonstrance was that "he who loved father, mother, wife or child more than his service was not worthy of it."[21]

There were other disputes with Friends that involved Fell and that were also litigated within the Friends' meetings. For instance, in 1655 Christopher Fell of Cumberland (who, as far as is known, was not related to Margaret Fell), was in prison and "gave a paper to bee read publickly in meeting," apparently against Margaret Fell. A letter from Margaret to Christopher Fell refers to him as having "sent a paper amongst Friends ag[ains]t a paper of M:ffs," the usual initials used to refer to Margaret Fell. The nature of the dispute is unknown. However, the Ellwood edition of Fox's *Journal* deleted his name, which is curious in light of the fact that his name had been included in Fox's original journal. This indicates that at some later date Christopher Fell fell into disfavor and/or may have left the Quaker community. It is possible that his opposition to Margaret Fell caused his eclipse within the community in much the same manner that Rawlinson was subsequently eliminated from the meeting, although in the latter case it was not permanent.[22]

Another instance of Quaker infighting with Margaret Fell occurred sometime in the 1650s when Isabel Gardner and Peter Moses were accused of witnessing against Fell both secretly and openly at meetings and of lying to her. Peter Moses had claimed that Margaret Fell had departed from the living God. He had, it seems, refused to acknowledge her reproof of them. Her alleged response was that she would laugh at their destruction. Such vignettes, although very sketchy, suggest that contemporaries' criticism of Fell was effectively resisted and that evidence of contention may have been suppressed. It also suggests that she, like George Fox, was capable of some petulance and lack of sympathy when thwarted.[23]

With this background in the dispute, let us return to the charges and countercharges in this case. Rawlinson accused Fell of wrongdoing in her business dealings and stated that he deeply regretted ever having done business for her, for through her he had "come to much harme and hurt."

[She] would needs compell me to speake and do things agat some . . . wch was not right in ye sight of god, [she] wo[uld] have laid on me to doe while I was in trust for [her] I cou[ld] not doe, and therefore being yn wearie of such like opression . . . I . . . wisht [they] all . . . fin[d] another to supplie my place yet [they] would not let me goe.[24]

In the first letter of 1663 to Fell and her daughters, Rawlinson alleged that Margaret accused him of skimming off £100 from her cash intake at the forge. To this he responded: "I have no money of yrs at all . . . but truth is yee are all much indebted and in arrears [to me] . . . but seeing ye ha[v]e accused me so falsely herein now I charge yu all to find when and wherein I [h]aue wronged y[o]u." He claimed that he "laid out of my purse for ye . . . wn [when] Iron would not sell wel as it had done." This states the crux of the dispute between them that escalated over the years. First, he repeatedly stated that he made his own personal doles of money toward running Force when money was tight. Second, he asserted that he was exact in his keeping of the accounts. He stood on the integrity of his account book to prove his honesty, concluding that throughout his years of stewardship neither Fell nor her daughters disagreed with his yearly accounting of the business and that they had had access to the accounts.

Rawlinson's accusation that Fell had not reimbursed him for the procuring and selling of iron ore, coal, and finished iron recurred in 1672. He wrote that he had "laid [ou]t of [his own] purse for her and her children in stake [stock] at ye fordge w[he]n they would [n]ot helpe to supply ye stocke yn wn iron fell from ye price it had been att . . . but it laid [on] my hand[s] in my custody unsould."[25] Following the purchase of a large stock of iron ore and coal, the price of iron subsequently fell, and it would not sell at his asking rate. The Fell women preferred that it "lay by us unsould" than reduce the sale price, which fell to about 50s or £3 per ton cheaper than earlier. They persuaded Rawlinson that it would rise to the old price again and they would wait for that to occur. In the interim he had "to lay out much of my owen money yn" to pay for a large stock of coal and ore rather than let the forge stand idle. When he left the forge, he delivered the unsold iron to Fell, hoping she would reimburse him for it. She did not give him his £100 upon demand; rather she began to "shiffle" with him and refused to pay him even though he followed her from time to time, asking for it.[26]

In a letter addressed to the men's monthly meeting in 1673, Rawlinson related an incident concerning his purchase of coal for the forge from some men of the county. He was later accused by these same men of not paying them a fair price, to which he responded that Margaret Fell had forbade him to pay them a higher amount. He had struck a bargain with the men to pay them half upon their first delivery of coal, and the other half when the remainder of the coal had been received. However, according to Rawlinson, they "brought ym [the coal deliveries] not in untill severall years after . . . longe after their bargan was expired." They wanted the rest of their money and Fell said to put them "in suit forth wth or she would looke to me [for] . . . ym so longe." She threatened Rawlinson that if he did not put them in civil suit, she would "fale [fall] out wth me." He allayed Fell's anger by suggesting to her that he ride from house to house warning the men to keep their bargain or they would be sued. Apparently, Fell was "pasified" when he told her that he had informed the men and everything was in order. However, when the time limit of their renewed bargain had expired,

> she asked me if they had all p[er]formed, and I tould her yt some of ym had not and wth this she began to be more angry with me, wherefore I had not put ym in suit before, and I tould her yt I was sore troubled they should p[e[r]forme no beter . . . [I] asked her if she pleased I would get ym to come over to her to see wt course they would take to see her satisfied, and she tould me yt she would not have ym to come att her attall, but yt I must goe to put ym in suit fort[h] wth I must not trouble her wth ym nor let ym come at her she said so wth this I went to ye fordge [and] sent for all those men . . . The men did not want a suit and were content with the payment of coales before ye rise in price.

They were paid according to the original bargain price and acknowledged themselves satisfied. He concluded with the note, "for all my end herein was nothing but . . . peace and good wil to ym all . . . and to make peace between ym." Rawlinson felt wrongly accused that these men later became witnesses for Margaret Fell against him for alleged unfair business dealings.[27]

Another of Fell's accusations against her former employee involved 300 quarters of iron ore taken from Adgarly. She claimed that this had been delivered to Rawlinson while he was steward at the forge and that he had failed to note it in his accounts. Rawlinson claimed that he had left the ore at Adgarly when he left the forge and that Fell had hired her own men to transport it to the forge. Later she had them swear to the commissioners investigating the dispute that they had delivered it all to Force while Rawlinson was still steward, thus making Rawlinson appear dishonest for not entering this in the accounts during his stewardship. Furthermore, he added, when he left the forge he gave her an exact account of "above 700 pounds in money, coals, iron and oare I left you in stock yn at ye fordge."[28]

Although the dispute was laid before the Friend's meetings for internal settlement, Fell seems to have taken the matter to lawyers outside the Quaker circle. This action renders this case unusual, for Quakers, as indicated above, sought to settle their differences among themselves, thus avoiding litigation in civil courts where the requirement of an oath conflicted with their beliefs. Fox's advice to Quaker meetings regarding the internal settlement of disputes was based on Paul's admonition to the Corinthians not to go to law before unbelievers (1 Cor. 6:1–8). Fox said, "it is a shame to the church of Christ, who had this power to judge, to go to law one with another before the unbelieving world." Thus Quakers from the beginning were rigorously counseled to judge one another's actions among themselves. However, occasionally matters could not be settled within the meeting, and in such cases resorting to civil litigation was not against the Quaker discipline if all internal negotiations had failed.[29] By 1667 Rawlinson wrote that Fell and her children, "who now sues me att ye law . . . are grown so churleish, impident and coveteous of harte . . . are so hardned in your sins . . . yt will not be content wth your owen, but thinks . . . yn shall maintaine thyself and thy children wth my wife and childrens right." He continued, "[It] is now almost three years since . . . yee first put me in suit and would not put ye busines to honest frends in ye truth though I [im]pplore[d] yee yn to chose two . . . faire frends . . . [and we] should end itt."[30]

Sometime prior to 1667 Fell called upon her non-Quaker brother-in-law Matthew Richardson, attorney of Dalton in Furness, and her Quaker son-in-law John Rous to be arbitrators. Rawlinson called upon his "father and brother" to be his arbitrators, with Colonel West, a non-Quaker, to act as the "umpior if need were . . . to end . . . ye arguemt betwixt us at Lancaster."[31] This early attempt to settle the dispute failed because Sarah Fell, acting in her mother's stead, refused the arrangement on grounds that Rawlinson must have "two strangers and no relations to him," while at the same time insisting that she would have West, Richardson, and Rous. Since Richardson and Rous were related by marriage to the Fells, Rawlinson retorted that this was "such unequall mesure . . . unjust and bad who durst not trust ye busines in ye ha[nds] of any honest men of no relation to neither of us . . . and so doe yee not now openly make your owen guiltiness and disho[n]esty exceedingly manifest unto all men who come but to understand how y[e] deale wth me." Meanwhile, the Fell women, according to Rawlinson, openly slandered him at their meetings at "Hawxhead, Swarthmore, Cartmel and Lancaster."[32]

In his second letter to the LQM of 24 June 1672, a copy of which was sent to Fell, he wrote that the arbitrators chosen for him by the Quaker meeting were Gervase Benson and John Wilson. Those chosen for Fell were Myles Halhead and Robert Widders. These four names are familiar ones in early Lancashire Quaker records, and Rawlinson noted that they were "quiet peaceable men" who "resorted much to her house." Nevertheless, Fell would not come to Friends' arbitration in spite of the fact that they sent for her. Rawlinson claimed that he was "stil ready and wileing to obay you in all things beha[v]eing [myself] . . . as a dutifull and obedient [man] . . . w[he]n they met on ye busines, he [Benson] saw me alwaise so tractable and wileing still to submit unto wt end so [t]hey pleased to make here in betwixt ye sd M:ff: [Margaret Fell] and me." Fell would apparently "not abide ye triall neither among frinds nor yet at ye law."[33] Although Rawlinson interpreted Fell's absence from the meeting as a sign of her obstinence, it may have been due in part to her frequent absence from Swarthmoor Hall. During the 1670s she traveled extensively for Quakerism. For instance, she traveled to Yorkshire in April 1672, and later in that year to an unknown destination.

In addition to the Quaker arbitrators, Benson and Wilson, Halhead and Widders, the two non-Quaker commissioners, Colonel West and Mr. Matthew Richardson, were concurrently involved in 1672. They called upon Margaret and Sarah Fell and Thomas Lower to meet in Ulverston to go over Rawlinson's Force Forge account book "onely between ourselves."[34] Rawlinson arrived at the appointed house on the designated day to meet the arbirtators and the Fell family. Instead he found a Quaker servant of the Fells awaiting him to take him to Swarthmoor Hall where they were expecting him. He recorded the following incident:

Sarah desired her [the servant] to tel me yt she would have me to come to her at . . . her house . . . not ye house ye comisioners appointed us to meet at. . . . At last she constrained me and I went alonge with her to her house . . . and yn wn we came to her house I found the[re] both Sarah, her mother and many frinds all there together in her house set. And so I sat down amonge ym and shortly after ye said M:ff: . . . began you to speake many lyeing words in this mater to frinds there present. . . . And when she had done I stood up and desired I might be sufered to speak and clear my selfe in this mater amonge ym yn. And as I began to let ym all see wt such a lyer she was there in how I could [say?] ye contrary to her face of wt she had said if

Feuding Friends 113

they pleased. And wth this frinds began yn to interrupt me and would not let me say no more but both she and they all began to tel me yt they would have me both to put ye mater in controversie . . . to frinds in ye truth and end ye busines, . . . and yt they would have no more such talk among us. And she tould me yt she was deserous now to put it to frinds in yet wth to end it and desired me to choose two or three frinds who I would and she and Sarah would choose as many. . . . She [told them] how [she] could never get me in minde to have put ye busines to frinds, so I did itt to ye stoping of all their mouthes.[35]

This phase of the Quaker arbitration process was sealed with a bond of security "according to her [Fell's] desire to abid[e] and obey their end order and award." In this maneuver, according to Rawlinson's view, Fell shrewdly moved the case back into the Quaker community while at the same time placing the blame on Rawlinson for not arbitrating among Friends sooner.[36]

Of the commissioners, Colonel West and Matthew Richardson, who were examining the case, it appears that they were not present at this meeting at the Hall. However, Rawlinson did comment that both Richardson and West were satisfied with his testimony at that time:

[They] all examined us both herein ouer and ouer againe yt [that] now knowes her wel enough [?] sees her clearly w[ha]t she is, and how she hath delt wth me . . . yt even of the [p]inciple man [Matthew Richardson], of her owen comissioners . . . upon examination [saw] how she had delt wth me acknowledged himselfe ashamed of her and her cause . . . he said yt it was a burthen to his conscience to heare it from ye mouth of such a one as she . . . he litle thought she had been such a woman to heare her speak and talke aboute religion he yt was her lawyer and counseller herein and neare [kin]sman and [in] ye flesh her brother in law who said to my comisioners he repented he should ever have had a hand in ye busines for her he saw it soe [v]ery bad.[37]

Rawlinson claimed that after Friends found Fell at fault and were ready to put an end to the business, she ceased relying on them. She would come no more "unto ym and yn after awhile, yee [Friends] know she . . . is now going to ye law againe agat me like a couerd [coward] and dare not stand ye treiall and abide ye end . . . she would not let ye said comisioners [Benson, Wilson, Halhead and Widders] meet on ye busines."[38] Rawlinson described this "shilly-shallying" thusly: "she yn [then] began to fret and cry out greatly w[he]n she saw yt she could not obtaine her maliceous, spitefull, lustfull, earthly ende on me by frinds ith truth no more yn she had got on me at ye law."[39] He ended his letter of June 1672 with the following self-perception *vis-à-vis* Fell's actions: "They saw me quit myselfe yn . . . like a man of a nouble and excelent spirit and understanding." He remarked further on the contrast between Fell's character and George Fox's: "Even for G:ff: [George Fox's] sake her now husband . . . would let her see w[ha]t she is who have lost ye sincerety and her virtue, soundness [and] sobriety." Finally, his musings include a rationale for Fell's going to the law. It was in order "to have terrified me so to save yourselves from restoreing me my owen." In this lengthy tangled process of Quaker and civil arbitration, Rawlinson mentioned that Fell had sued him in Chancery court, in "such a false bill of so many sheetes of pages in ye chancery suit agt me," accusing him of withholding more than £2000 worth of revenues in iron sales from her. Rawlinson countered this accusation with the response that such a quantity of iron was "more yn was made while I was steward att ye said fordge," adding that "many more such like lyes [were] exprest and declared in ye said bil and replication."[40]

THE SWARTHMOOR MEN'S MONTHLY MEETING vs. THOMAS RAWLINSON

Thomas Rawlinson's problems were not yet over. According to his account, Thomas experienced a hardening in Friends' attitudes toward him as the dispute dragged on, largely due, he felt, to Fell's studied attempts to denigrate him to his Quaker neighbors. Rawlinson witnessed what appeared to him to be Friends currying her favor:

> Some [Friends] yet are apt enough to doe for to plesure y[o]u all wth a kindness herein if they can now to gaine fauor wth y[o]u Mt [Margaret] yu are such a great woman that loves to be great in ye earth and amonge frinds; to sit as judges and to raigne as princes yt [that] site even as a queen upon ye people . . . and would to god yu did raigne wel in the truth and righteousness . . . for deceit and falsehood hath raigned longe enough.[41]

Elsewhere Rawlinson told Fell that she reigned over the Quaker church like the hated Anglican prelates, calling upon her underlings to do her bidding: "Thou assumes ye pl[ace] of supremacy so much . . . so I say even [as one] in ye great bishops chaire to judge ye brethren of ye lord and servantes of ye most high."[42]

Additionally, Rawlinson deeply resented her stirring up both the men's and women's meetings against him. Quakers, he claimed, "cannot be quiet and enjoy their meetings peaceb[ly] . . . [in the] mens monthly and quarterly meetings because of her and her children . . . alwaise puteing at ym [them] to stir them up into such a thing and troubleing ym on her behalf aga[ins]t me." He proclaimed she would endure her own torment for dealing so wickedly with him and his wife whom, he believed, Fell sought to exclude from meetings as well, despite their being unacquainted.

> They [the Fell's] lust so much after my blood also of my dear wife and children who never had any dealings wth her nor wth any of her children and yet she cries out agat ym and tould ye women at your womens meetings yt the[y] must not sufer my wife to come among you but that they must forbid her of their meetings.[43]

The Quaker fellowship was very important to Thomas Rawlinson, and his persistence in attempting to prove his right of membership is impressive. In spite of the fact that he was first dismissed from meetings in December 1668 and not fully restored to active membership until 1680–81, he never gave up his attempts at reconciliation although his bitterness grew over time. His persistence is evident in remarks written after he heard of the 1668 order suspending him from fellowship:

> I am willinge to subscribe and to give it under my hand in writing even as a certificate or a declaration to goe abroad to be read amonge all frinds if this wil give yu any satisfaction and content in this mater through my humility in abaseing myself to cut away all ye occasion of ofences yt hath been amonge us through . . . humbleing my selfe unto yu in the matter yt I am freely wileing even for ye lords sake to become anything unto yu all . . . and so be reconsiled together in ye lord.[44]

He claimed consistently that he was never guilty of any "wilful offense . . . [or] rebelion" toward the men's meeting and expressed anger that they never dealt "face to face" with him during their business meetings. Rawlinson also asserted the meeting had been unjust in conditionally expelling him before he settled his dispute with Fell without hearing his side of the story. Although the order against him read that he had refused to attend both the monthly and quarterly meetings to which he was called to defend himself, he claimed that his absence was due to the fact that no one had notified him to attend.[45]

Rawlinson's response to an order of 11 August 1668 included an account of his serious illness, which creates a vivid picture of the tenuousness of life in the seventeenth century. He wrote that he was "sicke nier unto death" and missed two monthly meetings, one held at Swarthmoor and one held at William Satterthwaite's in Cartmel.[46] He added, "I continued ye most pte of three quarters of a year before I was well and many neighbours about so died yn of ye same feavor . . . I [was] too sicke and weake in body since I can scarce sit straight in my chaire but for a litle untill I must lie down again . . . so I was disabled of all mine intentions herein." He also added that no one came to see him from the SMMM. However, the meeting claimed that two men were sent to him from the monthly meeting held on 8 September 1668 to request that he attend the next quarterly meeting at Lancaster. Rawlinson claimed that this also was a lie, for no one ever came to him either to attend him in his illness or to order his attendance at Lancaster.[47]

By December 1668, although still weak in body, Rawlinson was ready to attend the quarterly meeting at Lancaster. He set out early on the day of the meeting (28 December). Crossing the treacherous sands of Morecambe Bay when the tide was out, enroute to Lancaster, Rawlinson's horse, which had never crossed the sands before, suddenly shied away from the deep streams and pockets of quicksand and would go no further. The panicky horse was lamed by this incident, and Thomas lost the tide. He decided that the "hand of the Lord was in it" that he would not be in time for the meeting.[48] Rawlinson claimed that he told several friends, including James Taylor, of his misfortune, and they had appeared satisfied with his explanation. He sent a man to Swarthmoor the next day who explained to Thomas Salthouse, "how it happened to me," Salthouse acknowledging he had heard of his mishap already. As the order was dated 30 December 1668 and circulated at the meeting of 17 February 1668, it was apparent to Rawlinson that despite his explanation and their apparent acceptance of it at the time, the meeting had been two-faced with him and had decided to cast him out anyway.[49] Rawlinson's excuse seemingly did not arouse much sympathy on the part of the LQM members. This may be due to his leaving home the morning of the meeting day. Considering the distance to Lancaster and the difficulty in crossing the sands at low tide, it might have sounded like a halfhearted effort on the part of Rawlinson not to begin his journey a day earlier. Yet the very real possibility also existed that no one in his Quaker community would have been willing to put him up for the night enroute to Lancaster.

Rawlinson made numerous attempts to regain membership in the men's monthly and quarterly meetings. Finally, after repeated failures, he threatened that if the meeting did not rescind the order put out against him, he

Feuding Friends 117

would write an answer and send it out to be read among Quakers to explain how "even as ye said M:ff:, her children and yee all have delt with me." As his eighteen-page letter, written to the quarterly meeting and delivered on 30 March 1670, was never answered, he asserted that his copy of the letter would remain as "a testimony aga[ins]t ym."[50]

Rawlinson's second letter of 24 June 1672 to the Lancaster quarterly meeting was likewise never answered. The meeting, he claimed, in following the "unruly, disorderly, lyeing spirit . . . amonge you," forfeited the real spiritual power and unity of the true church:

> This darke wilfull spirit in yu all agat me wch seekes my hurt, though now for awhile I patiently heare it until the heat of your spirits be over . . . yee have . . . hereby lost your dominion and your sincerity and become as blind men herein . . . who cannot see, even as dead men that hath no cense and understanding . . . ye lord will arise [and] tread down his enemies under his feet.[51]

Rawlinson's invective rose to a crescendo as he described the meeting's unjust treatment of him in favor of Fell:

> Frinds verely I say yt it were more comendable for yu to put out ye [s]aid M:ff: and her children forth from amonge ye . . . and to have nothing to do wth ym . . . which now [would be] even becomeing ye truth to lay aside all their auld dirty, lust full, filthy, stincking strife . . . [they are] most fit for worldlings . . . and not for ye children of ye kingdom . . . seeming she is a woman and her children that will not be admonisht . . . and advised by truth and good reason . . . [who] stinks now even in ye nostrals of god and of sober, honest, just, good and godly peaceable men.[52]

Several incidents related by Rawlinson strongly suggest a studied plan of coercion within Quakerism, sometimes through silent ostracism, sometimes by rough verbal handling. The defendant expressed it thus: "I verely felt yt [that] . . . [you] stood all even as strangers to me in y[ou]r currige [and] behavior to wards me at ye mens meeting." He claimed that when he and Fell presented their case before the meeting, she spoke against his dealings so strongly that "wn they began first to meet and sit upon the busines . . . they began to speak very sharply and roughly to me." He insisted to the meeting that they misunderstood the whole business and their cold treatment of him was due to the fact that, "I had so litle to say for myself in this mater to ym yn [then] but let her go on wth her talk . . . she was so fule of words: onely I sp[a]ke a few words." He chose, in typical

Quaker style, to remain taciturn and "so let ye thing and truth . . . rest upon you after I had cleared ye busines."[53] Such behavior suggests that Rawlinson may well have exhibited his own brand of perversity as well.

Rawlinson attended the next monthly meeting despite his dismissal from the men's SMMM and the LQM in February 1668. When he began to speak in his own defense concerning the order, William Wilson, "drew all the others outside away from me and I followed out the door and asked him to see the order but he refused."[54] Rawlinson was very persistent, for he continued to appear at meetings after his suspension, while refusing to accept his condemnation in writing. The meeting of 28 March 1670 put yet another order out against Rawlinson, which was delivered to him by the same William Wilson. The order said he stood in rebellion against the meeting and it would therefore have no further fellowship with him.[55] He responded, "I stand in steadfastnes of faith and am now nothing terrified herein nor discouraged in this." He accused his erstwhile friend, James Taylor, in whose handwriting the order was written, of being an equivocator against him. Taylor expressed the exasperation of the men's meeting, as well as the frayed nerves of all, when he labeled Thomas Rawlinson the "faulty man in ye auld busin[es] between M:ff: . . . and I say for shame quit thy selfe like a man . . . keep from meetings for yu canot be sufered to come amonge us." The order against Rawlinson here points out that it was not a complete disownment. Rather it was a disciplinary order to take a moratorium from attending meetings until he could come to terms with Fell.[56]

Thomas Salthouse provides another example of how the membership treated the persistent foe of Fell. When Rawlinson appeared at a men's monthly meeting sometime in 1673, Salthouse treated him with disdain:

> There is a distance between yee and us said Thomas Salthouse to me. . . . [He] bad[e] me be gone and wth this I asked him a reason yn wherefore he would have me gone, wt evill I had done, and desired him to convince me of sin or of any disorderly walkeing amonge frinds . . . and he tould me yt they had other businesses in hand yn to stand talkeing wth me, and so bid me be gone.[57]

Moreover, on 8 April 1673 Rawlinson gave 5s to a Quaker collection for the poor, but the money was returned to him from the men's meeting with the following statement:

> Th. Rawlinson we understand yt yu gave 5 shillings to a colection at Hawkshead meetinge and yu being disobedient to ye Order of ye

quarterly meetings and so out of unety with friends, we do agree and judge it conveinent yt yu have thy money againe and yt yu do refraine comeing to meetings untill yu come into unety wth frinds wch canot be untill yu be reconsiled to ye quarterly meetinge or ye busines between M:ff: or her children and yee be agreed.[58]

According to Rawlinson, the meeting was firmly in the grip of Margaret Fell. For nearly two decades, internal pressure on Rawlinson to back down in this dispute within the local Quaker meeting was unyielding. Rawlinson was intractable, firmly resisting all forms of coercion, always reiterating that he was innocent of the accusations laid against him. He insisted that the meeting expelled him without having given him a fair chance to defend himself. After his unaccepted contribution for the poor, he accused the meeting of spiritual elitism in making exceptions as to whom could join the household of faith. He added, "yee . . . hath renderd me [?] redickolous amonge frinds."[59]

THE ACCOUNT BOOK OF FORCE FORGE

Thomas Rawlinson maintained that, over the years of his employment by Margaret Fell, he had stood on handshake honesty in respect to his dealings with her. He asserted, "I simply trusted you herein alwaise makeing no doubt of any need to have anything to show under any of your hands in writing between us, we all agreeing alwaise." During his years of stewardship, it was his annual custom to leave his ledger of the forge with the Fells at Swarthmoor, giving his year-end tally of income and disbursements. Sometimes he left the account book in their hands for a week or more, they always returning it to him, having "acknowledge[d] ym selves still fully satisfied, wel pleased and content."[60]

However, Fell and Reginald Walker, Rawlinson's successor at the forge in 1663, accused him of taking kickbacks. Fell, with the aid of Walker's testimony, claimed that Rawlinson had sold iron, labeling the sale as being in arrears and unpaid, when he had actually received all the payment from the smith before leaving Force. The defendant challenged his accusers to prove this, but according to Rawlinson "wth this he [Reginald Walker] put to his horse and road away." Rawlinson counterclaimed that Walker framed him for his own selfish ends for Walker was deeply in debt when he became Fell's steward and sought to enrich himself. Walker wished to make Fell and Rawlinson enemies ("to set us att a distance," in Rawlinson's words) because Rawlinson frequently revisited the forge after his tenure there was finished and Walker did not want him there for fear he would discover irregularities. Rawlinson claimed that Fell sent him there "to see how things were."[61]

Rawlinson's "Queries to Margaret Fell," written in 1672 contained an accusation of falsifying the accounts relating to the charges for the stock of iron ore cited above. His query read,

> whether you did not ad and deminish to ye same booke and altered ye figures of yt [that] book . . . behind my backe and neuer made me acquainted wth it w[he]n y[o]u gave me ye book . . . who therein charged me wth yt wch I had never received . . . y[o]u yt altered ye charge . . . from a litle sume to a greater so [sum].[62]

Rawlinson repeated this point in another querie of the same letter: "In money, Iron Oare, coales and stocke as alsoe did [alter?] by Sarahs owne hand writing in ye said originall booke of accounts did itt not . . . [and she] brags behind my back of many hundred pounds more." It was only at a much later date, Rawlinson claimed he discovered the changes. "Longe afterward . . . I accedentely [found] yt place and saw yt it was not my hand writeing and began to muse and to consider on itt awhile upon readeing of itt and caleing then to mind wel yt it was a false charge. I came to see you and asked yee wherefore yu had don so."[63]

Finally he accused Fell of bribing witnesses to speak against him. He wrote that she had enticed witnesses to swear in her behalf before the commissioners of the law and,

> awhile after when they had been before ym commissioners and examined but some of them chanct to say that they said as thou bad[e] ym. M[argaret] they were so only . . . wth aboundance of ye good [?] strange [strong] drinche [drink] and much money yee had bestowed on ym yt they cryed aha, aha, who but M[argare]t ffell for a brave and generous gentle woman . . . yee all joined hand in hand together those commisioners, lyers, druncherds wth yu such false witnesses and evill speakers.[64]

Rawlinson tried to prick Fell's conscience in referring to her recently-deceased son:

> Did you not counsall thy son Geo: ffell falcely to forswear himselfe agat me to things he never saw nor knew by me touching my accounts, disbursm[en]ts and charge[s] and receits. . . . [He] is now cut of[f]

and gone to his longe home and has a [reward?] according to his work no doubt. . . . Wherefore take heed his blood be not requiered at thy hands M[argare]t, for hereby ye may learn and hence observe, yt it is not good nor safe for any to be so hate[ful] and envious one agat another as yee have all been agat me herein.[65]

The single greatest problem Thomas Rawlinson had to overcome with the Fells and the men's meeting was his own somewhat mysterious affluence. When he became a Quaker as a young man, he was rejected by his father, an episode experienced by many Quaker converts. Because of this, Quakers assumed that young Rawlinson was in impecunious circumstances when he first came among them. When it became apparent to the local Quakers that he had some money of his own, this irritated Fell and made Rawlinson suspect to the Quaker fellowship. In 1667 he discussed his affluence and Margaret's actions against him:

Yee had heard yt I had lent out much money in ye country and I tould yu also yt it was true yn [then] w[he]n yee tould me did yea not M[argare]t [?] I also p[re]fere[d] . . . to tel yu then how I came by it all . . . seing I was [your] steward. If yee had any jealousie of me or had a mind to know [how] I came [b]y itt and yee tould me yn yt [that] yu had none but said yee did beleve yt I am honest . . . and so frinds wt is ye mater now yee begine . . . so longe after I have left yu to covit after this money . . . many years since; I say are yee come to such want and growen so poor [n]ow wth ye costly liveing yt yu now would have my estate and money . . . or is it more through envie and malice in ye harts towards me.[66]

Subsequently the Quaker meeting put the same question to Rawlinson. He explained that on becoming a Quaker, Friends offered to supply his wants and expenses. As Margaret Fell's steward, he traveled all over the south, to London and to many counties, and he took nothing from Friends, for he had enough of his own. He claimed, "I bought great store of linen wolen," before becoming Fell's steward. Before he knew her, he had accumulated some money by parsimonious living and by peddling textiles or investing in the textile market. He subsequently made more money of his own in the textile industry while traveling for Fell. According to his account, he had told the Fells at the beginning of his employment that he had loans in the county, and they were not upset about it at that time.[67]

The Rawlinson manuscript is unfortunately silent on the denouement of this lengthy dispute. Following a four-part letter to the men's monthly meeting and quarterly meetings, there is a gap in the correspondence from 1673 until 1680. A five-page epilogue written in 1680 expresses his joy at the happy ending to his problems with Margaret Fell. Mysteriously, Force Forge became Rawlinson's property in 1680. It is known that George Fox was in residence at the Hall from September 1678 until March 1680. A necessary question must therefore be posed: Could Margaret Fell's spouse have convinced her to relinquish her property? If so, on what grounds? Because Fox was at Swarthmoor, it is possible, that because of his status he may have been called upon to be the final arbitrator. Fox was probably the one man who could convince Fell to give up her iron business for the sake of internal unity lest the dispute become too widely known. Rawlinson, for unrecorded reasons, leaves the modern reader with the nagging question of what exactly happened in 1680. Did the meeting finally swing over to Rawlinson's side in the dispute, and, if so, did this occur through George Fox's personal influence? This will remain unanswered unless other evidence comes to light. The evidence that we do have, namely that of the SMMMM, the LQM, and the original Force Forge accounts, authenticates this dispute and records its outcome in favor of Rawlinson, a man who lacked Fell's social status and influence in the local Quaker community. Moreover, the litigation within the Quaker meeting finally vindicated a man who was completely excluded from the personal support network of the Quaker group during the entire protracted internecine feud. This raises a significant issue relating to the history of the early Friends.

The dispute was recorded in the SMMMM and the LQM over an extended period. Considering this, it is surprising that the Rawlinson document, which had originally appeared in several copies, is missing from the early Quaker records. Even more curious is the fact that the minutes of the SMMM, which are essentially intact from the beginning of the meetings' existence in 1668 into the modern period, are missing for the years 1674–91. These years span part of the contentious debate between Rawlinson and Fell and its aftermath. Some stray epistolary evidence indicates that the case was still discussed as late as the mid-1680s.[68] It might be thought that these minutes, which cover the last years of Fox's life, would have been deemed very valuable to the earliest Quaker record. Considering that Fox and his contemporaries manifested the deepest interest in preserving all documents dealing with their activities in the formation of the early Quaker meetings, the missing minutes may suggest an effort to suppress an unsavory affair that compromised Fell's reputation.[69]

Such a predicament, one may suppose, would have undermined the solidarity of Quakerism in a crucial moment of its development. Fell's alleged questionable business ethics and her apparent disruption of church discipline become more probable in light of this possible suppression of the early records. However, a caveat is necessary at this juncture. This theory is based on evidence that comes from one side of the dispute, and the source is obviously biased. Fell may have had her own unrecorded reasons for believing Rawlinson had defrauded her. Furthermore, it was the period of another quarrel, the Wilkinson-Story dispute, which led to open schism in the mid-1670s. It is possible that Rawlinson was part of this Wilkinson-Story group who opposed Fell and Fox, but there is no evidence to prove that. This notwithstanding, the document gives us splashes of new light on an early Quaker internecine argument of considerable proportions. The overall sense of this evidence is that Margaret Fell was a very powerful woman in Quakerism and her word was not easily overturned.

Although Rawlinson's alleged innocence and final victory might have been a blow to Fell and her family in terms of personal prestige, this setback was only temporary. Despite absence of the monthly meeting minutes of the dispute, there is no evidence to suggest that the meeting ever admonished or disciplined Fell. Rawlinson concluded in his "SALME of praise and thanksgiving to god after such a notable deliverance," that he was a despised man who suffered alone and finally won. He wrote, "My frinds fled fare from me and stood at a distance to see wt would become of me and [?] they looked so lightly on me." His song was for divine vindication of innocence as well as justice done to those who despised him:

> [God] yt hath been so merciefull to me herein and put al mine enemies to silence before me. . . . [The] man yt walkes with truth, righteousnes and inocency and speakes up rightly, yt despises ye gaine of opresion, shakes his hands from houldinge of bribes . . . and shuts his eyes from seeing of evil [?], his broad [board] is ful and his waters shall never faile.[70]

*　　*　　*

In reviewing the case of Margaret Fell versus Thomas Rawlinson, this chapter does not pretend to be a definitive study of this seventeen-year bitter battle, for there is insufficient evidence to accomplish that. Rather, this chapter has noted particular aspects of the dispute that directly or indirectly relate to Margaret Fell's role as a significant Quaker leader. Several important points have emerged to indicate the value of a further study of this internal Quaker litigation and, more broadly, of church discipline and business ethics in the formative period of Quakerism.[71]

A major theme emerging form the evidence is the importance of the Quaker meeting to people like Thomas Rawlinson. His identity, sense of security, and personal economic stability were closely linked to the Quaker community. Without it he apparently felt adrift in the world of political and religious turmoil of Restoration England. That he was so persistent in his efforts to regain full membership status in the group is remarkable. His insistence on appearing at meetings despite orders to stay away, his unwanted donation to a collection for the poor, the lengthy letters that he laborious recopied to clear himself, all denote a man deeply committed to, and in need of, his religious fellowship. Rawlinson's numerous *cris de coeur* were summed up in his assertion in 1673: "I said I had never done anyone any wrong but you had done me much wrong . . . I have no need to seek reconciliation, only to come and testify against you who have abused and slandered me. You should seek to be reconciled to me." But despite these flashes of anger against his religious family, he ended on a note of imperturbable faith in the possibility of reconciliation.

> I performed my duty to my parents in all things until the day my father turned me out of his house and shut his doere agat me . . . ye lord opend ye doer of life unto me . . . and brought me into his house . . . wherein he will never shut me out.[72]

Secondly, this long-term episode is highly suggestive that an elitism and covert favoritism contrary to Fox's theory of spiritual parity, operated within Quakerism. Social and economic factors did seemingly operate to exclude undesirables from the group. Rawlinson's resiliency to stand and fight the Fell family's attack also suggests that Rawlinson was not a poor, struggling workingman, but was a man of some independent means. The fact that his social background appears to have been of minor gentry or well-off yeoman stock places him on a rung probably not much inferior to Fell on the local social ladder. A servant of the "meaner sort" would not have had the financial backing or the social status to stand successfully against the power structure that was evidently a de facto reality in the Swarthmoor meeting.

It is also apparent from this document that Margaret Fell wielded enormous power in local and wider Quaker circles. Her methods of intimidation and ability to sway the local membership to reject Rawlinson indicate that her position in the Lancashire meetings was virtually unassailable. Rawlinson's predicament and his anguish are clear in this letter to the Swarthmoor men's meeting.

> I have tould yu [the meeting] ye truth herein and yet yee wil [not] hear me in this mater but has set me at nought because I am not as

[e]loquent in ye busines as ye said M:ff: and her confederates to set it fort[h] . . . and if ye said M:ff: spooke but [a] word to yu in this mater agat me, though never so false yee would hear [h]er and believe her, receive her sayeings and obey her voice and would doe [what?] she would have yu [do] . . . she is [s]uch a great, rich woman yt hath openally caried such a great sway [in?] all busines (?) amonge frinds and wil doe by her meanes yee are soe . . . under her all.[73]

This new evidence portrays Fell as a woman who held a hegemonic power within the early Quaker meetings that was probably without parallel anywhere in the wider English or American Quaker community. That she could effect a power conspiracy within the men's monthly and quarterly meetings for almost two decades, in stark contrast to the consensual and democratic ideal of discipline in the church laid down by Fox and other leaders, was a striking personal feat. That this long-term episode has remained virtually unmentioned in the annals of early Quaker history was another personal feat of major proportion.

It becomes clear that this ancient internecine struggle posed a major problem in primitive Quakerism's order and discipline. The internal order that Fox envisioned, as well as the ideal of spiritual egalitarianism so unique to Quaker worship and business meetings based on the "sense of the meeting" or group unity in the spirit of Christ, were fragile principles that were overturned when one individual became too powerful. Fox made his ideas concerning internal discipline very clear. He stressed that all differences should be settled quickly among Friends so that "justice may be speedily done, that no difference may rest or remain amongst any. . . . So that Friends may not be one another's sorrow and trouble, but one another's joy and crown in the Lord."[74] The early Quakers saw themselves as new men and women in Christ, establishing a new church order based on the purity of the primitive church. The fact that the SMMM failed to achieve the goal outlined by Fox, Barclay, and others, because of the apparent lapse in business ethics of Fox's closest companion was a real blow to the internal discipline and had significant ramifications.

The fact that no other copies of this document are known to exist points to a probable cover-up of Fell's suspicious business ethics accomplished by a pay-off of Rawlinson. It is very possible that Rawlinson became owner of the Forge in 1680 for the sake of maintaining Quaker unity and stability. The Fell sisters signed over Force Forge in 1680–81 to Rawlinson and, in so doing, may have silenced a malcontent who was doing damage to the Quakers by attempting to send abroad his letters against Margaret Fell, the SMMM, and the LQM. Moreover, it is possible that Fell bought

Rawlinson off simply to quiet him, even though she was not guilty of the accusations against her. However, his last-resort tactic of taking his case to "Friends everywhere" was quite possibly that of a man who felt deeply that he was innocent and wronged at the hands of his co-religionists.

Finally, Fell's alleged business dishonesty is an important issue. It cannot be proved from this document that Margaret was involved in dishonest business activities. However, if this were the case, we might conclude that Fox and Quakers as a whole would have disapproved of Fell's business practices or her resorting to civil litigation to win her case. Rawlinson's closing statement of personal vindication was written in April 1680, one month after George Fox left Swarthmoor. It may be significant that the final settlement occurred during Fox's last visit to Fell's home. Thereafter Fox lived and worked mostly in the London area until his death a decade later. It is logical to ask a second question closely related to the first: If Fox's last visit to the Hall at the moment of the settlement of this bitter dispute had anything to do with its outcome, then did this unhappy affair cause hard feelings, or even estrangement, between Fox and Fell? Could that too have been hushed up for the sake of outward unity in the Quaker church? There is no known internal evidence pointing to any rift between Fell and Fox. More information is needed to answer these questions.

*   *   *

It has been argued by one historian that the best way to locate the sense of identity of a religious group is in its doctrine of sin.[75] In their understanding of sin, the Quakers differed markedly from their "cousins," the Puritans. Bunyan, the traditional representative of the Baptist wing of Puritanism, wrote long and soul-searching descriptions of the oppression of sin in life and the consequent need for God's grace. Fox, Fell, and other early Quakers did not feel this same oppressive weight of sin upon their lives. The early leaders preached little about sin and grace but much about the divine light and "that of God in every man." Fox preached that a state of "perfection" in the individual was possible through Christ's indwelling spirit in the believer. He and his circle chose to withdraw from the sinful, decadent world of Restoration England and create their own community, which, when filled by the inner light, could live a spiritually perfected life. Fox endlessly reiterated the fundamentally optimistic statement, "that of God in every man," meaning that the divine seed existed in all humanity and that each person could be brought to the light of Christ, and perfected by it. The early Quaker doctrine of spiritual perfection was a significant component of their spiritualist faith that underwent modification after its initial definition by George Fox.[76]

This Quaker theological stance may shed some light on a specific difference that appears to have separated a Quaker, like Margaret Fell, from a Puritan such as Nehemiah Wallington, a contemporary artisan of London. Paul Seaver's study has described quite dramatically the life of an unobtrusive London Puritan artisan.[77] Wallington's life spanned the Puritan political rise to power of the 1640s and 1650s. His life was shaped by daily reading of Scripture, and he conscientiously put his religious thoughts in writing. Wallington's diary reveals a man who was often morose in his self-condemnation. He explained that in his youth he was enveloped in "a most vile and sinful condition," for he was "born in sin and came forth polluted into this wicked world." Wallington sought, during the better part of his long life, to find deliverance in God's mercy from the temptations of Satan. The striving for a life of virtue punctuated by self-scrutiny for lapses of sin and a need for God's grace marked this unknown artisan with all the features of rigorous Puritan theology.

Another characteristic of Nehemiah Wallington was his sense of calling and his business ethics. The process of getting money and spending it were, without exception, subordinate to his concern for personal salvation. This may well have been due in part to basic personality characteristics as well as religious belief. Upon one occasion, Wallington discovered that his journeyman had stolen upwards of £100 from him over a two-year period. Until the man's confession, Wallington had never noticed the missing money, possibly because he did not keep careful accounts. Rather than prosecuting the man, he called upon God to "bless and sanctify this my poverty unto me." Upon another occasion, when his house was robbed on a Sunday morning in 1641, he remarked, "the Lord doth see the world is likely to steal away my heart; therefore he doth it [allow this theft] in love to wean me from the world." According to Seaver's study, Nehemiah Wallington searched for the spiritual significance of events in his life as being more important than the loss of money itself.[78]

The difference between a man like Wallington and a woman like Fell is striking. Although they experienced a similar political and religious milieu, they came from different social-economic backgrounds, and their personalities were very different. Wallington's lack of financial exactness and wish to enrich himself may have been largely a personality trait. Margaret Fell and daughters showed an aggressive business acumen. Nonetheless, theology had a deep impact on both Fell's and Wallington's lives. Both persons attempted to lead pious lives, following the tenets of their faith. The Rawlinson document does at least raise the question whether, in Rawlinson's case, Fell rationalized or winked at her personal code of business ethics? It is possible that Fell's seeming lapse in integrity with Rawlinson could be explained, in part at least, by the Quaker principle of "perfectionism" and its concomitant deemphasis on the sinful human condition and the need for divine grace. Those who dwelt in the perfecting rays of inner light may at times have become oblivious to the possibility of their own error.

Part III
The Political and Religious World of Margaret Fell

6 "We Have Been a Suffering People under Every Power and Change": Margaret Fell and Politics, 1659–61

In the spectrum of Quaker women's writings of the mid- and late-seventeenth-century, Margaret Fell's writings stand out as more abundant than any other Quaker woman of her generation. Fell addressed political issues that went beyond the strictly religious sphere, although she was not alone in doing this. Other Quaker women, both well-known and lesser-known, published tracts addressing the government on its policies affecting religious sects. Like Fell, Dorothy White, a prolific Quaker writer, addressed Parliament, the nation, and the city of London on political and religious problems in the last years of the Commonwealth. Anne Downer Whitehead addressed the king and Parliament in 1670; Joan Whitrow addressed King William in 1689. The prominent London Quaker, Rebecca Travers, who was the most prolific Quaker woman next to Fell, wrote several tracts defending Quaker religious tenets against non-Quaker detractors. Quaker women wrote against what they saw as the false worship of the Church of England and against justices of the peace who imprisoned Quakers. They called on inhabitants of cities like London, Oxford, and Cambridge to repent, and they all generally defended Quakerism against its opponents.[1] However, Margaret Fell was singular in that she went further than other Quaker women in her political interests. Her writings reflect the widest range of issues addressed by a Quaker woman. She addressed the question of the Jewish reentry into England in the mid-1550s when it was a popular political issue, a subject to which we will turn in Part IV. She warned the military to change their ways the last days of the Commonwealth. More significantly, Fell was the first Quaker to inform the newly arrived Charles II of the peaceable ideals of the Quakers.

According to a recent bibliographic study of seventeenth-century women writers, Quaker women produced about 171 published documents between 1641 and 1700. Of this total, Margaret Fell's published output in this period represented thirteen percent. Of the 171 documents, thirty-one were calls to repentance, fifty-four were statements of religious doctrine, eighteen were polemical tracts against public officials and ministers, ten were epistles reporting on activities of women's meetings, and sixteen were complaints of persecution.[2]

The issue of persecution, which inspired so many Quaker writings, was a gnawing problem for the Quakers from the mid-1650s onward for more than a generation. As early as 1653 when Oliver Cromwell dissolved the Long Parliament and became Lord Protector of England, the Quakers were rapidly winning new converts. Oliver Cromwell was sympathetic to Quakers, especially those with whom he had personal contact in the early years. However, in his last years, Cromwell expressed some negative sentiments about Quakers. In 1655 Margaret Fell wrote two letters from Swarthmoor to Cromwell warning him not to support the priests of the Church of England, who were "without the Spirit of Truth." In that same year Cromwell showed a kindness to George Fox, who was charged with blasphemy in London. Cromwell had Fox released from this charge "to go whither he pleased." In 1656 George Fox spoke personally with Cromwell, telling him to "lay down his crown at the feet of Jesus." Two years earlier Cromwell had learned of two Quaker preachers, James Nayler and Francis Howgill, who were imprisoned in Appleby after a trial that found them guilty of preaching and refusing to doff their hats in deference to superiors. Cromwell instructed his magistrates to be "tender" when dealing with "such poor deluded persons" who refused proper deferential manners in court. In 1656 Fell penned a third letter to Cromwell reminding him of his promise of "liberty of Conscience."

> Thou art feasting and feeding with Riotous Persons, with Musick, and sporting with them; which Practices are abominable to God; as also thy upholding such an abominable Priesthood; who instead of allowing the Liberty of Conscience promised, do Persecute and Sue Men to Treble Damages for obeying Truth, and seize on their Goods, and Imprison their Persons.

In her fourth letter in 1657 Margaret informed the ailing Lord Protector of his continued persecutions of "the Innocent." She reiterated that "God had a suffering People in this Nation" and further that "God is just and will not be mocked; the Cry of the Innocent reaches to him, and he hears their cry." Fell rebuked the Protector for failing, according to Quaker perspective, to keep his promises.[3]

The friendly attitude of Oliver Cromwell toward Fox and the Quakers in the early and mid-1650s cooled gradually. Cromwell evinced a growing dislike of the Quaker's aggressive prosetylizing methods such as entering churches or "steeplehouses" and interrupting public worship services. Likewise, Fox and Fell grew weary of the disjunction between Cromwell's rhetoric and actions. With Quakers increasing in numbers in the 1650s, Cromwell applied repressive measures against various forms of nonconformity. For instance, in 1655 he subdued an Irish rebellion of Roman Catholics with determined force. However, this did not stop Quakers from continuing to seek Cromwell's help to relieve their own imprisoned flock. Richard Hubberthorne, Humphrey Norton, and a third Quaker wrote to Cromwell asking to replace with their own persons their leader George Fox, who lay in Launceston gaol. They were denied this request, but Fox later wrote in his *Journal* that Cromwell was moved by such petitions. According to Fox's account, Cromwell asked his council: "Which of you would do so much for me if I was in the same condition?"[4]

By late 1656 Richard Hubberthorne wrote Fell that Cromwell saw the Quakers as seditious. This was due, in part, to the infamous Nayler incident in Bristol. In 1656 James Nayler brought the wrath of Parliament upon himself by impersonating Christ's triumphal entry into Jerusalem by entering Bristol on a horse led by some women. After this episode, Cromwell moved rapidly away from his earlier tolerance. Thus when Fox encountered Cromwell in October 1656 and asked him to "lay down his crown at the feet of Jesus," the Protector was a man of a different mood toward Friends. Following Cromwell's death in September 1658, Quakers encountered an almost unrelenting whip of persecution from Richard Cromwell, the former Protector's son.[5]

Richard Cromwell's failed attempt as Protector ended in May 1659. One historian of the period has noted that at this moment England experienced a "vacuum of leadership," which culminated in the recall of monarchy in 1660. Barry Reay has argued that the last chaotic year of the Interregnum was a signal year for Quakerism. He maintains that the turn toward monarchy was intimately connected with a widespread fear of radical sectarianism. A Baptist colonel, William Packer, summarized the antisectarian feeling: "Before ye Quakers shoulde have their liberty he woulde draw his sworde to bring in Kinge Charles." Mrs. Lucy Hutchinson, wife of Colonel Hutchinson, wrote in her memoirs that "the whole nation began to set their eyes upon the king beyond the sea, and think a bad settlement under him better than none at all, the whole house [by December 1659] was divided into miserable factions." Likewise priests began praying from their pulpits "and began openly to desire the king; not for good will to him neither, but for destruction to all the fanatics.[6] The month-to-month political developments of 1659, as described by Reay, demonstrate the political chaos and sense of bewilderment undoubtedly experienced by many Englishmen and women. Quakers were active in calling for liberty of conscience and abolition of tithes. Leaders such as Fox, Edward Burrough, Samuel Fisher, and Richard Hubberthorne were actively preaching and holding Quaker meetings, some of which were attended by soldiers.

During these eventful last months before the Restoration, what was Margaret Fell thinking and doing so far removed from London? Fell is one leader of the Quakers who has been overlooked in Barry Reay's study of this eventful period. While at home in Lancashire, she wrote fervent millennialist messages to the army and Council of Officers during the recall of the Rump. Margaret and her daughters also headed the petition to Parliament that seven thousand Quaker women signed in July 1659 against the hated tithe.[7] Margaret's epistle to the Council of Officers, written in October 1659, exhorted them to believe in Christ and to be just to his innocent, suffering people. She admonished the officers of the army to act justly as the instruments of the Lord, lest they be overthrown as their predecessors had been overthrown:

> You have bin the Instruments of warre, and the Battle-axe in the hand of the Lord. . . . And now you are come into their place which they were in, with whom you warred . . . where you may execute that which ye formerly desired after, and suffered for. . . . And now that the Lord hath put it into your hands, even to try your faithfulness, and your integrity; Not that he hath need of you, more than them that went before you (as ye are men natural) but as you are righteous men, and guided by the just and righteous principle of God in you, so you may be serviceable unto the Lord . . . to bring to pass in these latter days . . . for the Almighty will show forth his righteousness . . . and his glory. . . . Therefore as ye love your soules, and your eternal peace . . . beware that ye turn not your hand against his work that he is working in the earth in this day.

Fell warned the officers to "keep out of strife, and contention one with another" and let the "just principle" lead and sustain them. She added, "let Christ Jesus . . . have his throne established, which is for ever and ever." Fell's statements conveyed her belief in an imminent second coming. She not only called the army the "Battle-axe in the hand of the Lord," but she signed her statement, "from a true lover of true Peace and Righteousness, which is coming to rule."[8] Whether these letters were ever read by her intended recipients is unknown.

With Charles II's return to the throne in May 1660, Fell's mood remained strident in its confrontation of institutional evil as she perceived it. In July 1660 she wrote *To the Citie of London*, a blood-and-thunder broadsheet exemplifying the apocalyptic imagery then current in religious propaganda:

> Repent thou Bloody City, who dust, and hath long Crucified the Just . . . whose *Pride* and *Hipocrisie, Deceit* and *Dissembly* hath reached unto Heaven. . . . But the Righteous God hath seen you all . . . and he will uncover you, and lay your Abominations open, and rend off all your false Coverings. . . . The day is come wherein you cannot hide your selves.[9]

Fell wrote a final call to convert the army leadership in 1660. She addressed Major General Harrison, a former Cromwellian soldier and preacher, whose career was soon to end at the scaffold:

> This is the day of your visitation, while ye have time prize it; God hath long holden forth a hand of love unto you, though you have for several years past blighted and rejected his council . . . come to the . . . Light of Christ Jesus in whome is Life . . . if ever you find peace and reconciliation with the Lord God, ye must have it in the Light.

The theme of the statement was that immortality was to be gained only through "partak[ing] with us, and tast[ing] of that love which is in Christ Jesus." She concluded, "I am a true lover of all your soules, and desires your Eternall peace and comfort of the Lord, who feeles the weight of your sufferings and travells."[10]

Although Margaret Fell did not arrive in London until May or June 1660, it is evident that she kept in close touch with political developments and wrote to leaders in that city prior to her arrival. She is one of the few, if only, Quaker women of the period who wrote expressively, if not with complete originality, to men in high places concerning the political issues of the moment. The rank-and-file Quaker woman usually confined her writings to prophetic messages to her own people. Fell's statements in this period express not only a sense of immediacy of the religious and political fervor that was abroad but also they address specific political issues. We know that Quaker women ministers, Fell among them, made numerous visits to King Charles II to make their personal requests. The "pretty Quaker woman" referred to by Pepys in his diary in 1664 who "thou'd" King Charles in addressing him was probably not Margaret Fell. Nonetheless Pepys's comment captures the image of persistent, serious-minded Quaker women confronting an initially indulgent king. For instance, in 1662 Elizabeth Hooten recalled her first visit to Charles II:

> My goeing to London hath not beene for my owne ends, but in obedience to the will of God, for it was layed before me when I were on the sea [coming from New England] . . . that I should goe before the King, to witness for God, whether he would hear or noe, and to lay downe my life as I did at Boston if it been required. . . . I followed the King with [MS torn] this cry: I waite for justice of thee, O King, for [MS torn] in the countrey I can have no justice among [MS torn] the magistrates . . . for they have taken away my goods contrary to the law.

Hooten continued to follow the king "which way so ever he went." She gave him her letters and the onlookers grumbled at her lack of deference, for she did not kneel, rather followed him, talking as she went, but apparently with little or no effect. Finally, Hooten gained an opportunity to address the king as he was approaching his coach at court. She recorded that the Lord's power witnessed in her at that moment, "till a souldier came and tooke me away, and said, It was the King's Court, and I might not preach there . . . and they put me forth at the gates."[11]

In July 1660 Ann Curtis of Reading accompanied Margaret Fell to Whitehall with another personal request for the king, the release of George Fox, then in prison in Lancaster Castle. Ann Curtis, nee Yeamans, was the daughter of the late Sheriff of Bristol, who died in 1643 at the hands of the parliamentarian army for his loyalty to Charles I. Curtis's pro-royalist family background was influential with Charles, for in Margaret's account of the interview, the king was deeply moved when he heard Ann's story. Margaret wrote:

> I went with her to Whitehall and brought her to the King, and she made known to him whose daughter she was and how that her father was executed for him or in his father's cause, whereupon he showed much love to her. And she said she had now a request to him. He asked her what it was. . . . [Upon hearing her request for Fox's release] he gave command to his Secretary to issue forth an order to that purpose, but the sublety of the Secretary gave out order to the Judge to [that George should] be brought up by Habeas Corpus and to appear before the Judges, so that she was disappointed of her request to him and of what

he had granted. If the Secretary had proceeded according to order, we might have had it.

Just a few days later Margaret and Ann went again to see the king concerning this aborted attempt for Fox's freedom. A letter of 24 July 1660 from Margaret to George describes their efforts:

> My dear eternal Love and Life: I gave thee an account the last week, how far Ann Curtis had gone in the business concerning thee, and according as I wrote we went the next morning, but before they would suffer us to go in to speak to him, General Monk did come, I believe on purpose to prevent us; and we were with him a pretty while before we were called in. And we were called in while he was there, and while we spoke to him, he stood by, and before we could get anything spoken to him of any purpose, they took him away from us, and the most he said to us was that he would speak to the Judges and they should set Friends at liberty.[12]

Fell's first meeting with Charles on 22 June 1660, less than a month after his arrival in London, was a noteworthy one. Fell has been credited as the first Quaker to write a declaration against the persecution of peaceful Quakers and against war and violence for any purpose. This self-delivered paper has been largely overlooked by students of early Quakerism, despite its political import. When placed in the broader picture of the recent historiography of early Quakerism, this omission suggests that Fell's political writings and ideas have been seriously underestimated. Her impact was not negligible according to her contemporaries. Alexander Parker's letter to George Fox described her first of several meetings with Charles:

> [She] had a full and large time to lay all things before him of Friends' sufferings. He was very moderate and promised fair things, if he performs them it will be good for him; he also desires a particular account of all Friends that are present sufferers here, and the cause of their sufferings and the names of the magistrates who sent them to prison and the time how long we have suffered.

Francis Howgill spoke also of Fell's work in London in this period: "[Her] service ... hath been good, and that which could not have been performed by many."[13]

Fell's most important statement addressed to Charles II, *A Declaration and an Information from the People of God Called Quakers to the Present Governors, the King and both Houses of Parliament and all whom it may concern*, was a defense of the Quaker principle of peaceful living, and it explained why Quaker opposed oath-taking, hat honor, and payment of tithes. It included a vehement protest against the persecution of Quakers:

> We who are the People of God called Quakers, who are hated and despised, and everywhere spoken against, as People not fit to live. . . . We have been a Suffering People under every Power and Change, and under every Profession of Religion, that hath been, and born the outward Power in the Nation these Twelve Years, since we were a People . . . not only . . . the profane People of the Nation [were against us], . . . but also in the highest Profession of Sorts and Sects of Religion, we have suffered under, and been persecuted by them all: Even some persecuted and prisoned till Death, others their Bodies bruised till Death, stigmatized, bored thorow the Tongue, gagged in the Mouth, flock'd, and whip'd thorow Towns and Cities, our Goods spoiled, our Bodies two or three Years imprisoned, with much more that might be said, which is well known to the Actors thereof; and this done, not for the wronging of any Man, nor for the breach of any just Law of the Nation, nor for Evil-doing, nor desiring any Evil, or wishing any hurt to any Man, but for Conscience sake towards God, because we could not bow to their Worship.

She assured Charles that

> our Intentions and Endeavours are and shall be Good, True, Honest, and Peaceable towards them; and that we do Love, Own, and Honour the King, and these present Governors, so far as they do Rule for God, and his Truth, and do not impose anything upon Peoples Consciences. . . . We do not desire any Liberty that may justly offend any one's Conscience; but the Liberty we do desire is, that we may keep our Consciences clear and void of Offence towards God and towards Men, and that we may enjoy our civil Rights and Liberties of Subjects as freeborn Englishmen.[14]

She ended her declaration with a second assurance: "We are a People that follow after those things that make for Peace, Love and unity . . . [we] do deny and bear our Testimony against all Strife and wars . . . . Our weapons are not Carnal, but Spiritual." Thirteen early leaders signed it, including George Fox, Richard Hubberthorne, and Ellis Hookes.

Seven months later George Fox and Richard Hubberthorne issued a *Declaration from the Harmless and Innocent People of God, called Quakers against all plotters and fighters*. Fox's statement on peace went through several reprintings, while Fell's was not reissued. Fox's statement was a specific declaration against wars and fighting in the wake of the Fifth Monarchist activity of January 1661. Meanwhile, Fell went to the king and reiterated her earlier statement that the Quakers were a "peaceable innocent people." In Fox's words, Fell told Charles II: "We must keep our meetings as we used to do and that it concerned him [the King] to see that peace was kept, that so no blood might be shed." She returned to the king and his council twice more to lay before them the account of "several thousands" of Quaker imprisonments. Fox commented in his *Journal* that the king and his council marveled at the amount of inside intelligence that Fell had about Quaker prisoners, for the government had ordered the intercepting and search of all letters. Fox then referred to his own peace statement: "We drew up another declaration and got it printed and sent some of them to the King and council."[15] Fox's and Hubberthorn's *Declaration*, which has subsequently been looked upon as the official Quaker peace testimony, was delivered to the king on 21 January 1661.[16]

In the face of the enormous and rapid political changes of 1659 through 1661, the Quakers, in Christopher Hill's words, "survived, prospered and rewrote their history." Quaker writing of this interval appears to have been not only revised in its encounter with later Quaker editorial work but also probably subjected to outright suppression of controversial statements as well. For instance, Edward Burrough, concurrent to Fell, wrote a statement about the peaceful Quakers that was edited after his death with portions deleted.[17] Fell's three fiery writings – *To the Generall Council of Officers of the English Army* (31 October 1659), *The Citie of London Reproved* (4 July 1660), and *This was given to Major Generall Harrison and the Rest* (1660) – were censored according to later Quaker literary tastes and not reprinted in her *Works* in 1702. They were undoubtedly found too politically confrontational and apocalyptic in tone.

RESTORATION PERSECUTION OF NONCONFORMITY

The political fears that inspired the legislation against nonconformity in the 1660s occurred in the aftermath of the flowering of Interregnum sectarianism, most of which withered and disappeared after 1660. We know that the Quakers were among the "fanatics," the popular label for the sectarian groups, that did not disappear by attrition after 1660. In fact, the

Quakers continued to expand rapidly, despite intermittent and sometimes heavy persecution. The civil authorities, representing the Cavalier element of the country, were alarmed by Quaker activity, not only by their refusal to pay tithes, but also by their refusal to take oaths and their persistence in meeting for private or conventicle worship. The Quakers appeared as potential subversives who engaged in civil disobedience and dissimulation to achieve their ends.

From March to July 1661, shortly after the Restoration of Charles II, the Savoy Conference of Anglican bishops met and demonstrated a belligerent mood. Anglican uniformity and the revised Book of Common Prayer were in place by May 1662 when Parliament passed the Act of Uniformity. Clergy ejected during the Civil War and Commonwealth were restored in September 1660; and ejections of Puritan and Presbyterian ministers commenced. By 1662 the ejected nonconforming clergy numbered approximately 2,000. Nonconformity became a way of life for all who refused to embrace the Anglican Church. This enforced conformity was implemented by a series of penal laws known as the Clarendon Code. The Restoration legislation thwarted the ideal of religious toleration expressed in the Declaration of Breda, which may have revealed the true intent of Charles II. When the Cavalier Parliament met in May 1661, its mood was essentially the antithesis of the notion of "liberty to tender consciences," and Charles II never achieved this idea in law. In November 1661 the Corporation Act eliminated any representation by nonconformists on the borough, county, or national government levels who refuse to take the oaths of supremacy and allegiance. The oaths stipulated nonresistance against the king and a rejection of the Scottish Solemn League and Covenant. In May 1662, with the Uniformity Act in effect, all clergy were obliged to use the revised Book of Common Prayer and take all the oaths and declarations imposed in the Corporation Act. Those who did not conform were ejected and their stipends presented to conforming clergy. Vestries were obliged to take the oaths by 1663. The Conventicle Act of 1664 prohibited meetings of five or more persons (who were not family) for any religious occasion.[18]

Under the Conventicle Act, Quakers, Margaret Fell included, paid heavy fines for continuing to meet openly. Persecution by this act came in waves, subsiding periodically due to the plague (1665), the Great Fire of London (1666), and the Dutch War (1664–67). In 1670, two years after the first Conventicle Act expired, Parliament passed the second Conventicle Act. The severity of this second act was seen in the heavy fining of nonconformist preachers at meetings. The first fine was £20 for the preacher, the second offense brought a £40 fine, and the person who hosted such a meeting was fined £20 as well. Quaker martyrologist Joseph Besse called the informers who told the authorities the location of Quaker meetings "beggarly rude informers." Besse recorded such an event involving Margaret Fell that took place in 1677 at the home of William Gandy in Frandley, Cheshire:

> One Midsummer-day this Year, Sir Peter Leicester, a Justice of the Peace, who also acted the part of an Informer, came personally to a meeting at the house of William Gandey, shut up the Doors, and placed a Guard of Soldiers at them, while he took a List of about two Hundred Names, and fined Margaret Fox and Thomas Dockra £20 each for Preaching; he also ordered £20 to be levied on several of the Assembly for the House they met at. . . . The distraint of £20 value levied on Cheshire Friends, was paid by Margaret via a Friend for their relief. She sent the money to Cheshire Friends yt suffered upon Acct of her being fined there but they would not take it, so . . . Returned it againe.[19]

The political events of 1659–61 saw deep involvement on the part of several Quaker leaders. Fell's involvement was notable for her energy, her political activities, and for some originality of ideas such as her peace statement. However, she was not alone among her own gender, for other women ministers wrote tracts touching on political issues. Likewise, women of less than gentry status who did not leave their political ideas in print were also active in "God's vineyard." Barbara Blaugdon of Bristol, a traveling preacher in the primitive period of Quakerism, was suspected of witchcraft and turned away from the homes of her erstwhile friends. Elizabeth Fletcher of Kendal, as a girl of sixteen, experienced the wrath of Oxford scholars who resented her public preaching in their town in 1654. She was physically abused by being dragged through a pool of filthy water and thrown against a gravestone with such force that she sustained permanent bodily injury. In 1659 in Colonial America, Quaker Mary Dyer confronted the Boston magistrates and insisted on returning to Boston until the Puritan oligarchy of Massachusetts Bay Colony hanged her in 1661. In London, women such as Rebecca Travers, Sarah Blackbury, Mary Elson, and Anne Whitehead, among others, were visible in their London ministries. Other women preachers such as Alice Curwen and Joan Vokins traveled abroad to spread the Quaker message, and left accounts of their journeys. Vokins, in her travels to New York and Long Island in 1681, urged her sisters in an epistle to the women's meeting there: "Be valiant for the Truth, that it may not be undervalued . . . quit yourselves like faithful Soldiers, fighting under the Banner of Love." Letters to Fell survive from many of these "she-soldiers" of Quakerism.[20]

Virtually all of the correspondence addressed to Fell looked to her as a special leader among both women and men. Fell's published tracts and epistles of the 1650s and 1660s reflect a high political awareness and participation in the polemical arguments relating to the problems of religion and politics. Her status, education, and reputation were such that gender did not compromise her leadership role, even when it involved volatile political questions. Although shaped by the patriarchal patronage society of their time, women and men still saw her as a patron worthy of deference.

Fell and Fox were both motivated by a religious zeal that heavily politicized their public religious activities. Fell's *Declaration* of peaceful intentions of Quakers, followed by her overt and verbally pugnacious resistance in her oath trial of 1664–5, to which we will turn shortly, were examples of a confrontational style toward institutional injustice as she and her contemporaries viewed it. Fell attempted to precipitate change while she was in the spotlight. The following two chapters will interpret the political implications of Fell's public religious work both within and outside the Quaker movement, first as co-founder with Fox of the women's meetings, and second as one who used her social status as an additional tool to augment her religious and political authority.

7 "Walk as Becomes Truth": Margaret Fell and Women's Meetings

Margaret Fell's home was a haven for George Fox, a clearing house for Quaker correspondence, and an important center for the organizational activities of men and women in the rise of Quakerism. George Fox wrote in his *Journal* that while he was at Margaret Fell's home, he took it upon himself to organize the growing number of Quakers into meetings, for, he observed, "The Lord's truth was finely planted over the nation and many thousands were turned to the Lord." According to one *Journal* entry, Fox recalled that in 1656, "I was moved of ye Lord to send for one or two out of a County to Swarthmoor to sett uppe ye mens meetinges where they was not.... And about this time I was moved to sett upp ye mens Quarterly meetinges throughout ye nation."[1]

It is generally agreed that George Fox's *Journal*, compiled between 1674 and 1676, while at Swarthmoor, has had an enduring, if not predominant influence, on the subsequent historiography of early Quakerism. According to his *Journal* the year 1656, when he began to organize Quakers into local meetings, was a landmark year in his religious pilgrimage. Although there is no specific reference to separate women's meetings in the above statement, by the time of Fox's death in 1691 women's meetings were an accepted part of Quakerism although they were not settled everywhere. What occurred between 1656 and 1691 to bring this remarkable, if not unique, seventeenth-century religious institution of separate female meetings into existence? In particular, the roles of Fell and Fox in setting up the female component of early Quaker polity need to be reinterpreted. The traditional view promoted by Fox's *Journal* must be reevaluated in the light of the early women's perspective, as best we can uncover it.

The standard description of the origin of Quaker men's and women's meetings throughout England and in the American colonies is summed up by the Quaker historian, William C. Braithwaite. Braithwaite, who was deeply indebted to Fox's account of the early movement, admits that the formative factors which converged to help create a religious organization out of a religious movement are not easily analyzed. The earliest Quakers, especially the itinerant ministers, saw themselves as carriers of the "fire of the Inward Light of Christ over the whole earth." They were the church of God gathered out of the world. Ostensibly, under Fox's leadership, Quakerism evolved into a multi-tiered organization where business was conducted in particular or local meetings, county-wide monthly meetings, and, subsequently, quarterly meetings which met semi-annually. Quaker church organization evolved first in the north and then elsewhere. The duties of the men's particular, monthly, and later quarterly business meetings evolved out of the necessities of the moment. The immediate needs of the poor of the respective areas were met by taking a collection at the meetings. The collection of monies also paid for books used by each meeting's itinerant ministers, as well as for aiding prisoners and sufferers for the "Truth's sake." Further, each particular meeting kept a record of local births, marriages, and burials. Finally, the monthly and later the quarterly meetings acted as mini-courts of justice where Quakers resolved internally any disputes among their own flock. By 1659, Braithwaite claims, the Quaker church organization had particular, monthly meetings and a general meeting which met two or three times a year.[2]

Braithwaite next explains the origin of the women's meetings. By the second decade of Quakerism, George Fox felt a deep concern to liberate both women's and men's gifts for God's service. Braithwaite attributed this novel idea to Fox's genius and claimed that Fox's various epistles to women's meetings thereafter were intended to stir them up to take their rightful place in church government. Braithwaite concluded that the venture was a daring one for Fox because "it taxed seventeenth-century feminine capacity to the utmost, but this only adds to its significance as a landmark in the movement for giving woman her true place of equal partnership with man."[3]

There are other historical commentaries on the early women's meetings in Quakerism. Arnold Lloyd, while following Braithwaite's view that Fox set up women's meetings, summed up his study with the statement that women's meetings "deserve fully study." Isabel Ross, writing on the work of Fell and her daughters for the women's meetings, concluded that Margaret and her daughters were among the "chief creators" of women's meetings but proceeded no further in defining their role. Michael Watts suggested, but did not pursue the idea, that along with Fox's genius for Quaker organization, Margaret Fell was also a chief architect of the Quaker "central administration." Richard Vann's work on Quaker organization and membership depicted, for the most part, the masculine perspective in the formation of early Quaker order.[4] In short, the secondary literature is consistent in its assessment of early Quaker church order and the origin of women's meetings. Fox called forth the women of emerging Quakerism to meet separately, and Margaret Fell, in a secondary role, supported Fox's efforts.

It is impossible to gain an exact picture of what occurred in the obscure early years of Quaker development. What part did Fell play *vis-à-vis* George Fox in the earliest phase of Quaker church development? If the record yields evidence that Fell was a "chief architect" of women's meetings, what does that mean and what are the implications? This chapter will reevaluate the early evidence and present a revised interpretation of the role Margaret Fell played in relationship to George Fox in the early development of Quaker organization.[5]

Margaret Fell's role must be seen in the broader picture of the early London women's activities on behalf of women. In the evangelizing fervor of the first decade of Quakerism, women Friends in London were the vanguard in distinguishing themselves by their public ministry. Although initially not encouraged to meet separately, by the late 1650s, when the men's business meetings were evolving in the North, the women Friends of London organized a distinctly separate female organization. That the public did take notice of Quaker women's activities proves they were sufficiently unusual. According to an older history of London Friends' meetings, the extraordinary nature of these London women Friends' separate meetings in the 1650s inspired the earliest drawings of London Quakers. The drawings depicted Quakers in meetings "at a point in time when a woman Friend . . . [was] addressing the assembly, not, however, from the ministers' gallery, but on a stool, tub, or bench."[6]

London women such as Anne Downer Whitehead, Rebecca Travers, Sarah Blackbury, and Mary Elson were among the activist Quakers who participated in London separate women's meetings. Two of these women left their ideas in print. Mary Elson, writing an epistle circa 1680, claimed that the women's meetings had been created "betwixt three or four and twenty years ago," which places the origins of the London women's separate meeting around 1656 or 1657. At that time, Elson recalled, the needs of the poor and sick were so intense that they attended Friends' meetings and "looked unto us for Charity; and we could not send them empty away." She concluded:

> So we considered, and had a Weekly Gathering for them, that it might be fulfilled, as it is written, *Do good unto all, but especially to the Houshold of Faith*. And after some time of our meeting, together, there came two of the Brethren from the Mens Meeting to us, when we were met together, expressing their Unity with us, and also did declare the mind

of the Mens Meeting. . . . *That they would be ready to help and assist us in anything we should desire of them for Truth's Service.* And so in some time it was agreed upon, in the Unity of the Truth, that the Men Friends should pay the poor Friends Rents, and find them Coles . . . and put out such poor Friends Children, as we should offer to them; And this has ben done almost from the beginning of our Meeting to this Day.

Elson's epistle on the London women's meeting mentioned George Fox as an instrumental figure in the setting up of the earliest women's meetings. Her account refers twice to Fox's role, noting: "Our dear George Fox, that man of God . . . when he came to behold their want [the poor of London] of things needful for them . . . he was moved of the Lord to advise to a Womans Meeting; and in order thereunto, he sent for such women as he knew in this city." On the other hand, Anne Whitehead's epistle neglected to mention George Fox or any one person as the founder or instrumental organizer of the women's meetings in London. Rather, she enumerated the reasons for supporting the women's meetings "for the future well-being in the Creation." Whitehead recalled, "about twenty six Years ago, in the South Parts we were a very little handful of People, gathered into the Belief of the light of Christ Jesus." Her epistle only referred to their "smallness of Number in the beginning." It conveys the sense of a community-recognized need that resulted in a women's meeting by common consent.[7]

Two nineteenth-century Quaker historians, William Beck and T. Frederick Ball, give an explanation of the origin of the earliest women's meetings in London that differs from that of the Foxian tradition. They explain that the early London women Friends only established their separate meetings to address the pressing issue of poverty among early Friends. One such meeting, known as the ancient London women's Box Meeting came to "possess considerable funded property and freehold estate . . . that enable them [the women] to make . . . disbursements." Further, the Box Meeting was not accountable to any beyond its own collective authority as to how and where the money should be spent.[8]

The origins of the Box Meeting and a second separate women's meeting in London, namely the Women's Two-Weeks Meeting (WTWM), are shrouded in obscurity. Although Fox, through the account given in his *Journal*, has traditionally been considered the originator of the WTWM, there is a second tradition, less well known, that offers another explanation, according to Beck and Ball. This second tradition of Gilbert Lately, a London Quaker who wrote a document describing the forming of the London women's meetings that the WTWM was organized only three or four years after the London men's meetings. Unlike the Box Meeting the WTWM received its funds from the men's meeting and from its own internal collections. In the earliest years the burden of caring for the urban poor, sick, aged, and imprisoned took all its resources and energies. Lately says nothing about Fox as the instigator or overseer of the WTWM in its earliest years.[9]

Foxian tradition also states that Sarah Blackbury, a leading London Quaker, approached Fox concerning the overwhelming problem of caring for the poor and sick. Fox told her to gather some sixty women to meet with him, and he directed these women to create a meeting that would oversee these human services: "to visit the sick, and for the relief of the fatherless and the widow, and to see that nothing was lacking among them." Beck and Ball point out that Fox gave considerable credit to himself, failing to mention the "important share others had had therein." The important distinction between the Lately version and the Fox version is that the former says the men's meeting first conceived of the idea and then approached Fox on the matter, while the latter version gives sole credit to Fox. Both accounts were written long after the actual events had taken place. Fox wrote in the mid-1670s, and Lately's account was written considerably after the Fox version had been published. Gilbert Lately, therefore, already knew of the Fox tradition, and as a member of that early London men's meeting, did not feel compelled until very late in life to write his version of the origins of the London women's meetings.[10]

Although it is risky to draw conclusions from this impressionistic evidence, the four differing versions of early Quaker history by Fox, Elson, Whitehead, and Lately suggest that of the various records on the origins of women's meetings, the Foxian perspective became the dominant version from which subsequent Quaker history has drawn its tradition. That an effort was made sometime later in the century by Gilbert Lately, to correct or perhaps counteract the Foxian point of view, may have been an eleventh-hour attempt to change the record, and thus prevent the elevation of Fox to the status of idealogue and originator of this creative aspect of Quakerism. The dearth of evidence precludes a decisive settlement of this issue. Whatever the circumstances of the origins of the two London women's meetings, at least one conclusion may be suggested: Sarah Blackbury, Mary Elson, Anne Whitehead, Rebecca Travers, and their cohort of London women were sufficiently concerned with the overwhelming needs of the poor, sick, and elderly in London that they became instrumental workers in the organization of separate women's meetings to specifically address these issues.[11]

Another significant influence in the birth of separate women's meetings was the dispute between the John Wilkinson and John Story faction and the larger Quaker movement in the mid-1670s. Margaret Fell and George Fox were deeply involved in this controversy. Wilkinson and Story, two Westmorland Quakers, were influential church leaders in their own territory as two early Publishers of Truth. They led a group of disaffected northern Quakers into outright rebellion against Fox and Fell and their policies for several reasons, one of which was a vigorous opposition to the establishment of separate women's meetings. The effects of this schism lasted well into the eighteenth century. The Wilkinson-Story opposition was the result of several factors, one of which was the probable jealousy of and competition among some of the primary northern leaders, notably Fox and Fell, and Wilkinson and Story. A second factor was persecution of Quakers. Persecutions in the 1660s and 1670s led some norther Quaker meetings such as the Kendal Monthly and Quarterly men's meetings to relax their discipline on some issues such as not requiring written statements from members avowing their non-payment of tithes. As a result, Margaret Fell and others reproved the apparent backsliding of these meetings. At the same time, Fell and Fox were trying to organize women's meetings in the north. Additionally, the dispute was based on another deeper issue, that of group versus individual spiritual authority in the church. The question was, was authority to rest, as Fox earlier had preached, in the individual's spiritual guidance by the inner light, at which moment the believer's direct revelations of the Spirit were valid? Or, was authority and disciplinary power to rest on group decisions based on the spiritual rule of the collective body? That is, was George Fox moving away from his earlier position of placing authority in the genuine first-hand religious experience, for a group discipline, while at the same time acting as the supreme leader of the newly-emerging Quaker organization? Wilkinson and Story labeled Fox a self-styled "Moses," and his new conception of church order as "Foxonian-unity." Further, the Wilkinson-Story circle looked upon Margaret Fell as a meddling woman close to Fox who took it upon herself to visit their meetings and tell them how to organize according to the Foxian model. They resented what appeared to be his and her unilateral decision to set up separate meetings under Fox's self-assumed and quasi-Pauline command,

> I was sent out by the Lord God, in his eternal light and power, to preach the word of life . . . that all might be reconciled to God by the word . . . which gospel I received not of man, nor by man, but of the Lord Jesus Christ, by his holy spirit . . . and so after I had received this and preached it . . . I was moved to advise the setting up of the men's Monthly and Quarterly Meetings, and the women's meetings.[12]

The Wilkinson-Story group believed that women's meetings were unnecessary and unscriptural, except in large cities like London and Bristol where the needs of the poor required women's special efforts. John Story believed that women's meetings were essentially worship meetings and therefore to meet separately for worship was both "monstrous and ridiculous." Moreover, Wilkinson and Story felt that marriages should not be laid before women's meetings for "clearness," or certification. They also believed that the women should not discipline unruly members or record condemnations for backsliders who had either paid tithes or strayed from the path of truth in other ways.[13] In response to the opponents of women's meetings, George Fox wrote, in the polemical style of the day, a very tart letter in October 1676 while at Swarthmoor. He warned all opposers of true religion that they would not prosper:

> Whosoever shall come, under what pretence soever, to alter these meetings [men's and women's meetings], it is in the spirit of confusion and opposition . . . its work is to disquiet the simple minds; . . . [and] some earthly loose spirits it may draw after them. But mark the end of it; . . . the seed reigns over the head of all such false spirits, . . . and it will grind them to pieces; that is the word of the Lord to you. For all our men and women's meetings, which are set up by the power and spirit of God, these meetings are for the practice of religion and to see that all that do profess truth, do practice it and walk in it.[14]

The enmity felt toward Fell and Fox by the Wilkinson-Story faction is expressed in William Rogers' book, *The Christian Quaker* (1680) which contains a vehement denunciation of both Margaret Fell and George Fox. Rogers accused Fell of reading a paper at a quarterly meeting in 1672 that made "grievous accusations against our Faithful Brother John Story," who had referred to Margaret as "the Female." William Rogers defended John Story's behavior at that meeting and other meetings. He denied the derogatory remarks made against Fox's wife and claimed Story did not refer to her as "the Female."[15]

According to *The Christian Quaker* Margaret Fell was praying at a northern meeting where John Story was present. Story allegedly mumbled words of reproof about her, to the effect that the meeting should "keep order" against the "noise of Deceit by one Person." Rogers denied this accusation.[16]

Three years later Fell's defenders, John Blaykling, John Pearson, and others, published a lively rebuttal to William Rogers' book, entitled *Anti-Christian Treachery Discovered and Its Way Blocked Up* (1683).

In defending Fell's actions regarding the paper read at the 1672 quarterly meeting, the authors blamed John Story for "whispering Reflections, and smiting Accusations against her, behinde her Back; and could not be got to speak thereof to her Face."[17] Blaykling's account described the Wilkinson-Story party as unjustified in their vendetta against Fell, who, as a gentle-born woman of moral stature, deserved more courtesy:

> [Margaret Fell] had been and was honourable in the Church of Christ for her Love and Lifes sake in the Truth . . . and of whose Care, Labour, and Service therein, from the Beginning, many were Witnesses of, and were comforted in, *having turned her Back of the Pleasures and Glory of the World, in self-denial for Christs sake*, and taken her part amongst the dispised and hated of the world . . . and we testify it was below the image of God, and the Spirit of Man for him to treat her, as he was resolved to do.[18]

According to Blaykling, the Westmorland Quarterly Meeting had, previous to this dispute, signed a paper of consent for the women's meetings, and likewise consented unanimously to the reading of a paper from Fox encouraging the women to meet. Margaret Fell visited the quarterly meeting around that time and answered questions concerning "church care" and "Gospel Order." At that meeting, John Wilkinson asked Fell, "What ground have we to practice things imposed upon us by man, or in the will of man, that there is no Scripture Proof or Example for?" According to Blaykling, Fell responded by asking who was to imply that church order was not also gospel order?[19]

These contemporary accounts reveal a depth of animosity that reverberated long after the deaths of the chief protagonists, Story and Wilkinson. Fox was very aware of this smoldering enmity for in a letter written to his wife from Oxfordshire in May 1674 he cautioned Margaret not to stir up the anger of their opponents:

> I desire that thou would do nothing to provoke to strife . . . I desire that thee would rather forbear and be over such things . . . *I desire you may be wise and if you do leave Westmorland Women's Meeting to themselves a while and let their spirits cool and not strive for the power*, life will arise over all. And so be still in the power and life and do as it moves.[20]

However, Margaret had already made up her mind, for two months later a letter dated 7 July 1674 from Fox, who was then in London, appears to concede to her will. His letter includes a terse one-line response: "If thou findeth it lieth upon thee thou may goe over to the wimenes meeting in west mor land."[21]

William Penn's letter to Fox at Swarthmoor, dated 4 March 1676, gives us another glimpse of Margaret's role in the matter of women's meetings:

> Poor Marg^t is so much smitt at, and run upon (as I believe never woman was, for wch God the righteous Judge will Judge, and plead wth them), as if she was the Cause; and of an Implacable temper; without bowels, or spirit of Reconciliation to show them that she can pass by all that past between them that Concerns herselfe; and to make the most of the good or the tender thing, if it arise never so little in either of them, this would be a deadly stroak upon that back biteing spirit, and so Confound them that smite, yea so stop their mouths.[22]

It appears that the degree of bitterness engendered in this dispute was due, in no small way to Margaret Fell's preemptive manner in dealing with the disaffected Westmorland Friends. Margaret's determination to see the women's meetings established reveals much about her leadership role therein. It is evident that Westmorland Quakers felt jealousy regarding Fox's assumed primary leadership, but resentment of Fell's power was at least as great. Fell was challenging patriarchy in this dispute and, considering the era, it is understandable that many Friends reacted negatively to such a powerful woman who was overtly stepping out of the traditional female role.

The open break between the Wilkinson-Story camp and the Foxians came in May 1675. In the same month the London Yearly Meeting of Ministers published an official statement on church authority and discipline. The rift was briefly healed by a meeting in April 1676 at Draw-well in Yorkshire. However, by the summer of 1676 the Wilkinson-Story faction had regrouped against the Foxians.[23]

The schism brought about the breakup of old friendships. One instance of this is seen in the longtime friendship of Fox with two prominent Quakers of Reading, Thomas and Ann Curtis. In 1660 Ann Curtis and Margaret Fell together visited King Charles on Fox's behalf because he was then in prison. Thomas Curtis had once described Fox as "one who should come, not born of Flesh, but of the Spirit." However, by January 1675, Thomas Curtis's opposition to the Foxians was so strong that he locked the doors to the meeting house in Reading so that the women could not gather there. A letter from Thomas Lower to Margaret in February 1675 described a contentious discussion that took place between Fox, Wilkinson, Story, Thomas Curtis, and Ann Curtis:

> My father once or twice bid me walk away, when I drew near to hearken to their discourse . . . I walked five hours upon the gravel walk expecting when to be called; and often took occasion to mind my father of going in lest he took cold; and A. C. [Ann Curtis] was as careful of dear John Story, as she caled him. . . . I hear [Fox] cut and hewed them to pieces and kept them at sword's point still; and told them if they continued in that spirit they were in he must bear as great a testimony against them as ever he did against the priests.[24]

These events give us a glimpse of the involvement of Fell and Fox in the schism of the 1670s brought on to a significant degree by the establishment of separate women's meetings. The attitudes expressed in the books and letters of the period are highly suggestive that Fell's involvement in the establishment of separate women's meetings was central in the 1670s.

Although the Wilkinson-Story schism was an unwelcome diversion in Fox's and Fell's plan for women's meetings, they did not desist in their efforts. In fact, George Fox wrote in favor of women in ministry right from the beginning of his public ministry and he never modified his idea despite controversy. While Fox asserted in a 1653 letter that in Christ all were one, male and female, he cautioned, "ye daughters, to whom it is given to prophesy, keep within your measure."[25] An examination of Fox's early works reveals that he neglected to be specific when writing on women's rules except in connection with London women's meetings. As for the establishment of women's meetings outside of London, Fox neglected to mention them in 1656 when he was moved to set up men's meetings.[26] In an April 1661 epistle, in which Fox gave advice to the county men's meetings, women were conspicuously absent. His statement read: "Be valiant for the Truth upon the earth, and of a noble mind, and keep your men's meetings in every county concerning the poor, and see that nothing be lacking." As late as 1667–68 Fox referred to those who had received the gospel, stating that "men's meetings are set up (as . . . in the days of the apostles) in the power of God and in the holy Ghost."[27]

Among Fox's collected Epistles, the first official exhortation to set up women's meetings was written in 1666 or 1667: "Concerning the women's meetings, encourage all the women of families that are convinced, and mind virtue, and love truth, and walk in it; that they may come up into God's service that they may be serviceable in their generation." In 1669 Fox claimed that the authority of these meetings was based on "the power of the Lord God," exhorting Friends that, "so many monthly Men's Meetings as you have in your country, you may have so many monthly Women's Meetings."[28] Fox apparently felt that women's separate meetings could better care for the needs of women and children in the new church order rather than men's meetings dealing with female care.

In June 1671, after recovering from a severe illness, Fox again wrote in reference to female meetings. He mentioned their existence elsewhere, then added that "the Women may come into ye practice of ye pure Religion, wch is to visit ye fatherless and widdows and to see yt all be kept in good order from all spots of ye World." In an epistle the following year, he wrote, "Friends keep your women's meetings in the power of God, which the devil is out of: and take your possession of that which you are heirs of; and keep the gospel-order."[29]

This was a period when Quakerism was aggressive, expanding, and still feeling its "nascent energy," when its ecclesiology was still in the fluid stage. It is apparent that in this phase Fox's thinking was changing from a highly individualistic stance to one where extreme individualism became subordinate to the unity of the group in spiritual decisions and where women's separate functions in publicly recognized deaconess style ministries appeared to become increasingly important to him.

Fox's theology of the equality of men and women in the spirit was rooted in the restorationist (or restitutionist) principle that recurs frequently in his, Fell's, and other early Quaker writings. Women were no longer under the curse but were restored to become the mutual companions Adam and Eve were before the Fall. Concerning the new man and new woman in Christ, Fox wrote:

> and some men may say, ye man must have ye power and superiority over ye woman, because God says, ye man must Rule over his wife, and yt man is not of ye woman, but ye woman is of ye Man: Indeed, after man fell yt command was; but before man fell, there was noe such Command, ffor they was both meett helpes, and they was both to have Dominion over all that God made.[30]

Margaret Fell's conception of Quaker order was also forming in these years. As early as 1653, Fell established the Kendal Fund from Swarthmoor and appointed two financial agents, George Taylor and Thomas Willan of Kendal, to collect and disburse the money to traveling missionaries and their families. It is highly probable that Fell contributed liberally to the Kendal Fund herself between 1654–57. Letters from missionaries in the field and in prison written to Margaret Fell reveal their gratitude to her both for her aid to them and their families as well as for her letters of encouragement to them. Several epistles in Fell's *Works* reveal how she and her fellow treasurers labored for the cause. An epistle dated 1654 exhorted Friends to give of themselves liberally for the service of truth: "So all my dear Brethren and Sisters . . . [The Lord] requires of you, in your particular measures, to be serviceable to the Body in your respective places." In an epistle of 1656 she rallied Friends again to service and giving, that all may be in unity

> with them in their Suffrings, in their Tryals, Travels, Troubles, Buffetings, Whippings, Stockings, Prisonings, Beatings in the Synagogues, Halings before Magistrates, who are sent as Lambs among Wolves, amongst whom there is no mercy, . . . for he shall have the Victory, who is King of Kings.[31]

Gradually the general meetings that formed in the northern counties, in London, and in Bristol took general collections for the traveling ministers abroad. The particular and monthly meetings began to assume responsibility for the expenses of outfitting their own traveling preachers and their own local poor.

By 1666 Fell was firm in her ideas about women's place in Quaker ministry. Her best-known tract, *Women's Speaking Justified*, was written in August 1666. As far as is known, it was the first written defense of female public preaching written by a Quaker woman. The document reassessed the Pauline injunctions against women speaking in church. She addressed those "clergy or ministers and others against women's speaking in the church," chastising them "for meddling in the things of God." She said they were "blind priests" who wronged the apostle in their interpretation of 1 Corinthians 14 and 1 Timothy 2. Her message was that "God hath put no such difference between the Male and Female as man would make" and all this "gainsaying of women's speaking hath risen out of the bottomless Pit and spirit of Darkness . . . in this night of apostacy."

Fell followed in the footsteps of Richard Farnsworth, who had written on this issue in 1655. Farnsworth believed that women who were in the Spirit were no longer in carnal wisdom and could speak as men did in the church, for it was the Spirit speaking through them and not the natural man or woman speaking. Hence a woman preaching in the spiritual state had equal authority, rather than equal power or status in the natural state.

Margaret Fell, as a woman of her own time, sometimes adopted the imagery used by her contemporaries of woman as the "weaker vessel" (1

Peter 3:7). This motif, so predominant in seventeenth-century thinking on women, acknowledged that God's "strength is made perfect in weakness" (2 Corinthians 12:9). It construed woman's role as authoritative only in the religious manifestation as prophet. On occasion Fell gave lip service to the weaker vessel image. She never exemplified it in any of her actions on behalf of Quakerism. However, Fell, like Fox, agreed that the Genesis story indicated that woman was the one tempted and the first to transgress in the Fall. Despite this, the stigma of the Fall was obliterated in Mary, the mother of Jesus, and in Christ both man and woman were renewed or restored to the image of God that they had before the Fall. Thus by 1666, and probably long before that, Fell was exceptionally outspoken in defense of female public ministry.[32]

Other events occurred in the 1660s that may have encouraged Fox to take innovative steps on behalf of women meeting separately. From February 1664 to June 1668, Margaret Fell was a prisoner in Lancaster Castle. For approximately the first fifteen months of her confinement, George Fox was also incarcerated in Lancaster gaol.[33] Shortly after his release in 1666, Fox wrote his earliest public exhortation to set up women's meetings. Their concurrent imprisonment at Lancaster Castle gave these two leaders an extended period of tangible contact when they undoubtedly formulated plans for group survival against their persecutors. Fox spoke out in favor of women's meetings again in 1669 the same year Fox and Fell were married. The formal cementing of their partnership at that time may well have had a strong impact on his actions.

Fox's strongest exhortation to organize women's meetings came in a circular letter during the summer of 1671. He had in mind the format of the London women's meetings with their visitation of the sick, the poor, prisoners, widows, and orphans, as well as the moral oversight of women.[34] Fox and Fell had these tasks in mind for the women of the church. They felt that separate women's meetings would tend better to these problems, and that would in turn strengthen group survival. It is noteworthy that Fox wrote that letter in the summer of 1671, when he and Fell were living together with Margaret's eldest daughter, Margaret Rous, at Kingston-on-Thames. By August 1671 Fox and his group of missionaries were prepared to sail for America. Margaret Fell and her daughter Susannah went to Lands-End to see Fox off and then immediately returned to Swarthmoor. Fell did not detour to Cornwall to visit her daughter, Mary Lower, who was ill at that time. Nor did she tarry long enough to visit her other daughter, Isabel Yeamans, who was also ill in Bristol. Instead, Susannah went to visit her sick sisters. Judging from Margaret's past religious motivation, it would not be out of character for her to put religious work first before family matters. Therefore, it would not be out of order to assume that only pressing business of a religious nature would have taken Fell north with such dispatch and without at least a brief visit to her two ailing daughters.[35] The minutes of the SWMM, the earliest women's meeting formed outside London, indicate that the first meeting took place at Margaret's home, Swarthmoor Hall, on 23 October 1671, just two months after George Fox set sail for America. Margaret lost no time in organizing her neighborhood women Friends. Their first order of business was to take a collection for the poor.[36]

The following spring Margaret made a three-week journey into Yorkshire, accompanied by her youngest daughter Rachel. They visited and held meetings at various homes before returning to Swarthmoor. We cannot be certain whether these were joint or separate meetings. However, a letter to Fell from one Jean Simcock of Yorkshire in 1672 states that she is arranging for Fell's letter to be read to the Women's Quarterly Meeting in Yorkshire. This suggests the possibility that Margaret's business in Yorkshire in 1672 was to encourage the setting up of women's meetings there. Simcock's letter also suggests that the Wilkinson-Story dispute against women's meetings was spreading into Yorkshire, for she writes that Quakers in Yorkshire object to collecting money for the "women's stock" (or treasury) and they are disturbed that this internal disagreement is becoming too public.[37]

Another glimpse of Fell's work in the 1670s comes to us in a letter from her friend Gulielma Penn, written in 1684, which makes reference to Fell's work a decade earlier. Gulielma Penn's comments on the nature of her friend's role in the Wilkinson-Story schism are revealing:

> In the unity of the Spirit of Truth . . . I see thee over all those bad spirits, that may touch thy heel but no further; and it rises in my mind as I am writing, something that I saw concerning thee in my sleep long ago, at the beginning of these bad spirits. I saw thee, and dear George and many Friends in a Meeting, and the power of the Lord was greatly manifested and methoughts there came dark wicked spirits into the meeting and they strove exceedingly against the Life that was in the meeting, and *their chief aim was first at thee and George, but mostly at thee*, and they strove to hurt thee, but only went to tear some little part of thy clothes, but thou escaped unhurt or untouched, and a sweet rejoicing and triumph was over them through the meeting.[38]

The sources strongly suggest Margaret Fell's deep involvement in the establishment of women's meetings, and that her role *vis-à-vis* the chief antagonists of women's meetings, the Wilkinson-Story circle, was widely known.

The long-term impact of women's meetings, as envisioned by Fell and her women friends, was problematic. Braithwaite reveals his uncritical stance when asserting that "the equality of men and women in spiritual privilege and responsibility has always been one of the glories of Quakerism." To suggest that their separate meetings gave women complete, or nearly complete, spiritual equality with Quaker men would be overdrawing the point. They were not recognized in Quaker polity until 1697 when the yearly meeting approved a separate concurrent meeting for women to take place annually. However, the yearly women's meeting only received full recognition in 1784 and was limited in its disciplinary power.[39]

Despite his statement above, Braithwaite points out that Quaker men's attitudes changed early, as revealed in minutes recorded in 1701. It seems that women were asked to leave their names with the clerk at public meetings if they desired the opportunity to "clear themselves [to speak in the meeting], and yet [they should] be careful not to interfere with their brethren in their public mixed meetings." Another minute of the same temper reads:

> This meeting finding that it is a hurt to Truth for women Friends to take up to much time . . . in our public meetings, when several public and serviceable men Friends are present and are by them prevented in their Serving, it's therefore advised that the women Friends should be tenderly cautioned against taking up too much time in our mixed public meetings.[40]

MARRIAGE PROCEDURE AND THE WOMEN'S MEETINGS

Fell's pivotal role in organizing the separate women's meetings was reflected in the internal ordering of those meetings. The development of the Quaker procedure for marriage "clearness," or certification, as handled by the men's and women's meetings, affords an example of Fell's authoritative role in the emerging organization and the actual example of co-authority concerning men's and women's meetings. Margaret Fell rigorously asserted that women's meetings held the prerogative in the marriage contractual procedure, especially insofar as it pertained to the clearing or certifying of women before marriage. She also stated her views on the legal aspects of marriage procedure that became the norm in Quakerism.

Very early in their existence the Quakers sought to regularize marriage in a way that would withstand legal assaults over the inheritance rights of heirs. In 1653 a civil marriage act was passed by the Barebones Parliament that made a marriage ceremony before a justice of the peace compulsory. The contract was duly recorded by the civil registrar for the public record. In the same year that the civil marriage law was passed, George Fox wrote on marriage procedures. Fox's procedure stipulated that marriage should take place in the presence of Friends at the local meeting, at which time one or more testimonies could be given as persons were so moved, and the witnesses were to sign a dated certificate of marriage. Hence, Fox ignored the obligatory civil procedure of recording the marriage in the records before a justice of the peace, as laid down by the parliamentary act.[41]

In 1656 Fell wrote an epistle giving advice to Quakers that differed in one important feature from that of Fox. Fell's epistle stipulated that a certificate first be signed by men Friends who wished to do so at the wedding, and then Fell deemed it mandatory for the married couple to appear before the justice of the peace the next day or as soon as possible to have it included in the civil record. This was an important innovation in the "regularizing" of marriage among Quakers, and it was Fell who was the first to call for an obligatory civil recording of Quaker marriages. Arnold Lloyd, however, claims that the catalogue of Fox's papers ascribes this innovation to Fox. Lloyd further points out that the original epistle docketed by Fox, entitled: "Epesell to frneds of mareges by Mff" (1656) was found circa 1950 in private hands and confirms Margaret Fell's innovation.[42] Although the SDMM did not organize until 1672, it seems curious that they did not preserve this epistle, which was of obvious importance for Quaker marriages, in their early collection of Quaker documents.

BUSINESS PROCEDURE OF SWARTHMORE WOMEN'S MONTHLY MEETING

The internal ordering of the earliest women's separate meeting outside London points to an informal hierarchy that operated in this women's fellowship. The women of the Swarthmoor Monthly Meeting, which drew together several smaller "particular" meetings of women in the Furness district, in the 1670s hammered out a discipline statement without any precedents to follow except the London women's meetings in their vision to "walk as becomes Truth." A relatively rare document describing the internal order and discipline of women's meetings was discovered by

Isabel Ross in 1949. Ross dated this epistle sometime between 1675 and 1680 and established its author as Sarah Fell, who probably wrote it in conjunction with the women of the Swarthmoor Women's Monthly Meeting. The document gives a glimpse into the Fell women's leadership in directing Quaker women away from complete reliance on inward spiritual authority to a group-oriented authority with stress on a deaconess-style discipleship through lives of outward service. The epistle was sent out from the Lancashire Women's Quarterly Meeting to "women's meetings everywhere."[43]

Under ten headings, the Swarthmoor Women's Meeting epistle documents the internal functions and discipline of the women's business meetings, and it describes in detail the responsibilities of those meetings as independent entities from the men's meetings. The document speaks with the authority of a mother meeting giving advice to the daughter monthly and quarterly meetings springing up and already functioning elsewhere. It is a beautifully worded epistle, heavy with biblical imagery, that rises to mystical eloquence in its introduction and ending. The minutes of the SWMM and the LWQM attest to the ideas of order in this epistle in terms of the women's day-to-day service to other women of the Quaker flock. It is useful to compare the content of this epistle with actual examples taken from the minutes of the SWMM and the LWQM. One can see a dovetailing of theory and practice in comparing the three sources.

The introduction preceding the clauses of advice on order and discipline is essentially a sermon. It includes a strong defense of women bearing their spiritual testimony in word and action. This defense against their adversaries, "the gainsayers against womens meetings," is based on every possible example of women leaders that the author(s) could cull from the Old and New Testaments.[44]

It opens with a bold statement of authority and purpose that yields mere lip-service to the "weaker-vessel" image:

> Dear Sisters, In the blessed unity in the Spirit of Grace our Souls Salute you who are sanctified in Christ Jesus, and called to be Saints, who are the true and Royal offspring of Christ Jesus, who is the root and offspring of David, and who is the resurrection, and life of all the Saints in light. . . . To you all every where, where this may come, is this written; that in this blessed seed, which hath the promise of the eternal God, that he should bruise the serpents head, and that in his seed you all live and dwell . . . in which seed all nations of the earth is blessed: that so we may be all helps meet for God in the restoration, and co-heires with Christ Jesus, who hath purchased us

with his precious bloud, and who hath washed us and loved us, who is no respecter of persons, but hath a care, and a regard unto all, the weak as well as the strong.[45]

The first heading for discipline calls for separate women's meetings to meet concurrently with the established men's monthly meetings, "to wait upon the lord, and to hearken what the lord will say unto them, and to know his mind, and will, and be ready to obey, and answer him in every motion of his eternal spirit and power." This suggests that the first portion of their business meeting was devoted to a silent worship. The Swarthmoor women expressed their experiences of renewal and fellowship in terms that express this spiritual orientation:

> wee ar well Refreshed in this ouer assemely to sarue [serve] ye lord finding his Exceptance and ye fauouerabell aspectt of his couentance [countenance] upon us: so yt ouer souells is more glad than if corne wine or oylle Encreses: and ouer Spiritts boued and broken beefore him: whoe is God worthy ouer all to bee magnifyed by those who has tasted of his ouercoming loue and of his hart breaken presence.[46]

Two headings of the document relate to the screening and disciplining of members who "walked disorderly." Public conduct was a very important aspect of Quaker discipline, and the women admonished and exhorted the disorderly to walk "as becomes Truth." Any serious public offense or immoral act required a paper of condemnation to be published by the offender before he or she could be reinstated into the group. According to the minutes of the SWMM, the unrepentant Hester Cooper is an example of long-term discipline. She had the uncomfortable experience in 1676 of being engaged to Thomas Wilson, who jilted her for another woman. That experience may help to explain her unhappy disposition in her subsequent encounters with the SWMM. Hester Cooper refused to write a letter of self-condemnation after she was reproved by women friends of her meeting for keeping and selling ale. A minute of 2 January 1679 reads:

> to the greife of ffriends; that shee will not hear reproofe, nor Regard Admonition: yet shee hath sent A paper in which Shee seems to bee sory, that Shee hath given occasion for the truth to bee Evill spoken off, and shee saith Shee will doe soe noe more: But wee haue litle Reason to give creditt to her words; knowing that Shee is led by an unstable false Spiritt, and covers herselfe with false coverings: Therefore our Testimony is: that if Shee doe not leave it off, that A

paper must bee written Against her and her evill practises, and made publiche, to cleare ye truth of her and her actions, yet wee are wilinge to waite a while longer; in hopes, that Shee may come to have some sence of the hard and unsubjected state that Shee is in; ffor wee know, that the Lord waites to bee gratious to All, and is A long-sufferinge God.[47]

According to subsequent minutes, her "stubborne Rebellious Spiritt" continued, and she slighted "ffriends Advice and councell from time to time" against selling ale and carrying "herselfe disorderly." The Swarthmoor women reached a point of exasperation in May 1680 "least Shee cause the Truth to suffer" and threatened to disown her for her continued refusal to write a paper of condemnation. By June 1680 she had written her paper of self-condemnation for ale peddling, just in time to forestall outright disownment. The SWMM accepted it but with little warmth toward Hester Cooper. Their advice to her was to come to true repentance and to "Rectify and Amend her life and conversation, which will be for her owne good; if not wee are cleare of her, haveinge been much Exercised for her good, and shee must beare her owne Burden."[48]

The Swarthmoor women's epistle also indicates that a chief order of business among the women was to gather from the members their testimonies against tithing or giving money to Anglican church-related work. It was the custom for someone to collect these testimonies which were written statements by each woman personally denying having paid any tithes to the local authorities. These statements were collected at the monthly meeting and carried to the women's quarterly meeting. The clerk of the quarterly meeting (which met biannually) duly recorded the testimonies. This was another area of dispute between the Wilkinson-Story people and the Fox party, for the separatists were more flexible about giving tithes. Every woman member was questioned, and the formulaic words recur many times in the minutes:

It is the desire of this meeting: that there bee a meeting in every ptuckler [particular] meeting within two weekes from this time to Examine and Enquier how ye woman ffrinds Stands as to ffaithfullnes in there testymonyes againestt tyths and in othr things where truth is Consarned: and an Acct caried to over Generall womans meeting wch is to bee at Lancaster ye next moth.[49]

The discipline of children and servants and the care of the poor, widows, and orphans were subjects addressed by the women's document. There was an abiding concern for Quaker household servants and children lest they be exposed to the world.

> It is our duty and care . . . to see that our Children are trained up in the feare of God. . . . For you know, that we are much in our families amongst our children maids, and servants, and may see more into their inclinations; and so see that none indulge any to looseness and evill, but restraine it.

The Quaker women of the Swarthmoor meeting fretted over this, and by the 1690s many comments appear in the minutes relating to the offspring of first-generation converted Quakers. It is apparent that these children lacked the fervent zeal of their convinced parents. A minute of August 1695 expresses parental concern:

> ffriends apoyentt to Enspectt the walking of ffriends makes Report to this meeting that things ar pretty well onlly Some ffriends of Swarthmore meeting gives an accouent that a young woman called Mary Dungley whoe have ffrequented there meeting for Sum yeares; have of latte keptt company with a younge mane of the world in order to maridge as its ffeared; Severall ffriends of that meeting haveing labroed much with her in the love of truth and admonidged her to bee wayer of that temptation and snare that Shee is in danger to ffall into.[50]

The procedure for marriage clearness or certification was explained in the women's epistle under the fourth and fifth headings. Great concern was expressed for orderly marriage procedure and the process of clearing a couple who wished to marry involved investigating each person's background to see if there were any unknown engagements, marriages, or children to others already in effect, as this had important legal implications. The clearing process also attempted to discover each person's past moral character, and to be sure that the young people were Quakers in good standing. The epistle also expressed a preoccupation with those women who went to the priest to be married.

> And dear sisters it is duely Incumbent upon us to look into our families, and to prevent our children of running into the world for husbands, or for wives, and so to the priests: for you know before the women's meetings were set up, Many have done so, which brought dishonour, both to God, and upon his truth and people.

Marriage procedure involved appearances before two consecutive monthly women's meetings so that members might examine the couple to see "that they be cleare and free from all other persons" and that the families consent. If all was in "gospel order," the marriage could take place at the following monthly worship meeting. The women insisted on keeping their prerogative, albeit shared, for the oversight of marriage clearing, stressing that "if anything be amiss concerning the woman, examine it, and look into it which may not be proper for the men."[51]

In the minutes of the SWMM, any of the disciplinary actions revolved around Quakers marrying non-Quakers without consent of the meeting. When women in the SWMM "married out," they could expect to be disciplined, and, if unrepentant, eventually disowned. The example of Sarah Cooper is typical. Sarah Cooper of Ulverston, sometime prior to February 1673, had married the non-Quaker, John Fell. Three women of the SWMM approached Sarah Cooper Fell and informed her "that ffriends cannott receive her, till shee have condemn [torn] her beinge Married with ye Priest." Apparently Sarah was not contrite, for "shee Answered friends corssly, and would not Acknowledge any wronge shee had done; so till shee come into a better sence of her Condition; shee cannott bee Received by ffriends." Four years elapsed before we encounter a minute of the SWMM on Mistress Fell's case. In April 1677 the SWMM received a paper of self-condemnation from Sarah Cooper Fell:

> of her sence and sorrow of the Transgression, that shee fell into: in being Married by A Priest, to one of the world; which wee are glad of and gives thankes to ye Lord in her behalfe, that hee hath broken her heart, and brought her to a sence of her Condition; wch paper is hereafter Recorded and the desire of our hearts are, that shee may walke soe wisely, and that shee may not give any occasion that truth bee evill spoken of hereafter; and soe as to gaine her Husband if possible.

However, Sarah was strong-willed, and we encounter her on a later occasion receiving reproach for her hostile behavior. A minute of the meeting at Swarthmoor in 4 May 1680 reads:

> A paper is read this day, from Sarah ffel, wife of John ffel of Ulverstone, wch signifies her trouble, for some words, that shee spoke at our last Women's Meetings with partiallity (without any cause given for it:) which shee spoke, in A forward hasty Uncircumcised Spiritt, – upon which shee owns Condemnation; which gives the Meettinge Satisfaction: – And its our desire, that shee, withall, yt makes A profession of

the pretious truth of our God may bee more cautious for the future, and not to runn into such things as brings grife upon themselves, and trouble upon others, but to keep in that meeke and quiett Spiritt, which is not easily stirred up, but keepes in Moderation upon all occasions.[52]

The procedure for marriage clearness was a mutually shared responsibility of the men's and women's meeting of Swarthmoor. The case of George Braithwaite and Ann Kirkby represent the normal procedure for marriage, both having appeared twice before the SWMM for clearness before May 1680. George was cleared by the SMMM, and Ann, according to the SWMM minutes, was not committed to any other man.

> A good Acct is given by the Men ffriends of him; and of his clearnesse from all other women; soe that there is nothinge appeares against Either of them, But that they may proceed, to A Conclusion in convenient time, according to the order settled Amongst Us, for the compleatinge of that Ordinance.[53]

The conclusion of the women's epistle expresses the rationale for existence of this women's religious order in spiritualized language. Although it follows a biblical form customary of the age, it reveals Sarah Fell's ability at literary expression on behalf of her sisters:

> And though wee be looked upon as the weaker vessels, yet strong and powerfull is God, whose strength is made perfect in weakness, he can make us good and bold, and valliant Souldiers of Jesus Christ, if he arm us with his Armour of Light, and give unto us the sword of his Eternal Spirit which is the word of the Eternal God, and cover our hearts with the breast-plate of righteousness, and crown us with the helmet of Salvation, and give unto us the Shield of Faith, with which we can quench all the firey darts of Sathan
> This is given forth for Information, Instruction, and Direction, that in the blessed unity of the spirit of grace, all friends may bee, and live in the practice of the holy order of the Gospell; if you know these things, happy are you if ye do them so.[54]

Speizman and Kronick rightly view this epistle as a "codification of the thought and experience of women's meetings, especially as systematized in the minds of Margaret Fell and her daughters." The document gives a summary of Quaker thought by a vanguard of women who were in the process of creating a Protestant female religious order with few, if any, established precedents to which they could turn for procedural advice. The SWMM and the LWQM were the models for other women's meetings. Margaret Fell and her daughters, Sarah, Rachel, Isabel (when at home), Mary, and Susannah dominated the SWMM and the LQWM from their establishment until Margaret's death in 1702. The women alternately met at three locations, one being Swarthmoor Hall, for more than thirty years. Margaret Fell's name always appeared first in larger, bolder letters on the list of signatures of women in attendance. Her daughters' signatures appear frequently. Fell attended these meetings faithfully (except when traveling) until about 1697, when she was eighty-three years old. Thereafter, she attended only those meetings that convened at her home. In the years of apparently lowest attendance (1688–92), her name sometimes appears alone at the end of the minutes. Her signature may have been representative of others at the meeting. However, by the 1680s a large number of Quakers were known to have immigrated to other parts of England and to America, and the low attendance may be a reflection of their depleted numbers in the Furness area. In August 1692, one such SWMM minutes appears with Margaret as sole signer. It conveys a certain wistfulness on the part of the clerk of the women's meeting (i.e., Rachel) and presumably her mother:

> Upon Enquiring wee finde all well In over Respective meetings which is mater of Joy to us; and those litle or noe busnes is preposed to this meeting, yett wee ar well Satisfyed with this over Assembling together; finding upon over Spiritts the answer of welldone, which is Encoridjment to us not to bee weary or groe Slacke in this Service.[55]

One final glimpse of Margaret and the oldest women of standing in the SWMM, such as Dorothy Beck, is seen in the minutes of December 1700 at Swarthmoor:

> It is the Desier of ye ones [?] and upright harted that the Anchantt Euenety [unity] in the love of god may continue amonges us and the Regarde and honer which is deue and Doth beelonge to the Elders and those that hath been as mothers in Issrall and have bore the burdine and Exercyes in the morning of the Day and have washed ther garmentts in the blod of the lame; pasing throug many treblations; that the younger Genration may not hurtt themselfs nor greeve the lord; by ther Disregard ungrattfullnes towards his worthyes whoe ar Deare and preshas in his Sightt and whoe ar knowne unto hime; Bllesed bee his holly nam ffor Evermor.[56]

The rationale and appeal of women's meetings in Fell's lifetime were based on several factors. It is apparent that women were easily overawed by the men in business or worship meetings where both were present and thus intimidated into silence. The posture of Ann Camm of Westmorland, a younger and respected contemporary of Margaret, who had endured imprisonment, portrays this situation. She was described as being properly deferential and knew not to be "too hasty, forward or unseasonable." In addressing the meeting, "she was a good example to her sex," for she spoke rarely.[57] Secondly, women's meetings appealed because they gave a sense of cohesion to women who sought not only renewal of strength in an era of sectarian persecution but friendship and fellowship among their own sex. Moreover, women's meetings acted as an outlet for women's energy and ideas, for women were barred, with rare exceptions, from important administrative groups such as the Meeting for Sufferings, the SDMM, and the Yearly Meetings.[58]

*   *   *

What conclusions might we draw concerning Margaret Fell's role *vis-à-vis* George Fox in the organization of the early women's meetings? The evidence marshalled in this chapter is incomplete and thus allows various possible interpretations. However, if my reading of the evidence is correct, then a revised interpretation of Fell's role is both plausible and in order.

We have seen Fell's importance in the SWMM and the LWQM over a period of thirty-one years. She was a woman whose name appeared in a respected position in the minutes, who organized and held the first meeting of 1671 at her home, who on occasion acted as arbitrator for internal disputes, who was innovative in her ideas on Quaker marriage and women's role in the certification process, and who was involved in every one of the activities of the meeting during the time under study. This signals a person of enormous personal esteem and authority within the local community. Secondly, Fell's daughters, Sarah and Rachel, acted as clerks of the meeting for more than thirty years. It should not be overlooked that the clerkship was a position of respect and power in the group. It is highly doubtful that Fell's daughters would have recorded for posterity anything in the minutes that would have cast any aspersions on their mother's leadership and reputation or that ran contrary to her wishes. This leads to the third and most important possible conclusion: that certain evidence was suppressed in the early years. We have seen in the Fell-Rawlinson feud that new evidence is highly suggestive that suppression of this internal strife, which would have placed Fell's leadership status and personal reputation in jeopardy, did occur. Such suppression was done in the larger interest of internal stability gained through concentrated leadership and for the survival of the Society beleaguered by sectarian disabilities in an age when tolerance was little valued in society.

The history of the origins of movements of a religious, social, or political nature is usually shrouded in obscurity and is first recorded much later by a member or sympathizer of the movement. The historical narrative often incorporates into the history a "myth of beginnings," an unintentional hagiographical approach that has developed from within and has been transmitted by the community. In Margaret Fell's generation of Quakerism, it appears that a myth of origins was already shaping its history. Some historians of early Quakerism, although certainly not all, begin with the assumption, handed down by Quaker tradition, that the founder and central leader of Quakerism was undisputedly George Fox. Winthrop S. Hudson has presented the argument that a close examination of the original sources reveals inconsistencies and discrepancies relating to the earliest phase of the movement. He points out, for instance, that William Penn's eulogy to Fox, contained in the preface of Fox's *Journal* published by the Society of Friends in 1694, called Fox the original founder and genius, indeed "no man's copy." Hudson has argued vigorously that on the contrary, Fox was not "no man's copy," but a man who assimilated ideas that were current at the time and made them his own, and that, subsequently, Quaker leaders piously suppressed evidence that went contrary to this tradition.[59]

Hudson's thesis is germane to this chapter because it encourages another question that deserves attention. What does a new reading of the evidence suggest about the origin and establishment of Quakerism, and what influences were brought to bear on that early development? In light of the evidence on early women's meetings, it seems reasonable to suggest the hypothesis that the record neglects to assess the full impact of Fell's role in relation to Fox. George Fox was an itinerant, charismatic preacher who spent most of his adult life wandering about England and America, while Margaret Fell maintained her home base at Swarthmoor and hammered out, year in and year out, with her daughters and neighborhood Quaker women, an organizational format that worked on the practical level and that borrowed from the experience and organization of the London Quaker women's meetings. It is reasonable to suggest that it was in keeping with Fell's personality, wealth, permanence of place, and organizational skills learned in the domestic sphere, that she, and not Fox, was the main ideologue and instigator of the women's meetings outside London. There is strong evidence that she and Fox were cofounders of equal importance in the Quaker administrative structure. Considering the images and conceptions of women in seventeenth-century England, it is highly probable that the actual pivotal leadership of Fell was subsequently modified in Quaker historiography to that of a nurturing and secondary role. In view of Fell's leadership and innovative qualities, it is reasonable to suggest that it was not only her idea to set up women's meetings, having been aware of how well they worked in London, but that she also consciously and methodically presented Fox as the innovator of this new gender role in Quaker church order. As cofounders, they worked closely together, and Fox, out of indebtedness to her friendship and patronage from the beginning, and in deference to her wishes as his co-worker and subsequently his wife, made the idea of women's meetings his own. He supported her work by publishing epistles to encourage the setting up of women's meetings wherever Quakerism spread. In mutual loyalty to him, Fell affirmed Fox's authority as the divinely-called founder of the Quakers in her travels, writings, and correspondence. Because of her unassailable position of authority in the early movement, she was able to insure Fox's preeminent position. For example, James Nayler had been an early rival leader to Fox prior to 1657. Fell, in 1656, after Fox and Nayler had a falling out, wrote to Nayler warning him not to attempt to supersede Fox as *primus inter pares* of the movement. She wrote:

> Since I have heard that thou [Nayler] would not be subject to him [Fox] to whom all nations shall bow, it hath grieved my spirit. Thou hath confessed him to be thy father, and thy life bound up in him, and when he sent for thee and thou would not come to him, where was thy life then. . . . I warn thee from the Lord God that thou beware of siding with unclean spirits [opposers of Fox], lest thou be cut off for ever.[60]

She alluded numerous times in writing to Fox's supernatural mandate, "an instrument raised up by God," thus helping to authenticate Fox's public ministry and leadership *vis-à-vis* other prominent Quaker leaders. In Hudson's summation, William Penn "was not entirely without guile when he wrote of George Fox . . . 'As to man he was an original being no man's copy.'" If Fox borrowed ideas from others, let us look to the woman closest to him.

8 Gender, Religion, and Class: A Seventeenth-Century Friendship[1]

The friendship of Margaret Fell, George Fox, and William Penn began with William Penn's conversion to Quakerism in 1667 or 1668. Fell, Fox, and Penn remained friends over many years despite a wide age difference and dissimilar social origins. Accounts of their personal contacts and the letters that passed between this triumvirate of early Quaker leaders reveal the significance of this interclass friendship and something of their various roles in the movement. Further, their friendship illuminates the emerging hierarchical church authority that came to characterize later Quakerism.[2]

The world of Fell, Fox, and Penn includes a deeply embedded system of social inequality. Hierarchical rank according to birth, wealth, land, and living standards formed the structure of English society. Social stratification seemed entirely natural to the seventeenth-century person. For instance, a contemporary of Fell, Mrs. Lucy Hutchinson, was very conscious of her gentle birth and station in life. She prefaced *The Life of Colonel Hutchinson*, her husband John, who died as an imprisoned regicide in 1664, by describing her maternal and paternal lineage. She then described how her indulgent parents at one point provided her with eight tutors in languages, music, dancing, writing, and needlework. She claimed she knew some Latin, Greek, Hebrew, and French and had read some classical and theological works. In preserving her husband's memory, she recorded a vivid account of the Civil War in Nottinghamshire. Referring to her Puritan family life, Hutchinson noted her social advantages:

> Herein I meet with so many special indulgences as require a distinct consideration . . . the parents by whom I received my life . . . the rank that was given me in my generation, and the advantages I received in my person . . . they give me infinite cause of glorifying God's goodness.[3]

Sectarian groups such as the Quakers introduced an unsettling egalitarianism that challenged the social hierarchy. Although the Quakers for the most part, accepted the legitimacy of the social order, they confronted its existing inequalities in their refusal to tip their hats, "to give hat honor," to social superiors or to recognize degree in dress, demeanor, or any public ordering of rank and status.[4] This Quaker radicalism gives rise to several questions relating to the Fell-Fox-Penn friendship. Did theory and practice of class and gender parity coincide within the spiritual fellowship of Quakers? Did Fell's collaboration with Fox and Penn upset the deeply inculcated conventions of class and gender? This chapter will explore these questions and propose the thesis that despite Quakerism's appeal to egalitarian principles, the correspondence of Fell, Fox, and Penn over more than twenty years suggests that the concept of "friendship" as rooted in social hierarchy and political patronage, was not eradicated within the movement. Secondly, Margaret Fell's roles constituted a new construction of gender and class.

Although the Fell-Fox-Penn friendship began in 1668, their correspondence spans 1674 to 1691. A total of twenty-six letters are extant, and the internal evidence indicates that more letters, now lost to us, passed between them. Sixteen letters were exchanged by Fox and Penn; nine letters were exchanged by Fell and William Penn and his first wife, Gulielma. The letters were for the most part mutually cordial and informative, yielding glimpses of their personalities and reflecting their attitudes toward their public and private roles. Warm salutations to family members and close friends were frequently included in the letters. Illnesses and pregnancies, two conditions fraught with danger, were mentioned with accompanying pleas for prayers. In spiritualized language they affirmed their mutual friendship and unity with God and each other. While Penn, Fox, and Fell exchanged news of Quaker missionary activities, Penn and the two women exchanged news of their personal lives as well.[5]

The story of this friendship with its implications for religious authority rooted in social status is better understood against the backdrop of two significant Quaker trials for nonconformity, one involving Fell and Fox in 1664 and the other involving Penn in 1670. The two trials illuminate the subtle but deeply rooted social differences between these leaders that were given nuanced expression in the correspondence.

Margaret Fell and George Fox had a highly publicized oath trial in Lancaster in 1664 during a wave of persecutions.[6] Justice Daniel Fleming, a staunch Royalist, investigated Fell in February 1664 in the company of other local justices. Their intent was to discourage Fell from holding further Quaker meetings at her home. When the justices threatened to offer Fell the oath if she refused to cooperate, she responded, "while it please the Lord to let me have a house, I would endeavour to worship him in it." Fell was issued a warrant of arrest and imprisoned in Lancaster Castle until the next quarter sessions. Although Fell and Fox were presented together at the assizes of March 1663–4, Fell, by virtue of her gender and class, may well have caused the greatest sensation. Her widowhood and ownership of Swarthmoor Hall, as well as her vulnerable position as the mother of six unmarried daughters, made it clear to all onlookers that she had more to lose than others, such as Fox, who was a man without real property or any apparent income.[7] At her trial Fell drew deference even from the Bench that was to try her:

> First: She was called to the Bar, Order was given to the gaoler by the Judge, to set a stool, and a Cushion for her to sit upon: And she had four of her daughters with her at the Bar: And the Judge said, Let not Mrs. *Fell's* Daughters stand at the Bar, but let them come up hither, they shall not stand at the Bar: So they plucked them up, and set them near where the Judge sate.[8]

Her trial was a display not only of her local standing among the county elite, it vaulted her into a position of wide notoriety.

Richard Bauman's ethnographic study of early Quaker speaking practices has demonstrated that the oath trials in the 1660s became the focus of a social drama where the defendants acted out in public conflicts over political and religious authority. Bauman has pointed out that the courtroom dialogues gave formal, symbolic expression to the opposing identities and their positions concerning the issue at stake. The Quaker, as the innocent defendant, was confronting the Bench, which was dishonest. Such episodes often vindicated those persons who publicly enacted the conflict and elevated them to a special leadership status.[9] At her Lancaster trial in March 1663–4, Fell expressed succinctly the symbolic conflict: She owed allegiance to the "King of Kings" before the "King of England." Fell was tried under the statute of praemunire and lost.[10] She remained in Lancaster prison for four years although her property was not confiscated by the Crown. Fox was tried under the same statute and imprisoned until 1668. Fell's role as a prominent local woman of some wealth and as a public speaker on behalf of a religious minority and its distinctive beliefs made her a cultural symbol far beyond the confines of Quakerism.[11]

Six years after the Fell-Fox trial, William Penn was tried in London on account of his Quaker activities. On 14 August 1670 the Grace Street Quaker meeting house was locked by the king's men. In defiance, Quakers assembled in the street in front of the barred meeting house, and William Penn spoke. Penn was arrested along with William Meade, another new Quaker, and committed to Newgate prison on the charge of seditiously creating a riot. The trial that followed would subsequently become a famous precedent case on jury coercion by a judge. Since Penn was not tried under the Conventicle Act, he argued in his own defense that in speaking to a crowd he had broken no law. The Bench refused to accept the jury's verdict that no law was broken and locked in the jury without food with the apparent intention of bullying it into changing the verdict. The jury became adamant in its stand, for in the subsequent court session the jury found both defendants "not guilty." News of the trial spread rapidly about London, giving Penn instant fame in the Quaker cause and in the cause of pleading for a jury's right to freedom from coercion by the Bench. The trial also was an important Quaker public tactic in the struggle against persecution of nonconformity under the Clarendon Code.[12]

A NEW FRIENDSHIP

In 1669 William Penn, as a newly converted Quaker, was in Bristol with Fell and Fox just prior to their marriage. He may have delivered a supporting testimony on behalf of the couple at a meeting for worship preceding the event. In August 1671, Penn, together with several Friends, including Gulielma Springett, his future wife, accompanied Fell and Fox to Land's End to see Fox off on his missionary venture to America. In 1673, Penn came from London with his wife Gulielma and other close friends to welcome Fox home from his new-world travels.[13] These were gestures of symbolic value among three "weighty" or public Friends.

Shortly after Fox's return to England, he was arrested for his Quaker activities in Worcestershire and imprisoned at Worcester gaol from mid-December 1673 until February 1675. This imprisonment was a great aggravation to Fell, and she criticized her spouse for not avoiding arrest. Fox wrote to his wife on 12 October 1673 from prison:

> der hart thou seemd to be much greeved when . . . j was taken thou began to fall vpon mee with blaming of mee and j tould thee that j was to bare it and why could not thee be and be content with the will of God and thou said som wordes and then was prety quiet.[14]

One year later Margaret Fell was a Worcester still awaiting Fox's release. She and her husband wrote a one-page letter to Penn on 9 November 1674, asking him for legal advice and for help in arranging Fox's release. Fell's letter conveyed her alarm over her spouse's declining health and her irritation due to the months of abortive efforts to gain his release. Dispensing with the usual social amenities, she went directly to the question at hand:

D[ear] William Penn – if thou thinke yt: my comeinge upp concerneinge this business of my husbands . . . If I coulde be any way serviceable In it: I am willinge to Come upp: although I have stayed heere soe longe yett I am willinge to doe my uptmost endeavour before I returne backe home if it be thought Convenient and requisite.[15]

Meanwhile Penn, in attempting to secure Fox's release, visited the Stuart Court for the first time since his conversion to Quakerism five years earlier. The king, who was kindly disposed to Penn, offered a pardon, but Fox rejected this on grounds of his innocence, for he had been arrested not at a meeting or conventicle, but in the private home of a Friend.[16]

Subsequently, Margaret Fell traveled to London and spoke personally with the king on Fox's behalf, apparently at the suggestion of a London Friend, Ellis Hookes. Hookes wrote to Fox: "I thought if Margaret had freedom to come up and speak with him [the king] it might have some good effect, he having a respect to her; she may come up in the coach." Her arrival by horse-drawn coach denoted a woman of some means and status. Hookes was not blind to the effect of outward appearances. The king did receive her, "spake kindly to her and referred her to the Lord-Keeper," but despite his personal friendliness, he was either disinterested or powerless to override his council.[17]

The next communication between Penn and Fell is a letter written from Swarthmoor on 26 June 1675, just after she and Fox arrived home. Fox was weak and ill after his long imprisonment. Fell's letter was once again gracious and solicitous, conveying a feeling of relief. In her salutation she talked about truth, everlasting joy, honor, peace and felicity, and unity of life in which, she added, "the Liveinge God has Ingaged us, for his owne Service and honour, and for our Etternall comfort." She acknowledged her indebtedness to Penn "upon many Accounts" and commented that her trip homeward was a "grasious and prosperous Journey." The same letter referred to the unfolding dispute within the ranks of Quakerism, the Wilkinson-Story feud. Her letter expressed alarm that the dispute had reached a serious impasse.[18]

The Wilkinson-Story schism was a factor that brought the Foxes and Penns into closer friendship against a common internal enemy. The dispute was the main subject of a letter written by Fox to Penn in September 1675 after Fox had spent the summer at Swarthmoor recuperating from illness. Penn had earlier informed Fox that London and Westmorland nonseparating Quakers were making an attempt to bridge the gulf between Fox and the separatists. Fox's letter of 30 September 1675 from Swarthmoor was a stern statement of self-defense, and he remained

174 *Part III – The Political and Religious World of Margaret Fell*

unyielding toward the two Johns. Fox told Penn in straightforward terms, "Lett them that dos complaine above, Come down, and not be frettinge and troubleinge themselves there." He said that it was up to the separatists to make overtures for a meeting:

> If they thinke others has not Judged Equally, they may come and mende the matter if they cann; But I doe not heare, that any of these dissatisfied ones [London and Westmorland arbitrating Quakers] doe mention anythinge that the two Johns has done amisse, but onely complaines of such as gave Judgemt.[19]

William Penn had close ties to the opposition party. Gulielma Penn's family were close friends of Thomas and Ann Curtis, two leading members of the Wilkinson-Story circle. It is therefore understandable that Fell and Fox at Swarthmoor may have had some lingering doubts as late as 1674–75 of Penn's loyalties. It is not surprising that Fell's letter to Penn in the month preceding Fox's terse letter, also conveyed, albeit in gracious language, an implied question as to Penn's loyalty to them. She greeted Penn, in the stance of a Quaker elder to a younger convert, advising him to remain constant to the truth:

> my Dearely beloued, keepe close and neere to the pretious and to the pure and liveing testimony of the liveing God in your owne hearte that never varyes nor Changes. You have seene tossings and Changes but the foundation of God stands sure, . . . Beware of men, that said they were Gods but they shall die like men, this is a time of tryall, . . . the serpent is very Cunneing and is close at worke and will appeare as an Angell of light, keepe cleare and all is safe . . . and the Lord and his truth and the Lamb shall have the victory and you will see that all things will worke together for good to those that love God.[20]

A year later, in April 1676, Penn made a visit to Swarthmoor enroute to a Quaker meeting in Yorkshire where he hoped to reach an accord with the Wilkinson-Story separatists. Prior to his arrival at Swarthmoor, Penn wrote a letter to Fox that glowed with admiration: "Eternally beloued, yea all Ages shall bless thee, and Magnify the Powr that . . . does guird and Crown thee ouer all envy, pride, darkness and thy kingdom is not of this world . . . and thy place is very neer the lamb." Penn went on to refer to the separatists: "and the[y] not seeing Thee in thy true place, nor haveing a sense of thy heavenly commission and Authority, folish shortsighted and pufft up spirits Intrude and smite and Rebell . . . well, my soul is in full

Confidence of its downfall, and I hope to behold it." He slipped in a one-line admonition to Fox saying, "some . . . wonder'd, . . . that thou shouldst be so sharp wth the J^s [Johns]." Then Penn referred to Fell's role in the dispute as well:

> Poor Margt is so much smitt at, and run upon (as I believe never woman was, . . . as if she was the Cause; and of an Implacable temper; without spirit of Reconciliation, to show them that she can pass by all that past between them that concerns herself; and to make the most of the good or the tender thing, if it arise never so little In either of them, this would be a deadly stroak upon that back biteing spirit, and so Confound them that smite, yea so stop their mouths.[21]

Penn's loyalty to Fox and Fell over the separatists was not without a price. He wrote Fox concerning this dispute: "I have suffer'd hard for it; yea more than ever I had from the greatest Princes of the world before I knew Truth; but the Lord blott out all; swe[e]ten all, and cement all, if it be his blessed will."[22]

Penn was also a great help to Fox in promoting publication of Fox's books. Printing was a Quaker defensive strategy that, in Penn's words to Fell, was not "unserviceable" for "divers are the weapons of our warfare. But a *time shall come*, when wars, in that sense [polemics] shall be no more: but those swords be beaten into ploughshares, and spears into pruning hooks. Oh! blessed Sabbath!" Penn commented in a letter to Fox, "Thy Book committed to me to gett Printed is out, a precious thing, as are all thy papers. Friends have great Regard to many of thy late papers and Books, deep and heavenly, openings and great variety of them.[23]

The epistolary evidence shows that Penn and Fell shared a common view concerning Fox's charismatic role. Both Fell and Penn and the pro-Foxian circle believed the solidarity of Quakerism would be enhanced by consolidating the entire movement behind Fox's leadership.

George Fox had an opportunity to defend Penn in May 1677. When Penn first became a Quaker, he continued for a time to dress according to his social station despite criticisms from other Friends. Samuel Pepys saw Penn in 1666 just prior to his conversion and claimed that he was "very smooth" and stylishly self-possessed.[24] When Penn was labeled a "perriwig man" by a Quaker critic, Fox argued that Penn, who suffered early in childhood from smallpox, had lost most of his hair, except for a thin border. He therefore found it necessary to wear a wig to keep from catching colds. Fox declared, "he wares them to keep his head and ears warm and not for pride; wch is manifest in that his perriwigs Cost him many Pounds a piece formerly, when of the world, now a ['civil border of a wig'] but a few shillings." Fox further retorted, "hee's more willing to fling it off if a little hair come, then ever he was to putt it on."[25] The fact that Fox, who spoke against fashion as a corruption of the worldly, should have defended Penn's wig-wearing seems curiously incongruous. Theory and practice may have diverged somewhat here for Fox, but operating within a certain range of freedom, he was willing to be politic, recognizing the value of Penn's friendship and patronage.

The friendship of the Foxes and Penns, although based on common religious beliefs, operated on another level as well. Friendship in its early modern range of meanings included a patron-client relationship.[26] George Fox's letters to Penn, written while Fox was in Worcester prison, illuminate this meaning of friendship. Fox resorted to Penn's social and political connections to gain his release. Penn informed Fox in September 1674 that he had, "stirred in [his] business, and . . . that a person of some quality," who preferred to remain anonymous, was working to obtain Fox's release "wth the K[ing]." Fox responded by suggesting that Penn acquaint himself with another person of quality, namely the Earl of Salisbury's younger son,

> whoe is much familiar with the duke of Mummuth [Monmouth] he was with mee heere to visitt mee In prison: and staied about two hours: hee took a coppy of the errors In my Indictment: and is Convinced In his Judgement of the truth: . . and [he] knew thee formerly. . . . I would hve thee acquainte thyselfe with him upon truths account: and what ever other service may occasionally offer. . . . And soe if thou canst effect my release: without the title of a [par]don: thou maist.[27]

It is apparent that Fox depended on Penn not only to write responses to the hostile Quaker literature and to give him legal advice but for his social connections as well.

Penn's stature as a public Quaker and close friend of the Fell family is evident in the fact that he was called upon to write the Preface to the first edition of Fox's *Journal*, published by the Society in 1694, three years after Fox's death. Penn's statement was glorifying in its loyalty to his leader. He eulogized Fox as one who stood in close parallel to Christ and who had reestablished a church in close approximation to primitive Christianity. Penn described Fox as one who was pure, grave, and pious in his youth, a shepherd of sheep, and introspective and solitary in that employment. Fox was untutored in the ways and education of the world that made him distinctive, indeed, "no man's copy." Penn pointed out that his speech may have sounded "uncouth and unfashionable to nice ears" and that his sentences sometimes tumbled out "abruptly and brokenly . . . about divine things." Despite this, Fox's words were profound, and he was "civil beyond all forms of breeding." It is probable that Penn's reiteration of Fox's unlettered ministry reveals, on Penn's part, an excuse for Fox's lesser social status, which, despite all his charismatic qualities, could not be passed over without comment. Nonetheless Penn was quick to add that Fox moved among social superiors with a presence that equaled "a religious majesty," and Penn could say of him: "I never saw him out of his place or not a match for every service or occasion." Fox's religious authority was inward not outward, and Penn wrote:

> Above all he excelled in prayer. The inwardness and weight of his spirit, the reverence and solemnity of his address and behavior and the fewness and fulness of his words, have often struck even strangers with admiration. . . . The most awful, living, reverent frame I ever felt or beheld . . . was his in prayer.

Fox's charisma as a religious leader allowed him to pierce the hard shell of aristocratic social circles for he was fascinating to those of higher birth and was thereby accepted and admired by some of them. Penn saw him as one given a special divine dispensation, for he wrote, "we may truly say, with a man of God of old, that 'being dead, he yet speaketh;' and though absent in body, he is present in spirit."[28]

Penn's veneration for Fox does not presuppose intimate friendship of an equal social nature. It is true that Fox and Penn, for more than twenty years, had shared many experiences and agreed on many theological issues although not all. The Penns regarded Fox as a great religious leader who initially may have played the role of "father confessor" to Penn to some degree. This notwithstanding, Fox and Penn were friends but at a certain social distance.[29]

William and Gulielma Penn found in Fell more than a co-leader in the cause. To them, Fell was a mother superior of Quakerism and an intimate adopted mother who showed them special affection and understanding. Penn wrote two letters to Fell. One was a wistful farewell letter just before his sailing to America. The second letter he wrote just three weeks after his return from Pennsylvania. Both reveal the warmth of their friendship. His farewell, dated 14 August 1682, opened:

> Dear Margaret, I am agoing. Remember me in the Fathers love and may God be with thee and bless thee. . . . Dear George I left yesterday at

Enfield. . . . My soul loves him beyond the love of women. . . . I have not else to add, but my wifes dear love, who is sweetly consenting and satisfied.

Upon his return to England, Penn wrote Fell on 29 October 1684 saluting her warmly and acknowledging their sweet fellowship in the spirit "above time and distance, floods, and many waters." He added joyfully, "I arrived well in my native land. It was within seven miles of myn own house, where I found my deare wife and poor children well." He wrote of establishing eighteen meetings in Pennsylvania by 1684 as well as the "many troubles in ye Settlement^s, of deaths from Ague and feaver." He thanked Margaret for her care of his wife, and then he boasted, "I have not missed a meals meat or nights rest since I went to the Country [Pennsylvania]."[30]

Likewise, in the 1680s the friendship between Gulielma Penn and Margaret Fell seemingly became a close one. Fell sent Gulielma gifts from Swarthmoor, such as clapp bread, a specialty of the Furness district. Three letters to Margaret Fell from Gulielma Penn address Margaret on domestic matters and give us a glimpse of the character of their friendship. In the first letter of 21 August 1683, while Penn was absent in America, Gulielma announced the birth and death of her three-week-old daughter, their seventh child. She wrote that George Fox had come to visit her after William had left for America, and she had found his company kind and refreshing. Margaret's eldest daughter and son-in-law, Margaret and John Rous, who lived in London, also visited Gulielma, and their daughter Bethia Rous stayed with her "a pretty while." The letter gave news of William Penn and the large Quaker meetings in Philadelphia, numbering up to 300 at a meeting.[31]

These letters reveal sufferings endured by both women. For instance, Gulielma Penn indicated that she was enduring some difficulties while her husband was in America. Not only did she bury her newborn daughter, but she was weak from illness which she called "St. Anthony's fire" – a fever that caused severe edema in her face and eyes. At the same time she was also caring for her dying mother. Moreover, Gulielma reported to Margaret that she was awaiting her husband's summons to sail to America with her children and hoped to hear from him before the 1683 sailing season was past. She appreciated her friendship with Margaret, referring to Margaret's "deare and tender lines. . . . I am truly glad when I heare of you and from you."[32]

One year later, on 24 August 1684, Gulielma Penn sent Fell a word of encouragement to endure unflinchingly the heavy persecutions of her family at Swarthmoor. The Fells were experiencing the whip of the Second

Gender, Religion, and Class

Conventicle Act, which sought to destroy the Quaker meetings in the Swarthmoor area. The local constables fined Margaret the considerable sum of £40 for holding a meeting at her home and £60 for preaching and praying. William Kirkby, a local justice of the peace and longtime enemy of Quakers, had killed one of Fell's oxen to supply his own table, while also distraining twenty-four of her cattle. Gulielma wrote, "I perceive th[ings?] are bad about you, and that thy Sufferings are large, but the Lord can and I believe will make it up, and in him, is thy great Reward for al thy many-fold Exercises." She further commented, "they doe begin to be troublesome, in this country [county], they have been pretty quiet heither to, . . . but threaten it."[33]

In this same letter, Penn's wife circumspectly counseled her older friend concerning her proper place beside her husband. George Fox had last left the Hall in 1680, and now lived in and about London, often staying with one of his wife's four married daughters. Gulielma Penn wrote, "methinkes if thou foundest a Clearness and Freedom in the Lord, it would be happie [if] thou wert nearer thy dear Husband and children, but that I leave [to] the Lords ordering, and thy Freedom." Whatever Fell may have thought, she did not follow Gulielma Penn's advice, for she remained at Swarthmoor Hall with her youngest daughter and son-in-law, Rachel and Daniel Abraham, for the rest of her life, making brief visits to London in the 1680s and 90s. It is evident that these two women were not alike in their marital attitudes. Penn's wife observed seventeenth-century norms for upperclass women and deferred to her spouse as patriarch of the family in much the same way that a recent study portrays an upper-gentry woman, Lady Verney of Claydon House.[34] Fell was atypical in her matriarchal role. She retained Swarthmoor and its financial resources in her own hands throughout her second marriage, a highly unusual if not unique arrangement. This placed her in an unassailable position as head of the Fell clan.

It was Penn's habit to report his foreign missions to Fell upon his return to England. After his trip to Europe in February 1677–78, accompanied by Fox and Fell's daughter Isabel Yeamans, Penn informed Fell of his journey. In this letter he spoke of his friendship with Fell in near rhapsodic terms.

> Truly sweet and precious is the holy fellowship that our dear Lord has given us together in his own pure eternall spirit, that hath . . . made us near, and very dear above the life, spirit and friendship of this corrupt world. . . . O! we are bound up together in everlasting love. Precious and sweet, is our Heavenly Kindred and relation. . . . O! we cannot forget one another, – nor time can wear out; . . . my love and due

regard to thee, and thy remembrance of me in thy two last letters, is very dear to me, as I hope they will always be. . . . I am in Endless bonds of heavenly friendship and fellowship.[35]

An interim of silence ensues in the communication between Fell and Penn from 1684 to 1690. William Penn fell from public favor at the Revolution of 1688. As a protagonist of toleration, he had become the close friend of James II. At the king's demise, Penn was labeled a Jacobite courtier and possible traitor. Penn retreated to solitary retirement from February 1691, following Fox's funeral, until November 1693. His relations with some Quakers suffered an eclipse. To the 1691 Yearly Meeting, Penn wrote in self-defense: "Receive no evil surmisings: neither suffer hard thoughts, through the insinuations of any, to enter your mind against me, your aflicted, but not forsaken friend and brother. . . . My privacy is not because men have sworn truly, but falsely, against me."[36]

In spite of the lack of correspondence, seemingly no break in friendship occurred between Fell and Penn. At Fox's death, Penn was the one designated to write the news to Fox's widow, a further indication of his prime leadership role and their enduring friendship. On 13 January 1690–91, Penn wrote from London to Swarthmoor:

> I am to be the teller to yee of sorrowful tydings . . . wch is this, that thy dear husband and my beloved and dear frd. George Fox has finist his glorious testimony. . . . O he is gone and has left us in a storm yt is over our heads, . . . a prince indeed is fallen in Israel today.[37]

It is a letter reflecting Penn's mood of gloom. He felt both political isolation and the loneliness of leadership in the wake of Fox's decease, for in several ways Penn was the foremost public Friend and logical successor of Fox.

Fell's son-in-law, William Meade, was among those Quakers dissatisfied with Penn in 1693. Meade, erstwhile codefendant with Penn in their trial of 1670, combined with others to prevent the printing of Penn's preface in some 1694 editions of Fox's *Journal*. Some of Penn's biographers have argued that Fell was aware of and sanctioned an investigation of Penn's personal politics in the 1690s. It has been suggested also that Meade's intervention to squelch Penn's preface to Fox's *Journal*, was in part based on the fear within the Fell family that the memory of Fox, as their founder of Quakerism would be eclipsed by Penn, the founder of Quaker Pennsylvania.[38]

A final undated letter from Penn exists, addressed to Rachel Fell Abraham at Swarthmoor for her widowed mother. It refers to Margaret with great warmth and respect as "ye ancient Standi[ng] enduring offspring of ye morning . . . who has weathered many storms these many years." The sense of this letter was religious, written by one who was "weary in spirit," who looked again to this indestructible mother superior as a source of inspiration and constancy. The letter reveals nothing that hints of animosity between these two longtime friends. Penn achieved, in his steady thirty-year friendship with Fell, a depth, warmth, and closeness to her that suggests natural congeniality and intimacy possible only between friends of the same social milieu.[39]

A study of the Fell-Fox-Penn correspondence indicates that the Wilkinson-Story controversy was the catalyst that brought the Penns and Foxes into a close working relationship. However, there was another factor operating in their shared alliance. Despite Fox's egalitarian theory of church order, Quakerism had assumed a degree of internal authoritarianism by the mid 1670s. Fell, Fox, and Penn operated as a de facto spiritual hierarchy in the emerging church. This authoritarianism came to the fore in the internal Quaker dispute over authority in church order. A recent study of Penn's writings by Melvin Endy points out that Quakers employed two means in singling out misguided revelations of individuals in the church. These two methods, in tension with one another, emerged during the Wilkinson-Story controversy. One was the strong leader concept, or "coercive rule of saints." Based on direct spiritual revelation, it tended toward an "unlimited personal rule by the apostolic leaders." This coercive force within Quakerism remained intact with the growing importance of eldership in the Quaker church.[40] By 1674 the Quaker theologian Robert Barclay claimed that not everyone was called, "in the same station, some rich, some poor, some Servants, some Masters. . . . Every Member has its place and station in the Body . . . yet is no member to assume another place in the Body then God has given it." Barclay further claimed that the true Church of Christ, "gathered into the Belief of Certain Principles, and united in the Joint Performance of the worship of God . . . still must [have] . . . a certain Order and Government."[41]

The second method of judging the authenticity of revelation, according to Endy, was the "consensual ideal." The chief hope of the early Quakers was to achieve, in the unity of the Spirit, a true voluntarism referred to by Quakers as the "sense of the meeting," in which decisions were based on loving relations between believers in spiritual unity with God. This perfect spiritual unity of purpose could be achieved only among men and women living and working in the divine or inner light. Robert Barclay used corporal imagery to express this unity of spirit and diversity of works: "The Variety of the Operations of the divers Members of the Body of Christ, working to one and the same End, as the divers Members of a Man's Body toward the Maintaining and upholding of the whole."[42]

Endy has argued that the Quakers were drawn to both these methods in church order, the coercive rule as well as the consensual ideal. The consensual ideal was expressed in Quaker social and political theory, such as the Quakers' stand on toleration, an issue that slowly gained ground after the Restoration. However, the coercive method never disappeared entirely and, under duress, such as in the Wilkinson-Story schism where contradictory revelations occurred, the secondary voices of the Spirit were rejected by the community through its leaders, in favor of the divine inspirations of the primary leaders.[43]

Fox's letters to Penn reveal a testy attitude toward Wilkinson and Story and their ilk, whom he regarded as disquieted spirits "jangling" out of the Truth. On the one hand, the Spirit was the vehicle by which all persons were drawn to the truth. On the other hand, those who claimed spiritual revelations that ran against those of the primary spiritual leaders, such as Fox, were considered to be listening to human will and not divine guidance. According to Endy, the fact that most early Quakers came from a Puritan background with its inclination toward the "rule of the saints," helped tip the balance in that direction when problems emerged in church discipline.[44]

The social-religious theory and practice of Fell, Penn, and Fox followed a similar pattern of attempting to incorporate these two opposing tendencies. The three leaders endorsed the consensual concept in Quakerism. Nevertheless, the discrepancy between theory and practice emerges clearly in these three leaders. It has been argued that Penn, the aristocrat, was criticized in his Pennsylvania experiment by those who disliked his order of government, which allegedly did not give people enough power, while "reserving too much for himself." His aristocratic self-assurance in his own "saintly rule" of the Pennsylvania colony was expressed indirectly in his last letter to Fell in the 1690s: "My life feels a work I have to do wth it [Pennsylvania colony] for the lord, yt his powr may be a top, but some are carnal, and others weak, and they would level all, and ye notion of freedom and commonweal yt I fear it will leave Truth but little powr over ye bad in ye end."[45] Fell's gentle-birth status was displayed in her authoritarianism as matriarch of her estate, in her self-assumed leadership role among Friends, and in her preemptive manner in dealing with the Wilkinson-Story faction. She also displayed this side of her nature in her business dealings. Fox's sense of chosenness arose from a deep inner conviction that he was called as a messenger of God to preach the gospel in its pristine form. Although Fox did stress the consensual principle in

Quakerism, he made exceptions when it diverged from his ideals, as seen in his manner of dealing with the Wilkinson-Story separatists.

It is apparent that religion, class, and gender made these three leaders anomalous in terms of the new social constructions they exemplified. The epistolary evidence suggests that both Penn and Fell, because of their rank, placed a high value on social order, despite the fact that they espoused radical egalitarian principles of religion. Fox was a charismatic anomaly as a religious leader, who bridged the gap between the ideal and the practical in primitive Quakerism. Fox's theory that judgment within the Quaker church had been given to all members through the guidance of the inner light broke down in practice. Fell, Fox, and Penn were innovators like Wilkinson and Story, and out of loyalty to Fox rejected the older consensual ideal. Fell and Penn were guided by an inherent instinct for hierarchical order inculcated by their social position in English society.

Because the Fell-Fox-Penn epistolary evidence is brief and intermittent, our assessment must be circumspect. This notwithstanding, an exploration of the correspondence and other accounts of their personal contacts suggest two possible conclusions. First, the combination of Fell's intraclass friendship with the aristocratic Penns as public Quaker ministers and her interclass marriage to socially inferior George Fox was highly unusual in seventeenth-century England when female stereotypes precluded a public role for women and where rank defined a person's public persona and private relationships. Fell's Quaker theology allowed her to break through the gender constructions of seventeenth-century English society, and that enabled her to disregard class restrictions as well. Second, in terms of modes of leadership in early Quakerism, Fell's relationship to Penn suggests that status superseded gender in Quaker leadership. This was a reflection of the social categories of seventeenth-century English society, despite the egalitarian theology of the early Quakers. Penn and Fell were analogues in the early movement. As public Quakers, along with sharing common religious beliefs, they demonstrated that religious authority rested to a significant degree on social status, despite the Quaker theory preached by Fox that nobility of the flesh gave way to nobility of the spirit.

Part IV
The Mental World of Margaret Fell

9 Fell's Worldview

Margaret Fell's life has been viewed from three axes: the social, political, and economic. We must focus on a fourth, all-embracing dimension of her life to flesh out the fullest possible picture of this woman's experience. The religious facet of her life comes to light not only in her public and private activities but in the structure of her thought as well. Fell's Quaker outlook, in a nuanced way, both overlapped and contrasted with a traditional Protestant outlook, and, more particularly, with the religious outlook of Puritans like Ralph Josselin and Nehemiah Wallington.[1] Fell's religious writings, published and unpublished, placed alongside her letters and other bits of evidence, manifest something of her mental world that was in most respects very alien to our own. As historians have warned, when taking an imaginative step into a past mental world, we run the risk of misinterpreting that world because we share certain common cultural traditions that can blur our own perception. Likewise we run a second hazard of distortion due to the comparative paucity of evidence. Fell's writings constitute a far smaller sample than do those of Josselin, Wallington, Fox, and Penn. Nonetheless, among literate women of her age, Margaret Fell stands out as a remarkably prolific writer.[2]

Fell's religious worldview is best reconstructed by examining her ideas about the material and spiritual worlds. On her deathbed she thanked God and committed to him all things enjoyed from him, "both in Spirituals and Naturals." Her recorded attitudes about those aspects of life that were beyond human control – pain and death – are minimal compared to two Puritan contemporaries, Ralph Josselin, or Nehemiah Wallington. From the few remarks she did write about suffering and dying, we can gather some clues about her attitudes.

Fell lived in a world that experienced deaths of friends, neighbors, and loved ones at very frequent intervals. However, the Quaker way of death gave only minimal outward observance to their own departed. In fact, customary outward observances of mourning the dead were eschewed by Quakers. We can only guess at the degree of emotional and mental tribulation that Fell experienced on the deaths of her parents, two husbands, an only son, and one daughter. One clue to her mental attitude toward death is contained in a letter written in December 1664 from Lancaster Castle at the impending death of her adult daughter Mary. She reassured her worried son-in-law John Rous, at whose house Mary lay seriously ill, "her Spiritt is neare and Deare, present with mee, whether in the Body or out. . . . To the Lord of heaven and Earth, she is given freely, and his heavenly and holy will, I freely submitt to." Her words were not unlike Josselin's in that Margaret too conveyed a certain resignation to death. Yet she did not express the depth of grief that Josselin expressed over his young childrens' successive deaths. Margaret remained composed throughout her letter and, after making the above comment, even changed the subject to more mundane matters. In so doing, she seemingly attempted to relieve the anxiety of her daughter and son-in-law who were responsible for Mary's care.[3] Although Mary subsequently recuperated, Fell did express a reasonably strong sense of loss concerning Mary. It is highly probable that the loss of daughter Bridget in 1663 and son George in 1670 were equally severe blows to her. Judging from Fell's concerns for her children expressed in her letters when she was away from home for more than a year in 1660–61, we may assume with some assurance that she felt deeply about them. She witnessed deaths of other kin who were not as close as her own children, that of parents and sister and brother-in-law, as well as her neighbors and friends. Although evidence is lacking, it is probable that she felt these losses less severely than those of her nuclear family, as was the case with Ralph Josselin, who felt an "unequal emotional response" to death of kin.[4]

The Account Book of Sarah Fell records the death and funeral of Margaret Fell's granddaughter, Margery Lower, who died as a relatively young child while visiting Swarthmoor. No emotional comment is given in the account book's terse entry of funeral expenses. However, the Fell letters of mother and daughters contain comments that show anxiety over death. For example, Margaret Rous was fearful throughout her pregnancies. Likewise, her sister Sarah Meade expressed fear of illness and death concerning her only child Nathaniel in 1685. When Nathaniel was ill, the Meades left London for their country home

> in hopes, when he came into the country, it might haue abated, but it yett continues upon him . . . wch keeps him pretty weake and low . . . we are fearful to give him things . . . least he should bee worse . . . I earnestly desire thy Prayers to the Lord for his preservation (if it bee his will) yt hee may bee an Instrumt in his hand to his glory; and may feare & serve him all his dayes.

William Caton, a close friend of the Fell family, was mourned at his untimely decease. Francis Howgill wrote Margaret Fell from Appleby gaol in 1665, "I heard . . . of the departure out off the bodie off dear

Wil Caton att Amsterdam for the which I am very sorey . . . he was a faythfull man." Howgill reflected on the many deaths among the Quaker leadership to Margaret: "Treuly when I consider of the taking off soe many faythfull men which could and would have done most service for the Lord in our generation makes my harte sad." Howgill was soon to join the list of departed faithful, for he died in Appleby three years after penning this letter.[5]

The Quakers opposed all outward symbols of grief or mourning. They erected no monuments as markers for the dead. Their burial custom probably included a meal served at the funeral for those in attendance.[6] Customary observance of the deceased took the forms of written Testimonies concerning faithfulness to ministry and death-bed words of expiring Quakers. We have seen that William Penn's testimony to George Fox, in a letter to Fell in January 1691, was adulatory in its reverence to his dead leader: "A prince indeed has fallen in Israel today." The testimonies to Fell, given by her daughters and a few of her surviving contemporaries in the preface of her *Works*, attest to her pure life and ministry as a great Quaker minister in the gospel.

Fell's personal thoughts of her own mortality also escape us for want of evidence. She did exhibit in her writings a typical Protestant and especially Puritan attitude of future blessing or affliction based on actions in this temporal life. She exhorted her daughters to "remain humble that the blessing of the Lord might be [their] portion." This humility required the cultivation of a godly, devout disposition and a pure lifestyle. Her dying words to her grandson John Abraham in 1702 were "stand for God, John, stand for God."[7]

Fell's public statements of woe or blessing in the after-life are recorded in her letters to Charles II. When his brother, the Duke of Gloucester died, she warned Charles II:

> The Lord is come very near thee, Oh that thou would'st consider it, and see his Hand . . . testifying that he doth not love Pride, Vanity and vain Glory; that now, in the very time of your Joy, he hath turned it into Mourning. . . . The God of Power give you to understand . . . who hath the Life and Breath of all Men in his hand.[8]

Fell gave further warning to the king that, "the plagues will come upon God's enemies." She sent several remonstrances to the later Stuarts that the Crown was the seat of justice and therefore it was incumbent upon kings to be merciful as "God is a god of mercy and forgiveness." She exhorted the King to stop persecutions before God would bring a divine curse upon the nation. Drawing on the Psalms, Fell wrote that God was "dreadful to the disobedient"; he was a God who would "break to piecies his enemies." Not only would the plagues come upon God's enemies, God would "laugh in derision" at their downfall. Moreover, she frequently reiterated in her pastoral epistles to her own circle, "wait low in the fear of God."[9] Her desire to serve God was to live in obedience to his will, which would bring blessing. Fell displayed less anxiety, however, than Puritan contemporaries like Ralph Josselin and Nehemiah Wallington who exhibited deep concern over their own sins, their hope of election, and their fear of damnation.

In her writing to the Ranters, Fell was most explicit in her thoughts of heaven and hell. In opposition to the libertine spirit of the Ranters, she envisioned the Ranters' end graphically: "thou beast, and thy spirit which is of the beast, that goes downward [in]to the earth . . . thou knows not the spirit of a [good] man that goes upward." Her belief in heaven and hell were couched n terms of God's blessing and God's vengeance, set at some future time. For example, to the Ranters she wrote, "you . . . are a generation whose hearts are turned away from God . . . your root beareth gall and wormwood, and ye are under the curse. . . . The Lord will not spare you when his anger kindles, . . . then all the curses that are written in the book shall lye upon you, and your name shall be blotted out from under heaven." It was evident that for her there was a distinct disjunction between those who both led a good life and held right belief and those who did not, with final rewards divinely adjusted accordingly. To the Ranters she also wrote: "Thou who sits at the Table of devils, art shut out from the liberty of the sons of God." The terms of reward and punishment are expressed in the Fell daughters' memorial to their mother after her death. "Her memory lives on [while] that of the wicked will rot." The overall theme in Fell's Quaker message was the desire to conform to the divine will, even though it meant undergoing a life beset with trials and tribulations.[10]

Fell stressed humility before the Lord, and she expressed her belief in the link between heaven and hell, reward and punishment. She held the conviction that nothing was sure and secure in life. Illness or accident could easily overcome one. Surely, the precariousness of life was deeply imprinted on seventeenth-century women, and Fell was no exception. Two of Margaret Fell's grandchildren had died while visiting Swarthmoor. It is possible that Fell may have sought God's mercy and attempted to understand why God displayed his wrathful side, but this is not expressed in words. The picture we receive of Fell and her daughters is that they did not experience fears of God's vengeance upon them in the same way as Puritans such as Bunyan, Josselin, and Wallington.

According to Macfarlane's study of Josselin, Calvinists tended to see

God as a "stern father figure who punished human failings on the human level."[11] It is true that to some extent Quakers shared this seventeenth-century religious outlook. Thomas Rawlinson described George Fell's untimely death in terms of divine punishment on his mother for her unjust machinations against him. As Fell reiterated the biblical woe oracles to those in power, it seems highly likely that she did not break entirely from this mode of thought. Nonetheless, whatever her level of anxiety may have been, her writings do not convey an excessive self-scrutiny and self-flagellation over sin. There is an absence of internal conflict and mental stress and also an absence of "fierce introspective battles against temptation" as Josselin experienced. Margaret's concern was to learn and live obedience with Christ as her model, for as Christ "learned obedience and was made perfect," she believed humans could receive Christ's perfecting spirit. This belief seemingly alleviated or at least lessened any sense of self-doubt that she may have had. Although this may have been due in part to her personality, nonetheless, her faith informed her simply that to live humbly and obediently before the Lord would bring blessing. Thomas Camm's testimony to her life and ministry expressed her positive faith, self-composure, and her optimistic sense of providential history that differed from that of Puritans like Josselin and Wallington.

> She never spared herself, nor doubted of good success in her manifold Labours on Truth's account . . . but approved her self such in Zeal that needed not to be ashamed of her work and service for the Lord . . . which she perform'd with all sincerity, and is now rewarded with the full fruition of eternal Life, and Peace with her God.[12]

Calamities of nature and consequent pain and suffering are a part of the human condition and were not alien to Margaret Fell despite her relatively high standard of living. It was an age of poor sanitation, chilly, damp homes, uncontrolled disease. Accidents and illnesses were often fatal due to ineffectual or primitive medical care. When Margaret jr. suffered from a chronic knee problem, she left home for London in search of medical treatment. Margaret sr. showed great concern for daughter Bridget's illness and fatigue in 1661 and wrote instructing her to take "jannes drink" as a cure. The account book is full of medicinal cures taken by the Fell household, including the application of butter as a salve in treating sick animals. Although Fell's family was unusual in that seven of eight children born lived to adulthood, still illness and death were very near them in the natural order of things.

Alongside the premature deaths of two of her children and grandchildren, and the deaths of her spouses, Margaret endured intermittently a particularly heavy burden of outward suffering in the form of Quaker persecution. She told her fellow sufferers that their faith would be tried by the fire. Margaret wrote, "the precious Faith . . . once delivered unto the Saints [they would] be called to contend for" in times of trial. However, in the "fight of Faith," Quaker believers would have the "Victory which is over the world, which is your Faith, and so know Faith working by Love; and so as you dwell and abide here, in your Measures, and as you learn of him who is Low and Meek, you will come to see Christ." Christ was the anchor in her existence to which she held fast throughout the vicissitudes of life. Whatever the affliction, "Christ Jesus the cornerstone," "the light," the "second Adam," the "Hope of Israel" had come and would continue to come and dwell in the hearts of believers. However stinging the persecutions, however devastating the loss of loved ones, she clung to Christ who was "the light, the way, the Truth, the Life." Faith in God was the ground and being of her existence and her acts and positive terminology used to express her faith conveyed this spiritual certitude in a compelling way.

But there was a duality to Fell's mental world, for along with "the spirituals" there were "the naturals" or the material world. Although she was of a strong spiritual orientation, she valued the material aspect of life and did not let go of it. Her attitudes on materialism surfaced in many ways. The Swarthmoor account book shows her this-worldly bent of mind and style of living. For example, Margaret sr. and Sarah invested in neighborly and commercial enterprises with the expectation of financial gain. She and her daughters invested in relatively costly wearing apparel. As farmers she and her family showed a business caution to maintain a solvent farm in an age when bad harvests and rising grain prices could plunge a farmer into poverty very easily. Her economic activities included the local coal and iron ore business. These activities were rooted in an attitude or frame of mind that reveals a strikingly materialistic and this-worldly orientation, despite her all-pervasive religious framework of living. This was not unlike her Puritan contemporaries, Josselin and Wallington, who also applied their service to God in the financial realm.

In expressing her own religious presuppositions, Fell resorted frequently to biblical images and to homey, domestic, and feminine metaphors in her writings. She also resorted to using familiar, well-loved biblical images. Her religious worldview was expressed in phrases such as, "God keeps his church and family as in the Hallow of his Hand." Such verbal imagery expressed security to her and to her Quaker friends buffeted by persecutions. She depicted Christ sometimes in material terms that gave force and comprehension, such as the "cornerstone," or the "Rock of

Ages." To those who desired shelter from the buffeting of persecutions, she wrote encouragingly: "Abide in the Cross," and let the "milk of the word" be received and believed as "newborn babes and grow thereby." She told her Quaker circle they had been "nursed up, and kept in, as living stones growing up in the Temple of the living God." To steel her compeers against outside derision, she wrote, "give your backs to the smiter," for the great of the earth were mere "grasshoppers" before the Lord. In face of persecution she wrote, "keep to few Words . . . wait in silence that you may come to know the . . . hidden Manna." For Fell these persecuted ones were the "salt of the earth," who were called upon to build a "city on hill," and the "armies in Heaven, clothed in fine white linen" would oversee their efforts.

To her enemies Fell did not mince words about their end. She frequently used domestic and natural imagery in castigating them. She warned the Ranters "The eye of God turns the wicked into hell." She warned the "blind and besotted professors" and priests that they would be seen for what they really were, for the "Fig Leaves of [their] Profession [would not] cover [their] nakedness [their] coverings are too narrow." The ecclesiastical structure was a "rotten house decayed [and] ready to fall," and the Lord "who makes clean the house at evening" would find theirs unswept and unclean and it would be "beaten down," unless the worldly clergy would "lay aside all superfluity of naughtiness."[13] In October 1659 she wrote to the *Generall Councill of Officers of the English Army*, "The Powers of darkness have long kept him [Christ] out of his Throne, and he hath holden his peace, and long bin still; but now he is crying *as a travelling [travailing] woman*; and now he will arise, and destroy at once, all those, that will not he should Reigne." Having been raised on the King James Bible, Fell's verbal pictures are quaint reminders that her world of thought was indeed an early modern one.

Fell's concept of Satan and sin is illustrated in her writings. To the Ranters she referred to Satan as "the serpent." The Ranters were "creature[s] under the bondage of corruption" to this serpent. Further, she told her Ranter enemies that they were messengers of the "Prince of Darkness"; they were thus "under the curse, separated from God, [and] under the power of the Prince of the Air, which rules in the children of disobedience." She foretold their impending doom, for they would soon be turned "into the bottomless pit." Although the Ranters used Scripture (Col. 3:11 and Eph. 1:11), to verify their ideas, she asserted that "neither [of these texts] belong to thee, nor none of thy generation, who are an enemy of God, and all righteousness. She warned both Quakers and non-Quakers not to let the serpent beguile them as it had beguiled Eve, for the diabolic was always lurking in one's life. She saw the "snares of satan" in the "outward preferment" of the clergy, and she explained it as the work of the Devil in disobedient men. Fell spoke of "hellfire and chaines" upon one occasion, and she warned all to stand against the wiles of the devil and to beware of the "subtility of the serpent." Although she did not use the words "hell" and "devil" excessively, hell as a place and the devil as a working agent of evil were real to Fell, as seen most clearly in her writing to the Ranters in 1656. She warned them that they would come to know the "wrath of the Lamb" and they would be "tormeneted day and night for ever" in the "bottomless pit . . . lake of fire and brimstone." Although mentioning witchcraft only once, she also linked this with the work of Satan.

> My Dear Hearts, God is Light . . . the work that he works, is in the Light . . . and all the deeds of Darkness, and all Earthliness, Lust, Pride . . . which is Idolatry, and in the Sorcery, under the dark Power and into Witchery, where the Devil hath Power.[14]

Another aspect of her religious thought that is striking to the modern reader in its alien tone is her millennialism, which is discussed in greater detail in a later chapter. Her belief in the imminent second coming of Christ and her apocalyptic anticipation of it, seen especially in her writings to the Jews, place her on an important spectrum in early modern Protestant thought. Those persons of apocalyptic bent shared the view that the world was an arena where angelic and demonic powers warred, a world that would ultimately meet a "definitive eschatological judgement." Margaret expressed her apocalyptic assumptions when she wrote, "The flying Angel is gone forth, who has the Everlasting Gospel to Preach unto them that dwell on the Earth unto every Nation and Kindred and Tongue, and People. . . . " Her apocalyptic language included frequent promises to the Quaker flock such as the "Lord will raise believers on the Last Day," and the Lord would gather the "dispersed of Judah from the four corners of the earth," for the "night of apostasy was drawing to an end." Fell ended one of her epistles with the plea, "it is the last time now . . . Jesus come quickly."[15]

Apocalyptic thought drew heavily on visions from Daniel and Revelation to see, as it were, through a glass darkly, the divine plan for the end of this world and for the new heaven and new earth to come. Visions and dreams assumed special import and were drawn upon to "exhort and console." Apocalyptic language and imagination frequently addressed the "issues of political and social liberation" without any specific program of action. Apocalyptic challenged the mind to view the world in different terms, to see human affairs determined by a supernatural power and that the end of the world was near at hand. In so doing it often caused discontent with the world, which led to a denouncing of worldliness in all its forms. According to one analyst, the apocalyptic mode of thought constructed a "symbolic world where the integrity of values [could] be maintained in the face of social and political powerlessness and even of the threat of death."[16]

Fell's mental and emotional stance about worldly things tended to follow this line of thinking. Although she did not record nearly the number of dreams her Puritan contemporary Josselin did during his strongest phase of millennial thinking, she did validate her statements by prophetlike utterances. She opened her 1660 epistle to the "Friends of God": "There is something upon me, as from the Lord, to write unto you concerning the precious Faith." In 1656 Fell addressed a pamphlet "To al the Professors of the World" in which she warned that "the Proud and the High and the Lofty" must repent quickly or they would be broken to pieces in their "rotten Profession." For those who refused to heed the call and walk in the light, the chaff would be burned, when "deceitful Hypocracy shall be as Stubble." She closed her epistle with this validation of her prophetic authority:

Before this Paper was written, as I lay upon my Bed, I saw a Vision of all the Professions in the World, and it appeared unto me, as a long, torn, rotten House, so shatter'd, and so like to fall, as I thought, I never saw a thing like it in all my Life, so miserable Old and Decay'd. . . . And a Pity and a Tenderness rose in my Heart to the People; and so in the Motion of God's Spirit I writ this Paper aforesaid.[17]

Margaret Fell's daughter Mary, as a child of eight, recorded on a scrap of paper her dream that foretold the doom of local anglican priest, William Lampitt. Fox thought it significant enough to record the child's message in his *Journal*:

'Lampitt,
  The plaiges [plagues] of god shall fall upon thee and the seven viols shall bee poweered upon thee and the milstone shall fall upon thee and crush thee as dust under the Lords feete how can thou escape the damnation of hell.'
  This did the Lord give mee as I lay in bed.
                                            Mary Fell[18]

This attitude on the part of a young child portrays a family deeply imbued with apocalyptic imagination and language. The Fells, like Josselin, saw a nexus between the spiritual and temporal world. The Lord would crush the wicked of the world and ultimately reward the good. Affliction was the fire by which believers would be tried in their faith; it was part and parcel of temporal existence. Those who persevered in the faith would be rewarded in the end. In the moral universe God was the great spiritual force that linked these things and gave them meaning. Thus visions, dreams, and apocalyptic imagery filled the mental world of Fell and those around her.[19]

Margaret Fell's Quaker faith relieved her of some of the Calvinist anxieties experienced by Puritan contemporaries such as Josselin, Wallington, and Bunyan. She shed their excessive concern over sin and guilt, and replaced it with a positive Quaker belief in the possibility of perfection through obedience to the inner leadings of the Spirit. Margaret felt secure in her own election and she in turn called upon her Quaker co-religionists to wait in the faith of the Elect, for the antichrists and false prophets of the world would meet their doom. The mistress of Swarthmoor scolded England as an elect nation for persecuting God's true prophets, but she added, "Christ [was] bringing his work to pass, [that] day; building up the ruins of Sion." She displayed a deep and uninterrupted confidence in living according to her measure of the Light, at least in the early years of her ministry. Although some changes seemingly occurred in Fell by 1681 as discussed in Chapter 5, she continued to reiterate her experience of immediate and continual revelation that ran counter to Puritan beliefs on revelation as contained only in Scripture. She considered herself and her Quaker brothers and sisters to be New Testament Christians and the Quaker flock to be the true church of apostolic root.[20]

Although we are given mere scraps of evidence of Fell's interior thoughts and assumptions about her world, a reader of Fell's writings does receive a picture of Fell as one who was "in" her seventeenth-century thought world but not entirely "of" it. That is, she held many of the predispositions, assumptions, and attitudes of her generation, but her thinking was altered in a peculiarly Quaker way. Her belief in a gracious God, perfectionism and universal salvation allowed her to overcome excessive Calvinist fears of human sin, election, and damnation. Her theology gave her a broadened vision of gender and race: it included women, Jews, and heathen as converts among God's elect. Further scrutiny of her theology in the next chapters will yield a still clearer picture of how her Quaker mode of thinking resonated in her life.

10 "Let Us Be of One Spirit . . . ": Fell's Spiritualist Theology

As a fervent and talented Quaker apologist, Margaret Fell was not reticent in participating in some of the timely doctrinal debates of her generation. Quaker theology underwent changes during the first and second generations, generally moving from a vigorous focus on the Holy Spirit in the 1650s to a more orthodox trinitarian stance by the time of the Toleration Act in 1689. Fell came from a Puritan background and first became radicalized in her religious views under the influence of George Fox after 1652. Fell's staunch Quaker position was reflected in her writings to her own community of believers. But to the outside world as a savvy controversialist she moderated her theological statements over time to better convince the non-Quaker audiences whom she addressed. To put this in context, her own theological stand *vis-à-vis* late seventeenth-century Puritan theology is the focus of this chapter.

Quakerism has often been viewed by historians of early modern English sectarianism as a variant strain of Puritanism. When Perry Miller first gave Puritan studies a respectable reputation as a field of academic investigation in the 1930s, he did not include sectarian enthusiasm. The succeeding generation of historians, however, tended to see Puritans and Quakers as religious "cousins," albeit warring cousins, who shared a "continuity of experience" and a dissatisfaction with the established church. More recently, Melvin Endy's study of Puritan and Quaker theology has reconsidered their similarities and differences.[1] Endy points out that Quaker beliefs, evolving over time, were sometimes similar to Puritan beliefs, and sometimes overlapping Puritan beliefs, but they also diverged at crucial points. Puritans deemphasized outward mediatory religion in terms of clergy, sacraments, and saints, while Quakers rejected all three. Both groups shared to some degree a similar apocalyptic concept of history, in which a cosmic battle was to be fought between the forces of good and evil, culminating in the establishment of the kingdom of God on earth. Just how Christ's kingdom would be established on earth received a range of speculative answers that incorporated both spiritual and historical ideas.

However, the breach between the two groups occurred when Quakers accepted, among other differing points of view, two significant ideas that diverged radically from Puritan and Protestant theology in general. One was the belief in continuing revelation in the present apart from Scripture. The certainty of salvation held by the believer through the indwelling light of Christ gained for the believer a sense of direct divine knowledge of purity of life and worship. Quakers repudiated sacraments and ordained clergy and all outward forms of religion because they hindered the workings of the divine inner light or Spirit. Quakers believed that the work of the Spirit in the believer had greater authority than the authority of Scripture or sacraments. Second, the groups differed on the christological issue of the two-fold nature of Christ. This rendered Quaker theology sufficiently aberrant to be considered heretical by contemporary opponents.

In view of the continuing scholarly interest in the delineation of Puritan and Quaker theology on some of the historic Protestant issues such as christology, election, revelation, and millenarianism, Margaret Fell's writings take on a new significance. Fell's life spans the years when Quaker theological development was fluid. Her published and unpublished works, including a recently discovered theological tract, amount to approximately seven hundred pages. This documentary evidence gives us one of the few in-depth feminine perspectives available on the first-generation Quaker theological debates. At times Fell repeated Fox's views; at other times she took the lead in articulating Quaker theology. Because she was so close to George Fox, William Penn, and the entire early leadership, and because her theological writings have heretofore been largely ignored by her biographers and the historians of early Quakerism, it is appropriate to take a fresh look at Fell's religious writings.

The doctrinal debates that took place between Puritans and Quakers offer substantial evidence that the contenders regarded themselves as divided by a deep gulf of difference. The Puritan movement was comprised largely of Anglicans, Presbyterians, Congregationalists, Reformed, and the Particular Baptists. These groups within the Puritan spectrum espoused certain central doctrines: First, that human beings were sinful creatures after the Fall and in need of redemption, and that Christ entered human history, preached the gospel, was crucified, and rose again as humanity's redeemer. Second, Scriptures contained the record of Christ's atoning work in history. True knowledge of Scripture, which was infused in the believer through the power of the Holy Spirit, contained the message of salvation whereby human beings could receive God's grace through Christ's atoning sacrifice.[2]

Various shades of Puritanism emerged in England by the middle of the seventeenth century. The Civil War and the Cromwellian years of experimentation in Puritan parliamentary government were characterized by the emergence of a wide range of religious ideas from moderate to that which went beyond the Puritan pale and held greater kinship to spiritualized christology. Spiritualist ideas, as defined above, were expressed by Quakers, Fifth Monarchists, Seekers, and Ranters who seemingly pushed antinomian-style ideas to their ultimate conclusion. However, the behavioral manifestations accompanying their ideas varied radically among these groups. The amount of vitriolic verbiage spilled over a half century of debate between these groups was mountainous. Each group came to look upon the other with antipathy and fear.[3]

Although students of the early English Quakers have often seen them as radical Puritans, Endy's study attempts to prove that most Quakers were neither Puritan nor strictly orthodox in the essential trinitarian doctrine of the Godhead. Endy concedes George Fox is often cited for Quaker orthodoxy concerning the incarnation and resurrection. For example, Fox opened his treatise "The People of God, in scorn called Quakers," by stating the Quakers believed in the Trinity and that Christ was "crucified without the gates at Jerusalem, and rose the third day, and sits at the right hand of God." However, Endy also points out that Fox was not consistent in this interpretation. He and other early Quakers exhibited a strong tendency to deemphasize the historical Christ and stress the divine guidance of the inner light. Fox wrote concerning the Spirit: "[It] draws off and weans you from all things, that are created and external (which fade and pass away), up to God, the fountain of Life and head of all things." It is apparent that the early Quakers were testing the limits of Scripture in the christological debate. Fox, Fell, Barclay, Fisher, among others, stressed Christ as the eternal Spirit within rather than as the historical Jesus.[4]

In the 1650s the Quaker minister, Edward Burrough, a friend and near neighbor of Margaret Fell, wrote on this subject. When Burrough's opponents confronted him on his docetic tendency, Burrough conceded that Quakers, like all Protestants, believed that the incarnate Christ who died and rose again in Jerusalem and that Christ was the central figure and event in Christian history. Endy sees Burrough as a "typical" Quaker in the way he handled this issue, for he avoided being specific about the problem unless pressed by his opponents to do so.

Margaret Fell confronted this issue in much the same manner as Edward Burrough. She wrote in 1660 of Christ as the Spirit of life who nourished the soul and then added:

we witness and bear our Testimony of the same Christ (and not of another) which witnessed a good Confession before Pontious Pilate, and was crucified upon the Cross at Jerusalem; This Christ which all the Christians in Christendom profess in words, do we hear Testimony of in the Spirit of Life and Power, according to the Scriptures.[5]

Although this particular statement confesses the "faith of Christ crucified," it was a belief infrequently expressed by Margaret Fell. By far the majority of her statements emphasized the inward spiritual Christ of faith. The apparent studied ambiguity of this Quaker position masked a range of interpretations varying from a very spiritualist approach (Fox and Fell) to a more conservative approach that reflected Calvinist influence, such as seen in Robert Barclay's *Apology*. Barclay claimed in Proposition Six of his *Apology*, that the doctrine of salvation freely available to all, was very evident in scriptural testimony. He wrote,

> The angel who declared the birth and coming of Christ to the shepherds said not that his news was for a few men but (Luke 2:10): 'Behold, I bring you good news of a great joy which will come to all the people' . . . if most of mankind had been excluded, the angels would have had no reason to sing: 'Peace on earth and good will towards men.'

By the time of the Toleration Act, the Quaker position as seen in Barclay's statement had moved closer to an orthodox one, for the Act of 1689 granted toleration only to trinitarian Christians.[6]

To observe Fell's position more precisely, let us return to the pre-Barclay and pre-Toleration phase of Quaker theology concerning the incarnate and postincarnate Christ. Although Fell stood in the spiritualist camp, she spoke moderately when addressing orthodox entities of power such as the Crown. In a letter to the king and his Privy Council in June 1660 about the persecutions and imprisonments of Quakers, she assured him that Quakers too were Christians:

> We are not Heathens, but Christians; and it is for Christ's sake that we suffer, although our Sufferings are more severe than other People's professing Christianity; in that so many Thousands should be cast into Prison only for their consciences towards God.[7]

Likewise she expressed herself to her Quaker compeers. "I write to you, as unto dear and near Members of this Body, the Church, whereof Christ

Jesus is the Head, and hath given himself for it." She described her own conversion in 1659 in terms that were not unlike Puritan conversion experiences, by alluding to Christ's redemptive act.

> I went through the vale of misery . . . when no light appeared, then was I following men, but darkness was over the whole earth in me, . . . all my acting and all my prayers returned in vain . . . then the Son of Righteousness appeared to me, within healing under his wings, then he rent the vail, and uncovered all that I had done . . . but I desired that I might be spared . . . for I was condemned, but the Blood of the Son of God ransomed me and brought me into freedom . . . then he shewed me his doctrine . . . and placed his word in my heart, . . . and caused all my bones to tremble and stagger like a drunken man, but the remainder of himself could not be shaken, for he was stricken, and by his stripes were we healed, and by his light we see all hirelings, and persecutors to be out of his doctrine.[8]

Among her scattered triniterian references, Fell did proclaim the apostolic root of Quakerism. The Quaker church was not only the true church, but its ministers were ordained by the Spirit as the apostles had been, not needing an education in terms of university training: "The holy Apostles off Xt [Christ] went nott to the university forr Xt endued them with this Holy Spiritt . . . By his spiritt he endued them with more knowledge than they could have gotten at the universitys, and soe they need not go thither."[9]

In her polemical writings, Fell seemingly accepted the orthodox incarnate and postincarnate Christ. In her debate with Anglican clergyman Allan Smallwood in 1668, she accused him of not preaching Christ "risen from the deade as the Apostles did . . . and soe the cross of Xt which is the power off God thou art ignorant off. The gospell wch is saluation to everyone that believes and surely those that preaches this gospell worshipps God in soe doinge." Elsewhere she wrote in 1664 that Christ "took upon him the form of a Servant, and in that he humbled himself to the Death of the Cross." In 1659 she also expressed her belief of Christ's redemptive act: "The eternal God keep you who brought again our Lord Jesus Christ from the Dead, through the Blood of the Everlasting Covenant; and by his Blood wash you, and cleanse you from all Sin." Several years later she wrote, "Your redemption and Perfection hath been purchased at a dear Rate; no corruptible Perishing Thing could purchase it; but the precious Blood of Christ Jesus, who hath given himself as a Ransom for you."[10]

Fell's view of Christ may be found also in what appears to be a possible inclination toward an adoptionist christology. She used the metaphor of adoption in at least one instance:

> The Whole Creation hath long groaned under the Burthen of Oppression and Corruption; the Redemption out of which, is thorough C. J., by the operation of his spirit are his sons; And the Redemption of the Body is through waiting in the Spirit of Adoption, whereby every Member may call God Father.

and she quoted from Galatians:

> God sent forth his Son, made of a Woman, made under the Law, to redeem them that were under the Law, that we might receive the Adoption of Sons. God hath sent forth the Spirit of his Son into your hearts, crying Abba Father.[11]

These two statements do not give enough theological content to ascertain whether she was adoptionist in her thinking. What is more important to see here is that Fell drew many of her christological ideas straight from Paul. She did refer numerous times to Christ's perfect obedience that had made believers free. What is evident also is that Fell and other early Quakers were most at home with language that described Christ in spiritual terms and this was congenial with Pauline theology. Fell frequently used metaphors of spiritual unity with the divine will. She exhorted disciples, "Let us be of one spirit," and expressed the mystical Quaker unity in Christ as follows:

> Christ [is] the Pillar and Ground of Truth. And when the Wind blows, the Storms and the Tempests beats, this House will stand; for they that are joined to the Lord, are one Spirit with Him, and of one Heart and Soul; if they were ten thousand, they are all one, and brought forth, and sanctified by one Spirit; and here is the Mystery of the Fellowship of the Gospel, which is in the Unity of the Spirit.[12]

Central to Fell's thinking was the idea that those who were without the light of Christ were living, teaching, and ministering in apostasy. She wrote, "come to the soundness of the Truth of the Gospel of Christ, who is the beginning, the first Born from the dead, that you being rooted and grounded in him, and that through the Spirit you may wait for the hope of Righteousness by Faith, and so grow up as . . . the Planting of the Lord." Elsewhere in her writings she made statements such as the following: "Oh Friends [stand in the] Faith of Christ Jesus, . . . that we may be justified by the Faith of Christ, For whatsoever is born of God overcometh the World" and, "the first Adam was made a Living Soul, but the second is the Lord of Heaven, a Quickning Spirit, which doth quicken and raise up the Just unto the Resurrection of Life," and "those that deny the Light of the Spirit of Life, they know little of the Pourings forth of the Spirit, or of the Manifestation of the Spirit of God, which he hath given to everyone."[13] This statement is a near quotation from the chapter in I Corinthians 15:47ff on the resurrection. It does define the power of the spirit in Christian faith.

Fell's spiritualism emerges in a pattern of religious statements written over a lifetime. Her frequent references to Christ as the light, the spirit of Christ, Christ the cornerstone, the seed, the shepherd, the image of God, the great prophet, priest, and king, the light of the gospel, and the cup of the New Testament do not deny the two-fold nature of Christ; neither do they convey a strong sense of it. It simply is a subject that receives minimal attention from Fell. In effect Fell had moved from the Puritan theology of her younger years into an oppositional theological camp that represented a rupture in the seventeenth century continuum of Puritan thought. Her spiritualism was the overarching theme of her writings. Only when pressed or when writing to non-Quakers did she clearly state that the Christ she spoke of was the same Christ who died and rose in Jerusalem. What was important to Fell was the concept of an inner spiritual Christ with whom she and others could have continual communion. In such a state, all outward performances would occur through Christ's quickening Spirit and as such one could become perfected through Christ who indwelt the believer and was perfect.

It is possible that Fell was genuinely confused and thus avoided a methodical interpretation of orthodox soteriology. It is possible that she didn't care about or recognize all the fine distinctions contemporary theologians were making on this debate. However, it is this writer's contention that when her writings are taken as a whole, they indicate that she had an astute understanding of the theological debate and that she pursued a studied avoidance of the natures of Christ and the trinitarian arguments because she no longer accepted them as central beliefs as a Quaker. Fell was *au courant* in Quaker-Puritan polemics and simply equivocated purposefully, depending on the need. To the king she wrote about Christ in typical Protestant terminology; to the Jews she avoided (initially) using the term Christ; to her own circle she wrote with great fervor about Christ's Spirit that led to Truth. Fell was sincere in her Quaker beliefs, but she was a shrewd controversialist rather than a purist theologian. Whether consciously or unconsciously, she tailored her theological remarks to her audiences. Her style of studied ambiguity enabled her to write to widely different religious and political audiences – namely, Oliver Cromwell, the

Stuart court, Jews, Protestant and Catholic Christendom, and to her own circle – without engendering internal Quaker hostility. It was a useful political approach for Fell; her strategy being to convince the world to tolerate Quakers, and to evangelize for Quakerism.

Margaret Fell spoke out on other religious matters as well. Her stance changed from a defensive posture to an assertive one when sparring on the subject of clerical abuses. She wrote that the real potential threat to the king's power was not the Quakers but his own Anglican prelates. In criticizing the wealth and power of Anglican bishops, she posed the question in 1668, "why may there not be a Kinge in Great Brittan without bishops?" Her answer did not rise above the angry verbiage that characterized the polemics of the period. She blamed the problems of the Stuart monarchy on the power of the clergy:

> The bishops acknowledge him [the King] in words, but robs him both by there power and there greate benefits and liveings that they keepe from him to maintaine there prid[e] and haughtiness wch the kinge may justely and righteously [take] from them, [to] . . . ease the heavie burdens and oppreshions of the people, that he is forced to lye upon them by tacssations [taxations] and other wages, for if the kinge had that wch you usurpers and devorors [the clergy] have, the poore people of the nation would find ease thereby.

Fell stated that if the king were ever to be "prosperous," he would first have to reduce the power and revenue of Anglican prelates. However, Fell's radicalism stopped short of any political theory of revolution. In this she was entirely representative of the Quaker position on the validity of royal authority in the civil sphere. A recurring theme in her apologetics was that of the suffering people of God, faithful subjects who endured persecution for refusal to pay tithes to the hypocritical and power-hungry clergy who meddled in civil affairs and manifested ignorance in the spiritual things of God. Although the clergy had "beene at Oxforde and Cambridge," their dark and ignorant minds were full of "fallible totteringe judgement." Fell was representative of a wide-ranging sectarian backlash, brought on in part by the severe persecutions after 1660 and in part by traditional English Protestant fear of clerical power and abuse. In Fell's mind, her spiritual faith was superior to the pretended faith of an educated clergy: "It is manifest that thou and thy brethren knowes noe other Holy Ghoast; but what you knowe by hearinge and readinge the Scripture ffor all your goeinge to the universitys. Yes, thousands know more of the Holy Ghoast than ever you did, that neuer went to the university."[14]

Margaret Fell's Theology

Like her Quaker brothers and sisters she held outward, sacramental religion in low esteem. She opposed pedobaptism, calling it unscriptural. In 1668 she wrote to the Anglican clergy, "Your practice in . . . sprinkleing infants, . . . [about which] thou makes such a pitifull mingle mangell . . . you have no ground nor foundation [for]." Baptism, according to Fell, did not insure the beginnings of sanctification. On the contrary, she wrote, "What if the infants that you baptise be children of wrath? The watter wch you throw upon them makes them noe better. . . . And the children [of faithless parents] when they come to riper years manifests the fruts of your throwing watters on them by the[ir] wicked ungodly lives."[15] Fell moved beyond this level of apologetic to define the Quaker stand on the meaning of the waters of baptism. It was not to be understood as literal water but spiritual water according to her interpretation of the Johannine passage, "Except a man be borne of water and the Spirit . . . he cannot enter into the kingdom of God." She argued that Christ's words should be read "watter of the spirit . . . in the spirit of Christ, who saith that wch is borne of the spirite, is spirit." The catechism, she reminded the clergy, asserted that "baptisme is an outward forme of an inward and spirituall grace." Therefore Baptism was an inward "watter of the word, as no materiall watter were to be used." She concluded:

> Dost thou belieue that X waseth and purgeth his church with outward watter, whose baptisme is with the Holy Ghost and with fire? Surely, they are blind that see [your] profenation and blasphemie in perverting and mishaping of the Scriptures, altering if thou could possably the very intent of the words of Christ and the Apoistells about this baptising infants.[16]

Fell told the Anglicans and Allen Smallwood that they were also too close to Roman Catholicism in their doctrine of the Eucharist. Drawing from Paul's letter to the Corinthians she wrote, "Christ instituted a supper but he never mentioned a sacrament . . . thy brethren and the pope your father . . . instituted sacraments." Fell had no use for "carnal ordinances," claiming that a person who eats and drinks "unworthily, eats and drinks [to] there owne damnation." The sacraments were simply a dead and empty show for those who knew not the inward Spirit of Christ. Her rejoinder to Smallwood's treatise against her was:

> Thou makes it a greater crime that M. F. should put the Spirit in the roume of your sacraments. . . . St. Paule did not put the letter for the Spirit as thou and the brethren doth, nor he put not watter for the Spirit

as you doe, nor he put not bread and wine for the boody and blood of Christ as you does. St. Paule was a minister of ye Spirit and not of the letter as you are.

On the issue of the authority of Scripture and continuing revelation, she expressed the Quaker position in an epistle of 1660. In agreement with Fox, Fell answered her opponents that the Spirit bore out the message of Scripture:

> We received not this Spirit . . . by Man, but by the immediate Power and Revelation of Jesus Christ, according to the work and operation of it in us; . . . for no Man could teach it us, but who had the Revelation of Jesus Christ. . . . Now though these Scriptures bear testimony to the Truth of this, yet these Scriptures are not our Testimony only, for we have our Testimony in the same Spirit as was in them that spoke forth the Scriptures; and these Scriptures bears Testimony with us and we to them, and so are in unity with the same Spirit which gave them forth.[17]

Fell, like other Quakers of her generation, repudiated Scripture as the only source of revelation, yet used it to undergird her theological arguments against any detractors of their special form of Protestantism. When writing about true and false ministry she claimed, "let the Reader . . . seriously search, examine, and try by the Scriptures and the Measure of the Spirit of God, who are the true and false prophets . . . and who are the true Teachers, and who are the false." For Fell, as for Fox, Scripture interpreted by ministers who were not in the spirit simply held no authority. To Smallwood, who maintained that the believer sought the council of the Lord only in Scripture, Fell spoke bluntly:

> This is thy way now without the light of Christ, in the apostasy. . . . Does not the Scriptures say if any man want wisdome lett him aske it off God. And doth nott the Holy Ghoast say I counsell thee to be off me, goulde tryed in the fire. And does not the Apostle say that Xt is made unto us wisdome and righteousnesse; and woulde thou have the Scriptures to bee this? O Miserable man, consider the darknesse thou art in.[18]

She accused Smallwood of perverting and corrupting the true biblical message when he rejected on-going revelation. Smallwood claimed that all revelation was "compleated and there shall nothinge ever be added."

Her rejoinder revealed the depth of her antipathy: "If God did add as many plauges unto you as you haue made additions to the Scriptures you will have a heavy loade upon you." She added: "Thou canst hardely speake any Scripture or any truth, butt it will be against thyselfe, who art of another spiritt and nature." Fell was reiterating in these statements the standard Quaker spiritualist argument for ongoing revelation. However, while affirming it for Quakers, she denied it to others. She saw the Quakers as returning to the pure and true understanding of Scripture. What the Quakers came to learn over time was that to rely on individual interpretation by the Spirit often brought chaos and schism.

Another clerical tradition that had a long history in English Protestantism was the ecclesiastical discipline of the laity. Fell decried ecclesiastical courts that persecuted Quakers. Smallwood defended them as spiritual courts to punish disorderly laymen. She retorted: "The spiritt that rules in amongst thee is blacke, as well as yor coate and girdle, ffor the spiritt off God and his truth never erected nor sett upp such courts, neither . . . was there ever such courts in the Apostles' days."

Probably the most hated aspect of outward formal religion in England was the tithe tax. The Quakers suffered severely for refusing to pay tithes. Fell explained why Quakers bore witness against these "ungodly covetious practices." In response to Smallwood's argument that tithes were scriptural for an "aloowance for the ministry" as it was under the law when Abraham paid tithes to the priest Melchizedek, she wrote:

Abraham gave the tenth of the spoyles to Melchesadecke. But this thou and the brethren doth not like. You must have it by force and compultion or else you know that if it come ever to giueing, you [a]re not soe well beloved but that your shaires will be small . . . the sober reader . . . if they compare thy words with the Scriptures [they will see], there is noe ground att all in the Scriptures for the[e] and thy brethren to lye, chalence [challenge] or claim to tythes.[19]

Fell's polemics were sometimes biting, sometimes reasoned. But on the heated subject of the tithe tax, to which she had borne personal testimony in the form of heavy distraining penalties, she did not rise above a vituperative response.

Margaret Fell is a feminine exemplar of Endy's definition of Quaker thought where distinctive theological ideas placed Quakerism outside the purview of Puritanism. Fell's spiritualized approach freed her from her Puritan past, giving her a new belief in the ultimate power of the inward Spirit of Christ. More significantly, her theology led her to incorporate other ideas that were considered entirely aberrant to Puritans. Fell's newfound spiritual faith allowed her to reinterpret gender images of female ministry. Fell opened her best known treatise, *Women Speaking Justified* with scriptural arguments to substantiate gender equality. She wrote, "Justified, Proved, and Allowed of by the Scriptures, All such as speak by the Spirit and Power of the Lord Jesus. And how Women were the first that Preached the Tidings of the Resurrection of Jesus, and were sent by Christ's own Command, before he Ascended to the Father." She went on to argue the point:

> And how are the Men of this Generation blinded that bring these Scriptures, and pervert the Apostles Words, and corrupt his Intent in speaking of them [women]? And by these Scriptures, endeavour to stop the Message and Word of the Lord God in Women by contemning and despising of them. If the Apostle [Paul] would have had Womens speaking stop'd, and did not allow of them; why did he intreat his true Yoak-Fellow[s] to help those Women who laboured with him in the Gospel? Phil. 4:3. And why did the Apostles join together in Prayer and Supplications with the Women. . . . [20]

What merits attention here is that separate women's meetings never did give women spiritual equality as Fell hoped it would. Thus, on the issue of women's meetings, Fell insisted that women should conduct their own separate business meetings, and, as we have seen, she worked unflinchingly for their establishment. She based this belief on the examples of New Testament ministers like Priscilla and Aquilla, that women in the Spirit could and should hold religious authority beyond a limited and impermanent prophetic authority. Her sense of the perfecting experience of the inward light of Truth gave her a self-confidence, and diminished sense of personal sin and self doubt. It also seemingly influenced her social and economic behavior in terms of breaking not only with the prohibitions of gender and class, but also in her long-term interest in and economic support of poor and laboring women. Moreover, her self-confidence deeply influenced people who came into contact with her, and this was a factor in her personal success.

There is another factor in Fell's Quaker theology that must be considered: her millenarianism. This aspect of her theology gave her insight into overcoming the endemic ethnic attitudes of her day in her belief that Christ intended all people, including heathens, to be saved.

Millenarian or apocalyptic beliefs based on certain books of Scripture were not the preoccupation of English Protestants or Puritans alone.

A wide-ranging interest in apocalypticism characterized most religious groups throughout Europe in the early modern period. However, apocalyptic ideas and imagery gained an ascendancy and respectability in England in the sixteenth and seventeenth centuries that it lacked in Europe.[21] Embedded in English millenarian thought of the mid-seventeenth century was the question of Jewish reentry into England. Jews had been expelled from England in 1290. English philo-semitism and the Jewish reentry question became a significant political-religious issue during the Cromwellian period. English philo-semitism was in part rooted in an interpretation of texts in Daniel and Revelation that stipulated that there would be a scattering of the Hebrew nation throughout the earth. As history approaches the end of time, the chosen, Hebrews and believing Christians, will be gathered from all corners of the earth prior to the Second Coming. For some English millenarians the land of the gathering of God's chosen people and the fulfillment of the promise was to be England.

This thinking was evident in the tracts that poured forth from English presses especially in the Interregnum. Multiple factors influenced the Christian millennial interest in the readmission of Jews into England. Economic and political factors certainly played a part in that interest, which peaked with the Whitehall Conference summoned by Cromwell in December 1655. English philo-semitists who attended included merchants and lawyers, some who were interested in Hebrew as a possible universal language, religious persons who wished to discover and understand cabalist knowledge, Puritan divines concerned with Old Testament themes, and some who searched for the ten lost tribes, all hoped that they could end the Diaspora and usher in the last days of the world. This widespread English interest, both in favor of and opposed to Anglo-Jewish rapprochement, focused both on the Whitehall Conference and on the person of Menasseh ben Israel, whose connection to the Quakers and Margaret Fell will be discussed later.[22]

Margaret Fell and George Fox were representative of many English Protestants, especially of a Puritan strain, who expressed an intense interest in millenarian ideas and Old Testament imagery. Fell and later Fox carried this interest to the point of establishing contact with the Jews in the Netherlands. Fell published a letter to Menasseh ben Israel and three other epistles to the Jews in Holland between 1656 and 1668. Fox addressed at least six tracts to the Jews. One way to understand Margaret Fell's compelling interest in Menasseh ben Israel, the Jews, and their reestablishment in England is to examine her religious thought and compare it to Fox's on this subject, I hope to demonstrate that one organizing principle of her theology, especially prior to 1680, was apocalyptic, like that of George

Fox. Moreover, her understanding of the role of Jews as well as women in salvation history was rooted in this theological outlook.[23]

Even though Quakers stood beyond the pale of Puritan theology, the theological background from which Margaret Fell drew in her apocalyptic writings was Protestant and specifically Puritan in origin. English Protestantism from the Reformation onward had viewed England as the place where God's divine acts in history were occurring and would continue to take place. John Foxe's *Book of Martyrs* was ordered to be placed in all parish churches during Elizabeth's reign, and English children were raised on the story of English martyrs according to his chronicle. Fell owned a copy of Foxe's book in her library at Swarthmoor. Many religious people within English Protestantism saw the English revolution as a prelude to the millennium, and therefore the role of the Jew in England took on an immediate significance. Christian divines interested in placing a date on the coming cosmic events searched Jewish sources. A flurry of tracts both pro- and anti-Jewish were published in this era, and the Dutch rabbi, Menasseh ben Israel, became the personality that especially attracted the attention of English millenarians.[24] Let us turn to Fell's interest in Menasseh ben Israel and to her ideas on salvation available to all and the Jews' role therein.

11 Fell's Work to Convert the Jews

Menasseh ben Israel wrote from Amsterdam to Oliver Cromwell in 1651 to request legal admission of Jews to England:

> The opinion of many Christians and mine concurre herein; that we both believe that the restoring time of our Nation into their Native Country, is very near at hand . . . and therefore this remains only in my judgment, before the Messia come and restore our Nation, that first we must have our seat here [in England] likewise.

Margaret Fell's writings to Menasseh ben Israel and the Jews are a typical expression of the fervor felt both by Jewish messianists and Christian millenarians for the cosmic significance of the decade of the 1650s as well in anticipation of the hoped-for "second-coming" year 1666. Menasseh's arrival in England in October 1655 to plead his case for the Jews touched off widespread public curiosity and interest in the Whitehall Conference. Margaret Fell kept close watch on these developments in London from her home at Swarthmoor. Two issues held the public's interest. The first involved speculations on the ten lost tribes of Israel. The second was Sabbatai Sevi of Smyrna, a Jewish false Messiah. Until his apostasy in 1666, Sabbatai Sevi's Jewish followers considered him the Messiah. It is noteworthy that Fell's writings to the Jews ceased after 1668.[1]

The most prolific period of Margaret Fell's intellectual life were the 1650s and 1660s, during which time she wrote thirty-nine of the forty-five published tracts that were collected posthumously into her one-volume *Works*. Her four epistles to the Jews were written in this period. She was the first Quaker whose message to world Jewry was translated into Dutch and Hebrew and exported to Holland. Her first book, entitled *For Manasseth-ben-Israel: The Call of the Jews out of Babylon* (1656), probably was never read by Menasseh himself. He was deluged by books such as this one when he was in London in 1656–57. Her second book to the Jews was also published in 1656 or 1657, *A Loving Salutation to the Seed of Abraham*. Then a third tract, *Certain Queries to the Teachers and Rabbi's Among the Jews*, appeared in 1656 or 1657, followed by her book *A Call to the Seed of Israel* in 1668.[2]

The exact printings as well as translations of Fell's books to the Jews is a subject in itself. In 1657 Fell's friend John Stubbs, prior to his trip to Holland, prepared her tracts for publication in London, according to a letter from Stubbs to Fell dated 7 September 1657.[3] At the same time, the Quaker scholar Samuel Fisher was asked to translate *For Menasseth* into Hebrew in preparation for his and Stubbs' Quaker mission to the continent. Stubbs' letter to Fell mentioned two tracts, "Thy Book called the Second Call to the Seed of Israel" and another epistle, both of which were translated into Latin in London. The reference to the "Second Call" may have been her thirty-seven-page tract, *A Loving Salutation*, printed by Thomas Simonds in London in 1656. However, *A Loving Salutation* does not agree with the title given in the Stubbs letter. The possibility exists that another tract was being readied by Stubbs and was appended to *A Loving Salutation*. Because Stubbs had referred to Fell's first *Call* as *For Menasseth*, it is also possible that his reference to the "Second Call" simply referred to her second book, *A Loving Salutation* (1656). At least we know that both books were printed and circulating in England when Stubbs wrote to Margaret, and that Samuel Fisher had been asked to translate *For Menasseth* into Hebrew. When Stubbs and Fisher departed for the Netherlands in 1657, Stubbs sent a letter to Fell from Gravesend dated 19 October 1657 saying, "I hope thy business concerning they Bookes will be gotten finished tymely."[4]

The proximity of the Jewish community in the Netherlands caused it to become the natural focus for English millenarian writers such as Fell and Fox. Jews in Holland enjoyed some limited privileges. By 1657 the Dutch Republic recognized Jews as full citizens and asked that no restrictions be placed on them when traveling abroad. Jews in England after 1656 were granted some legal protection and permitted to open a house as a synagogue and to have a cemetery. Their legal residence was not confirmed until 1664.[5]

A Quaker missionary effort to the Jews of Holland commenced in earnest by 1657 under the leadership of William Ames. It is believed that while in Holland Ames met Baruch de Spinoza shortly after Spinoza had been rejected from the Jewish synagogue. Ames wrote a letter to Margaret Fell from Utrecht on 17 April 1657 describing the following encounter:

Theare is a Jew at amsterdam that by the Jews is Cast out (as he himself and others sayeth) because he oweneth no other teacher but the light and he sent for me and I spoke toe him and he was pretty tender and doth owne all that is spoken; and he sayde tow read of moses and the prophets without was nothing tow him except he came toe know it within: and soe the name of Christ it is like he doth owne: I gave order that one of the duch Copyes of thy book should be given toe him and he sent me word he would Come toe oure meeting but in the mean time I was imprisoned.[6]

Several Quaker historians including Helen Crosfield, who first suggested the connection of Fell and Spinoza in 1913, and Henry J. Cadbury, William Hull, Isabel Ross, and the Dutch historian Jan van den Berg, have given support to the Spinoza-Quaker connection. Richard Popkin, however, has presented the most convincing argument delineating the relationship of Spinoza to the early Quaker mission in Amsterdam. Popkin builds his case on circumstantial evidence. By a process of elimination, he has presented a vigorous argument that the most probable Jew referred to in the above quoted letter was Spinoza. As Popkin notes, studies show that only four known persons were excommunicated from the Amsterdam synagogue around the date of Spinoza's excommunication in July 1656. The other three appear to have either recanted and been reinstated into the congregation or to have left Amsterdam and could not therefore be "the Jew at Amsterdam . . . cast out."[7]

Popkin has traced the liaison between the Quakers and Jews in Amsterdam to Peter Serrarius (1600–69), a Dutch millenarian who became friendly with the Quakers and who may have introduced Spinoza to Ames. Popkin further suggests that it was Spinoza who sought out Ames in an effort to discover more about Quaker thought. According to Popkin's study, Spinoza, Serrarius, the Collegiants, and Quakers in Amsterdam were expressing some overlapping views at that time, emphasizing inner conviction and inner light as well as repudiating external creeds and church polity, so that a temporary affinity appeared to exist between them.[8]

Fell's first tract, *For Manasseth-ben-Israel*, charged Menasseh with the divinely ordained responsibility of bringing the Jews to England and converting them to Christianity in order to fulfill biblical prophecies. She also asked that he have her book read among his compeers wherever they were dwelling. Shortly after completing this tract, Fell wrote to the Oxford-educated Samuel Fisher and asked him to translate her tract into Hebrew or find someone in Holland to do that work for her. As stated earlier, Fisher left for the continent with John Stubbs and apparently was unable to accommodate Fell in her request before leaving London. However, Fisher and Stubbs probably took Fell's work to Holland with them, and it may have been translated into Hebrew by Spinoza through the efforts of Fisher and Stubbs. In the September 1657 letter of John Stubbs cited above, Stubbs indicated that her two works, *The Second Call to the Seed of Israel* and *For Manasseth*, had been translated into Latin.

The epistles were most likely translated into Dutch by Ames before he gave them to Spinoza.[9]

The youthful William Caton, who had spent several years at Swarthmoor Hall and was Margaret's erstwhile amanuensis and adopted son in Quakerism, wrote several letters to her from Amsterdam. Caton reported faithfully to her from Holland from the late 1650s until his sudden death in 1665. Between 1655 and 1665, Caton made several visits to the Jews either at their synagogue or in their homes. Although he spoke no Dutch or Hebrew, he communicated with them probably through the help of either Fisher or Ames.[10] On one occasion, Caton described the following scene after Sabbath worship among the Amsterdam Jews: "After their worship was ended, I and another friend had some pretty good Service with some of them in their Houses: but they are very hard, obstinate and conceited People in their way." He further commented to Margaret in October 1656 that

> in one of their houses, we had 3 of 4 hours' discourse, with some of them, and a good principle of conscience they owned in words, but they could scarce endure to hear Christ mentioned. They are a high lofty, proud and conceited people, far from the Truth, yet there is a seed among them, which God in the fullness of Time will gather into his garner.[11]

In an October 1656 letter Caton informed Margaret that the Dutch Jews were "hungering" for her book to be published in Hebrew; presumably this referred to her epistle *To Menasseth*.

In June 1657 Caton reported to Margaret that her epistle *To Manasseth* had been translated first into Dutch by Ames and then into Hebrew, followed by distribution "at their Synagogue, some to the Rabbyes, and some to the Doctors." Caton's comment on its reception was "I cannot understand that they have anything against it, but only they apprehend that the Author doth judge that the Mesias is come already and they looke for him yet."[12]

Sometime prior to November 1657, Margaret asked Caton to negotiate a translation of *A Loving Salutation* into Dutch and Hebrew as well. Caton responded in a letter dated 18 November 1657 that he had contacted

> a Jew and have showed him thy booke, I have asked him what language would be the fittest for them he tole me Portugees or Hebrew; for if it were in Hebrew they might understand it at Jerusalem or in almost any other place in the world. And he hath undertaken to translate it for us, he being expert in several languages.

Caton also wrote that this Jew "remains friendly in his way." Popkin concludes that no other Jews expressed any interest in or befriended the Quakers at that time other than Spinoza.[13]

The Dutch mission became disorganized after the arrest and deportation of William Ames in 1657. Also, the credibility of the Quaker missionaries was eclipsed by the well-known Nayler incident of 1656 whereby Parliament punished Nayler for entering Bristol on a donkey in the company of other Quakers. The Naylerite followers in Holland also disrupted Quaker worship there. Not until October 1658 did William Ames have any further contact with Spinoza. By that time Ames was again back in Amsterdam. In the meantime, George Fox had written Ames requesting that his tract, *A Visitation to the Jews* (1656), be translated into Hebrew. Ames responded to Fox in a letter of 14 October 1658: "I have spoke with one who hath been a Jew toe translate it intoe Hebrew." Ames had Fox's piece translated first into Dutch and then approached the "Jew toe translate it intoe Hebrew . . . because he who is to translate it into Hebrew cannot understand English." However, Fox's work seems never to have been translated. Efforts to have his work published came a year or more after Fell's work first appeared in Hebrew.[14] Thus, it was Fell who was the initial and primary Quaker spokesperson for the Jewish international mission.

When Samuel Fisher and John Stubbs left Holland in 1660 for Rome and Constantinople in hope of converting the Pope and the Sultan, they may have taken with them the Hebrew version of Fell's *For Manasseth* and *A Loving Salutation* to pass out to Jews along their route. William Caton, the other Dutch missionary who took Fell's interests seriously and promoted her missionary endeavors, was back in London by May 1658. He wrote Margaret from London that he had peddled some 170 copies of her book in Hebrew "among the Jews who willingly and greedily received them," some of which were given out in Zeeland to Jews there.[15] Three years later Caton was in Frankfurt, Germany, where he distributed one of Fell's pamphlets. A letter from Caton to Fell with the heading "near wormes," read,

> Many of the Jews have gotten of thy Hebrew bookes, when I was Among them. [at] frramfrod [Frankfort], in their Synagougue I had of thine and of J. Penningtons, which were in the Germane Languadge, but thine in Hebrew they had more mind to them to the other; they are wonderfull darke and ignorant of any Spirituall good but zealous in their way, for twice a day they go to their Synagogue the whole year through as I have been informed, but exceedingly given to cozening and cheating not esteeming it a sin to defraude one of them who are called Christians, and of the like opinion are some of the Christians of this country who

count it no sin if they can cozen A Jew, so that the Truth reproveth them both.[16]

The above translation in Hebrew also included Fell's *Queries to the Rabbis* and a short call to conversion to the Jews, written by Samuel Fisher. Sometime after Fox's effort in late 1658 to have his works translated through Ames, the latter apparently lost interest in the mission to the Dutch Jews and turned his interest to edifying the general Quaker community in Amsterdam.

The early Quakers believed that the moment of confrontation and decision was at hand for the kingdom of God to break into human history. They believed that God would come not just within the convinced individual's soul but through a "visitation" that would, by degrees, become a worldwide salvific process. It is not clear if "visitation" was to be simply a "second coming" of Jesus as an historical event, or as a spiritualized event for believers only. What they did make clear, however, was that the earliest phase of this eschaton was God's judgment. The apocalyptic vision included a turning upside down of the worldly order. The wicked and worldly would be punished before the meek and righteous would be rewarded. The righteous ones of God were to do their part in helping to usher in the kingdom. Fox's and Fell's writings to the Jews both express a deep knowledge of and personal immersion in the biblical prophecies of Daniel, Revelation, Isaiah, and Zechariah. Fell and Fox were committed to an apocalyptic frame of reference, and they viewed themselves as workers doing their part at that juncture of divine history. Other leaders besides Fell and Fox felt this new rising strength of the Quaker message. Christ was come, said James Nayler, not to save "a few raging Quakers only, but with ten thousand of his saints is he come."[17] Fell and Fox, during the first two decades of Quakerism, expressed the sustaining conception of the "Lamb's War," that the eschaton was about to break into human history and their message to Jews everywhere was to heed the apocalyptic vision.

Fell's ideas of judgment, incorporated in *For Manasseth-ben-Israel*, include a stress on the "day of wrath" and its consequences. She quotes Isaiah frequently, using his imagery of judgment as a plumb line and righteousness as the plummet. She warned the famous rabbi: "Therefore to the pure Light and Law in the inward Parts turn your Minds, which shews you your evil Deeds, and makes manifest Sin and Evil." This first phase of reproving would be followed by a newfound sense of inward peace. "This is the Way, walk in it. Here you will come to drink the still Waters of Shiloh, which go softly, which the Lord spoke of to Isaiah." Despite this call to the Jews to repent, Fell did not dwell excessively on their sin or on the day of darkness and wrath, as some other Quaker writers did. Instead, her more positive theme was that of the "Light" that would first rip one up and tear one open, to be followed by blessing and a sense of personal peace. Typical of the apocalyptic writers of her day, she equated the Jews with the ancient Hebrews, alluding to them as the "Trees of Righteousness" and the "Planting of the Lord."[18]

There was, however, a certain ambiguity in Fell's and Fox's apocalyptic message. On the one hand, they said that the kingdom would come within the heart of the believer – a spiritualized approach. On the other hand, they called for repentance, for divine judgment would come as an historical event with its consequences for those who resist the apocalyptic message. At points there was an urgency to their call to the Jews to repent and follow the Light before it was too late. Fell quoted the Scriptures, "Seek the Lord while he may be found," and "call upon him while he is near." The kingdom would come, but time was running out. In her last writing to the Jews in 1668, her use of the words "Zion" and "Jerusalem" appear symbolic. "For out of Sion the Law is going forth, and the word of the Lord from Jerusalem; and Sion is redeemed through Judgement, and her Converts with Righteousness." She refers to the circumcision "which is of the Heart . . . now this is the Seal of the Covenant." Her later theme stressed Fox's belief: "Jesus the unchangeable is risen to teach his People. . . . That is the Light, by which the Lord teacheth his People, by which all the Seed of Israel, Jews and Gentiles will be saved." This message is more tentative in that it does not pinpoint any place or date of the Second Coming, only the belief that it would occur. The predominant theme running through Fell's and Fox's books to the Jews was expressed in the belief that the Quakers were the true church, a replica of the primitive church existing in the last days of the world.[19]

Fell's tracts to the Jews are laced with biblical quotations combined with her own exhortation. In contrast, Fox's early tract, *A Visitation to the Jews*, written in 1656, seemingly is more original in expression and style but less gentle and positive in tone than Fell's work. For example, Fell's treatise *For Menasseth* is diplomatic in that she does not once use New Testament references or the word Christ or Son of God but confines herself entirely to familiar Old Testament passages and images. She does this in order to emphasize the religious affinity between the Jews and her special type of Christianity. Instead of using the word "Christ" in her two earliest epistles, she employs the words "Light," "Living Water," "Lord," "Spirit," "Trees of Righteousness," "Planting of the Lord," "Fountain of Gardens," "Shepherd," "David's righteous branch," "Ancient of Days," and the "purging of the Daughter of Sion." Fox uses the word Christ" throughout and reiterates numerous times in the *Visitation* that the Jews knew not Christ and because they did not believe the prophet Moses they were in transgression, for "the builders [had] rejected the corner-stone."[20]

Fell admonished Menasseh to throw off outward ritual worship and become a "Jew inward." This phrase was undoubtedly Fell's best-loved quotation, and she borrowed it directly from Fox. Fell first heard Fox's exhortation to become a "Jew inward" in St. Mary's Church, Ulverston, on his first visit there in 1652. In her testimony to Fox years later she wrote: "And the first words that he spoke were as followeth. 'He is not a Jew that is one outward; neither is that circumcision which is outward: but he is a Jew that is one inward; and that is circumcision which is of the heart.'" She commented elsewhere that those of the true church had the Law written in the "inward parts": "We are the Circumcision who worship God in the Spirit, and have no Confidence in the Flesh, and this is the Seal of the Covenant, which stands forever, unto all the Seed of Abraham, to which the Promise is."[21]

Twelve years lapsed between Fell's first and last exhortation to the Jews. In her last tract to the Jews, *A Call to the Seed of Israel* (1668), she fervently implored Jewish conversion in positive biblical language:

Awake awake, put on thy Strength, O Zion! put on thy Beautiful Garments, O Jerusalem, the holy City! . . . unloose thyself from the Bands of thy Neck, O Captive Daughter of Zion; . . . you have sold yourselves for naught, and ye shall be redeemed without Money. Now is the free covenant of Everlasting Life set open to you, which brings Salvation unto the Jews and Gentiles . . . Happy art thou, O Israel![22]

Another difference in approach to the Jewish question between Fell and Fox is seen in her lack of emphasis on the guilt of the Jews for Christ's death. Although she mentioned that the Jews were a captive Israel, "like sheep gone astray" who had "disallow[ed] the Corner Stone, Christ Jesus the Light," she also stressed the return of Israel to rebuild anew. Fell wrote, "god destroys the chaff but the shepard is finding his sheep and will save his flock." Her emphasis in *A Call to the Seed of Israel* was centered on Christ's death which atoned for all human sin. However, just as she despised the Anglicans for their outward forms, she did not spare the Jews a stern indictment concerning Jewish outward worship and observances, calling them "the abominations to the Lord." She reiterated the standard Quaker theme but in a positive tone, stressing that the Lord would make a new covenant with Israel and put the law and testimony within their hearts, "Israel arise, shine for thy light is come. The visitation day is here. The

Redeemer is come to Zion . . . he brings salvation to the Jews." She told them not to keep their outward Sabbaths, for "the Lord is weary of sacrifice and outward observances . . . [instead] obedience is the best sacrifice."[23]

In contrast, Fox's *Visitation* repeatedly mentioned the Jews' guilt for Christ's death and their need for redemption. His opening statement to the "seed of Abraham" expressed this forcefully:

> To all scattered Jews according to the flesh, who have the law of God, the form of it, but being found out of the life of it, have not possessed that which Moses saw, who received the law from God . . . and your fathers in ages past had, who put Christ to death, and slew him that is the end of the law, God's righteousness, Christ Jesus . . . to keep out of all unrighteousness.

A third more nuanced contrast between Fell and Fox occurs in their attitude toward the fulfillment of the law. In several of his writings Fox stressed that the gospel had abrogated the Jewish law. In *Gospel Liberty*, Fox wrote that "Christ is come, who has ended the Jews law, and their weapons, and their religion and worship and has set up the gospel, the worship in spirit and truth; and Christ did not give forth any law, nor did his disciples after him give forth any law to put men to death."[24]

Fell stated that Christ came to fulfill the law and sought to de-emphasize nullifying it. She combined the Deuteronomic and Matthian texts in her 1668 treatise to argue the positive theology of fulfillment of the law through the incarnation: "Jesus Christ, who is the End of the Law of Righteousness, witnesseth to the Righteous Law of God, which Moses taught Israel; who said the Lord our God is one Lord." She immediately reiterated the biblical words of Christ, "Think not that I am come to destroy the Law; I am not come to destroy, but to fulfill." Taken as a whole, her writings to the Jews make clear that her approach to the Jewish-Christian question reflected Fox's theology only to a point. Her interpretation was a subtle disagreement with Fox on this issue of fulfillment of the Law. Further, her interpretation reflected a greater sensitivity and diplomacy through her careful choice of words.[25]

Seventeenth-century Puritan hermeneutics stressed a deep concern with the Old Testament and encouraged the study of Hebrew. George Fox attempted some study of Hebrew and spoke a few words of it to great effect in his preaching. One historian of the period suggests that this preoccupation with Hebrew and the Old Testament prophecies may have contributed to a slight improvement in the understanding of the Jew as no longer a scapegoat for Christian antipathy.[26] Although Fell's books expressed a more affirming note than Fox's, Fox's books stressed the idea that the "Jew inward" included all persons who came to live by the spiritual guidance of the inner light, while the "Jew outward" applied to many so-called Christians. Fell's openness to the Jewish question is further seen in the following encouraging opening statement of *A Loving Salutation*:

> Come, Oye House of Jacob, and let us walk in the light of the Lord. This is the day of your visitation, therefore sek the Lord while he may be found, call upon him, while he is near; let the wicked forsake his way, and the unrighteous Man his thought, and slight not the Mercies of the Lord, and his gracious visitation unto you: for he hath brought Salvation near, and his covenant is he performing with you; if you do turn to that measure of Light which you have received from him, who is the Light of Israel . . . [it] will lead you unto Justice and Equity, which Light will teach you to fulfill the Law of Moses.[27]

Later in the same treatise she elaborated the dark side of the visitation. The first phase of Christ's coming to earth would involved judgment of sin. This accords more closely with Fox's overall tone and theme in his tracts to the Jews: "The day of the Lord discovers them and will be sorrow and distress unto them, the great day of the Lord is near . . . and haste greatly . . . the day is a day of wrath . . . trouble . . . and distress . . . of darkness . . . a day of the Trumpet and Allarm."[28]

Fell believed that if the Jews would just read and listen to the noncoercive Quaker message, they would see the truth of Quaker principles and embrace the inner Christ. She tended to deemphasize the image of the scapegoat and shift the focus to a spiritual inward religion that had the effect of being less threatening or frontal in approach.[29] Fox was less willing to overlook Jewish "guilt," and this reflected the traditional Christian attitude of accusation. Fox saw the Jews as a blind people for not receiving the Messiah to whom the prophets in the Old Testament had testified. One statement is representative of this theme:

> So consider now, you whose eyes ar blind, and are yet prisoners in prison, and sit in darkenss and do not behold the new thing that the prophet declared . . . and now ye that will not receive the Messiah, the Immanuel, born of the virgin according to Daniel's number. . . . And so is not this he that was born of a virgin, the Messiah; that when he came among you, and to this day, you could never see any form or comeliness or beauty, but spit upon him, and buffeted him and put your

crown of thorns on him; yea, says Isaiah, 'he was despised and rejected of men,' and did not you despise him? did not you fulfill the prophets words?[30]

In spite of these subtle differences in written expression to the Jews, Fell and Fox, in common with other Puritans and Quakers of similar interest, looked upon the story of the Jew as one that was fundamentally tragic. Their history of chosenness, apostasy, and unrepentance had been followed by rejection. This notwithstanding, the Quaker evangelists did continue to believe that rejection could be turned into reconciliation and that the message of the Inner Light of Christ applied to all peoples. The new covenant of which Fox and Fell spoke included all believers who responded to God through repentance and obedience and who turned to an inward religious experience. Fell expressed it thus: "Now if ye will know your Redeemer . . . the mighty one of Jacob, turn to his Light, join your Minds into the Light which is within you and there will you see your Savior and Redeemer." But because the Hebrews had been a stiff-necked people, this would now happen through the divine instrumentality of the Gentiles, for Fell wrote, "ye shall suck the milk of the Gentiles, . . . thus saith the Lord God, behold I will lift up my hand unto the Gentiles, and set up my standard to the People, and they shall bring thy sons in their arms, and thy daughters shall be carried on their shoulders." The apostate church in Christendom, as well as Jews and heathen, held no different status in God's plan, for all had broken the covenant. The Quaker vision included a totally converted heathen world and reconverted Christian society and it made exceptions for no one.[31] Fell addressed the Jews as one final group or nation that she longed to see in the Quaker fold:

> Is there not even a bowing down [of Gentiles] unto thee in this living invitation, and even a licking of the dust of thy feet, . . . who have the standard of the Lord set amongst us, which is for the gathering of all Nations together. . . . Therefore arise and shine for thy Light is come and the Gentiles are come to thy Light. . . . This we desire to see fulfilled, and this we wait for. . . . The Redeemer is come to Zion, and unto them that turn from Transgression in Jacob.[32]

Fell told Menasseh that the gathering of the nations together was about to take place for "the good of all souls . . . and the bringing back out of captivity of the whole seed; which the Promise and Covenant of the Lord is." In her *Loving Salutation*, she continued the same theme:

The Lord . . . shall assemble the outcasts of Israel, and gather together the dispersed of Judah, from the four corners of the Earth; then shall the Lord utterly destroy . . . all the Power of Darkness; and with his mighty Wind, which is the Spirit . . . there is the high way for the Remnant of his People, who learn the Voice which doth cry . . . I will make an Everlasting Covenant with you. . . . Now if you will mind the Word, which the Lord hath sent unto Jacob, this will lighten and open your blind eyes, and the Lord's Promise will be fulfilled unto you.[33]

With only one or two exceptions, neither Fell nor Fox stressed, as some Puritan writers were wont to do, that England was the specified place for the promise to unfold. Fell resorted instead to the symbolic use of Jerusalem. She used such allusions and phrases as: "The Redeemer is come to Zion," the "gathering of all Nations' together," and "he hath Redeemed us out of every Kindred, Tongue and Nation, and hath made us unto our God, Kings and Priests, and we shall Reign on the earth." Only once, in her first address to Menasseh, did she mention England: "Thou who are called Manasseth-Ben-Israel (who art come into the English Nation, with all the rest of thy Brethren), which is a Land of gathering, where the Lord God is fulfilling his Promise. . . . "[34] Likewise, George Fox avoided mentioning the geographical locus of the future eschaton, or divine intervention of God in history, either in his *Visitation* or his *Declaration*. At one point in the *Visitation*, however, Fox alluded to Jerusalem as the possible geographical place: "[The Lord] will gather you from all the nations, and from all the places where I have driven you, and will bring you again to this place . . . whence I caused you to be carried away captive. . . . Now wait in the light, that you may come to witness this promise fulfilled."[35] Rather than emphasizing the place and date of the cosmic event, both writers expressed the hope of its imminence.

Although millennialists looked upon the readmission of Jews to England as a blessing to the church, other more prosaic factors operated as well that are beyond the scope of this study. Suffice it to say, a Jewish presence in England would be an economic benefit in fostering international trade connections and protecting English commerce, as the exiled Spanish and Portuguese Jews had done for Amsterdam. Undoubtedly, these factors came into play in the Cromwellian government's interest in the readmission of the Jews. A small community of Jewish merchants already lived in London by the 1650s. Aside from these factors, however, in the theological world of Fell's and Fox's day, the famous mission of Menasseh ben Israel brought the whole issue into sharp public focus and encouraged English apocalypticists to express their hopes.[36] That the

Quakers assumed their plan had even the slightest likelihood of success was highly presumptuous.

Fell's writings to the Jews in Holland not only predated Fox's tracts, but because she was more successful in getting her tracts translated, they were more influential in the international Quaker mission in the 1650s and 1660s than were Fox's. There is an additional reason why her writings were more widely circulated. One historian has commented that Fell's statement to Menasseh was a rehearsal of "Old Testament quotations best loved by Christians and unlikely to convince a Jew."[37] This is true, but even less likely to convince Jews was Fox's approach that was scolding in tone. An example demonstrates this attitude. Concerning the Jews' return to Jerusalem and their outward worship, Fox wrote in his *Journal*:

> The Jews must never hope, believe nor expect that ever they, shall go again into the land of Canaan, to set up an outward worship at Jerusalem, and there . . . to offer outward sacrifices . . . for Christ, the one offering, hath offered himself once for all, and by this one offering he hath perfected forever them that are sanctified.

Likewise, Fox's stern indictment is expressed in the following statement:

> [Think not] that the Messiah should come . . . and restore you to your sacrifices, do not look for that now, for that will never be, for by the offering of himself he ended your offerings . . . no prophecy you have in all the prophets to bring you back again to outward Jerusalem, to set up a temple and sacrifices and altars, but if you have any gathering to God it must be by the power of the Messiah up into his kingdom, and you must never look to be gathered to these outward things again.[38]

Fox recorded in his *Journal* that the Jews in Holland were "very dark and do not understand their own prophets." He also commented, "while I was here [Rotterdam] I gave forth a book for the Jews; with whom when I was in Amsterdam, I had a desire to have some discourse, but they would not."[39] It is apparent that Fox used very straightforward language to address the Jewish problem. It is true that Fell's more gracious and tactful writings probably made no converts, but it is more likely that Fox's approach was simply irritating in effect.

As noted earlier, Fell's and Fox's emphasis on Jews as the people of Israel who would fulfill biblical prophecy was representative of Protestant and, more especially, of the Puritan strain of eschatology. What was not representative of Puritan eschatological thought, however, was the more radical Quaker focus on the active inclusion of women evangelists in preparing for the cosmic drama that would redeem the world.[40] Fell stressed throughout her work, and especially in *Daughter of Zion Awakened*, that women as well as Jews were heirs in the process of worldwide salvation. This was a distinctive feature of Quakerism and expressed consistently in Fox's and Fell's, *Works*. Fell was a mouthpiece for Quaker women ministers and a female model for this emerging woman's role. She emphatically and repeatedly linked the Jews and women in the salvific process: "[Christ] hath perfected forever them that are sanctified, of what Nation or People soever, Jew or Gentle, Barbarian, Scythian, Bond or Free, Male or Female, all that ever comes to God by him, he is able to save to the utmost."[41] It is understandable that both Fell and Fox, as apocalyptic writers, espoused the radical idea of inclusion of the Jews in the true church of Christ in the latter days. It follows that their conception of those who go forth to evangelize the gospel must include all and any laborers. Further, it was Fell in her hopeful approach for Jewish inclusion in her writings, that enabled her to go further than Fox and others piercing through the racial and religious questions of her day. It appears that at least intellectually, Fell was able to overcome the endemic ethnic attitudes of her generation.

Quakerism appealed to Fell and many women because it allowed them to exercise their gifts in public ministry. At the same time this female experience must have heightened Quaker women's awareness of and sensitivity to other groups against which there had been traditional discrimination.[42] Fell's feminine interpretation of Scripture softened the harshness of the dominant Christian theme of Jews in transgression and tended to replace it with the themes and metaphors of promise to the seed of Abraham and the awakening Daughters of Zion. Her more positive apocalyptic prose and milder tone was a female expression of hope and openness to those groups excluded from the church, as women traditionally had also experienced exclusion from the inner circle of church policy and preachment.

*   *   *

A recent study of George Fox's thought and ministry maintains that the entire interpretative framework of his theology was consistently apocalyptic in understanding Christian revelation and prophetic in preaching his message about a new order in human organization.[43] Fox reiterated many times in his writings two statements that undergirded his apocalyptic thought, "that of God in everyone," and "Christ has come to teach his people himself." The first meant for Fox and his followers that each person must "sound deep to the witness of God in every man." This phrase was a euphemism for the light of Christ or the inner light. The second phrase stressed the apocalyptic message of Revelation 19–21 that foretold Christ's return to earth to make war against Satan. It conveyed the message of the victory of Christ's rule with the saints over Satan and death, which was to precede the beginning of a new heaven and a new earth. Once again, the language was not decisive in explaining Christ's return. This notwithstanding, Fox preached with unyielding confidence the belief that the Lamb shall have the victory in the end.[44]

It is apparent that Margaret Fell's theology was influenced by Fox's apocalyptic interpretation and by Protestant millenarianism in general. Fox and Fell saw that in the restoration of an apocalyptic new order of creation all things would take a radically new position. For instance, men and women in the light of Christ became as "help meets" or equal partners not only in marriage but in all relationships of the Christian life. This called for a rejection of all manifestations of social deference and inequality for those in the unity of the Spirit. Corporate authority of the church in the new order in theory, at least, embodied equality in the spiritual life, for God's word may be revealed in one of the least of them. Moreover, loyalty to Christ over loyalty to the state was fundamental, and passive resistance to the tithe tax and oath-taking were notable examples of Quaker resistance on account of religious conscience. The older order of church and state, both Protestant and Catholic, stood against Christ's direct rule of the saints within the Quaker community in its earliest phase. This iconoclastic and prophetic stance was all part and parcel of Fox's and other early Quaker preaching. The early Quaker ministers sounded "the day of the Lord," not only in their own circle, but in the public arena, "in the steeplehouses, synagogues and in the presence of magistrates" in the first two decades of Quakerism.[45]

Although the Pauline theological principle of Christ coming soon and the Johannine apocalypse and prophecy theme inspired Fox's and Fell's writings, those of Fell pose a problem. She was a prolific writer in the first two decades of Quakerism, but then she apparently ceased writing for a period of nine years between 1668 and 1677. Of the forty-five published writings by Fell in her *Works*, thirty-nine were written prior to 1668, far the greater part of her total recorded thought over a fifty-year period. It is also noteworthy that her writings ceased after the high point of interest in Jewish messianism and Christian millenarianism of the 1660s and shortly after the year 1666. She wrote four tracts between 1666 and 1668, then nine years of silence ensued, followed by six more tracts printed between 1677 and 1698. It is well known that she traveled extensively during the 1670s, which helps to account for her silence. However, considering her earlier prolific writing in defense of Quakerism, and to her own circle's need for her spiritual support as a key minister, some questions impose themselves. Why did Fell's writings significantly diminish in number after the surge in millenarian interest of the 1650s and 1660s? Does her silence suggest that Fell's earlier apocalyptic vision left her disappointed after 1668, causing a reorientation of her thinking and a temporary crisis in faith?

Christopher Hill has pointed out that the long hoped-for mid-century millennial phenomenon, which failed to materialize, had a significant impact on some of Fell's and Fox's contemporaries. Hill contends that political and religious sects were all fighting for what they conceived to be God's cause in that era of crisis. Groups like the Levellers and Fifth Monarchists believed that the Second Coming would achieve the political goals they had failed to win in opposing Charles I, the bishops, and the state. Political discord and Civil War gradually led to the defeat of political utopianism and religious hopes. Hill has written that "the radical sects in their desperation first became wilder and more millenarian and then gradually concluded that Christ's kingdom is not of this world." Milton wrestled with this dilemma as well. His masterpiece, *Paradise Lost*, reflected the ending of millennial hope and the resignation that God's ways were beyond human knowing.[46]

It is also noteworthy that Fox recorded in his *Journal* the experience of several depressions in this period. Specifically, Fox indicated that three severe illnesses or depressions occurred between the years 1659 and 1671.[47] It is true that Fox's statements are ambiguous, open to a number of interpretations. However, in light of the wider political-religious picture, it is plausible to interpret Fox's depressions as periods of crisis in faith for Fox as well as Fell. They, and other apocalyptic writers who experienced the religious enthusiasm of the early years of Quakerism, were forced to reconsider the timing of the imminent fulfillment of the promise of a Second Coming. It became necessary to reinterpret this breaking into history of Christ's kingdom in much the same way that the early church evangelists were forced to reassess their theology of an imminent Second Coming.

Fell restated her faith in 1677 in *Daughters of Zion Awakened*, which was an important summary statement of her theology. In it she spoke of Christ's Second Coming, but embodied in this statement was the idea that he was resurrected in the hearts of his saints "in these last Days [which were the] Night of Darkness; . . . Night of Apostasy." She reasserted her belief in the Second Coming: "the Holy City, new Jerusalem, is coming down from God out of Heaven," but her emphasis had shifted to a spiritualized one. She wrote:

And the Lord J. C. is building the Temple, which is his church, which is made up of living Stones . . . he is building a Spiritual Sion . . . his Word is at work, his Spirit is at work his Power and Light is at work, his Grace and Truth is at work . . . in the Hearts of Men and Women, he is at work.[48]

It is probable that Fell, like Milton in *Paradise Lost*, attained a "calm of mind, all passion spent," only after she had resigned herself to the failure of an imminent apocalypse. It took time to reorient her thinking to envision God's cause as an internal moral victory, not an external political and religious one; or a "paradise within thee, happier far."[49]

Throughout his adult life George Fox stated his apocalyptic faith in Christian revelation, but parallel to the development in Fell's thought, there is a theme in Fox's writing that denotes victory over sin, the Devil, temptation – or, in Fox's words, "to him that overcometh." Throughout his *Journal* Fox reiterated constantly, "the Lord's power was over all." In the latter days of his life Fox wrote, "Dear Friends everywhere, have power over your own spirits" . . . "be above" and "come over" all that is evil. After one of his depressions, he wrote that he had been "brought through the very ocean of darkness and death and through and over the power of Satan." For instance, at Enfield in 1671, Fox was deeply troubled by a vision:

in my deepe misery I saw things beyonde wordes to utter and I saw a blacke Coffin but I past over it. And at last I overcame these spiritts and men eaters (although many times I was so weake yt people knew not whether I was in ye body or out) . . . and towards ye springe I began to recover: . . . I came from under my travailes and sufferings.[50]

This suggests a man sorely troubled by external calamities such as persecution of Friends, but also seemingly by an internal struggle as well. Fox seems to have overcome these depressions when he resorted only to his "inner ear," not minding the voices of the external world. He expressed this despair as an alienation from God, in which one becomes confounded and presumptuous: "The cause of desperation is going from the light, for that which leads to presume, will lead to despair, . . . and that is that which wanders to and from, up and down and hunts abroad, and builds that which God confounds." His advice to Friends, learned from his own bouts of depression was "Quench not the spirit of God in you, but live in the authority of the Son of God and His power whereby ye may be kept on top of the world."[51]

Fell and Fox were like other millenialists in that they were influenced in this by "Judaizing conceptions of messianism" that called for an "historical realization of messianic utopia."[52] Fell and Fox were caught between one traditional Christian view of the kingdom of heaven present within and the belief in the imminent coming of the kingdom in an historical dimension. The Gospel writers themselves were divided on this issue. The Jewish conceptions, which deeply influenced them in the early phase of Quaker witness, were toned down as time wore on. Neither Fell nor Fox completely abandoned their apocalyptic approach, but Fell became disappointed and ceased to be fired by its message. This is seen in a falling off of enthusiast writings and a loss of chiliast vitality in those writings that did come forth from her pen. The process of "de-eschatologizing" or of de-spiritualizing chiliast conceptions may well have contributed to Fox's physical illness and psychological trauma, as seen in his severe illness and dark visions at Enfield in 1671.

A further question arises concerning Fell's and Fox's relationship based on their radical theological stance. If Fell began to doubt the apocalyptic message after 1668, did she and Fox begin subtly to part ways theologically? They who had felt a deep spiritual kinship in the first bloom of Quakerism in the 1650s and 1660s may have gradually ceased to feel that spiritual closeness if Fell lost her earlier apocalyptic enthusiasm. It must be noted that the 1670s was a decade of strife for Fell. Along with her public Quaker life of persecutions and imprisonments, her personal life was burdened by alienation from and the premature death of her only son, frequent separations from her husband, the Wilkinson-Story dispute, and the Rawlinson feud. All of these factors may have combined and contributed to the possible cooling of some of the earlier driving vision that she had shared with George Fox. To some degree Fell may have become somewhat disillusioned and materialistic, as seen in her dealings with Thomas Rawlinson. Additionally, if Fox silenced Fell in the Rawlinson dispute perhaps she was willing to see him leave Swarthmoor. The fact that Fox did not return to Swarthmoor Hall after 1681 might be explained, in part at least, by an awareness of her changed outlook. Her possible disillusionment may have caused him to feel uncomfortable in her presence and to feel a need to remove himself from close association with one who could cause or increase his spiritual depressions. By 1681 Margaret may have ceased to be George's closest spiritual companion and mentor. This possibility must, however, remain speculative until further evidence comes to light.

12 Conclusion

The goal of this study has been a reconsideration of Margaret Fell and her role in radical religion in seventeenth-century England, as seen in her social, religious, political, and family life. The results of this effort call for a reassessment of Fell in at least two respects. First, it is time to set aside the hagiographic approach to Margaret Fell as "nursing mother of Quakerism," seen in ample example in the secondary literature, and to try, as much as the old and new evidence permits, to see her as she really was. Granted, if one wishes either to condemn Fell or to put her on a pedestal, there is material to support either approach. If one wishes to understand Fell, it is first necessary to reconcile some contradictory material. We have seen that Margaret Fell was capable of courage; she was also capable of peccancy, pettiness, and belligerence. She was given to generosity; she was likewise given to self-serving materialism. She was endowed with the skill of giving spiritual inspiration to beleaguered Quaker missionaries; she was also endowed with an ability to exhibit highly authoritarian behavior. As matriarch of Swarthmoor Hall, she demonstrated *noblesse oblige* to those who were loyal to Quakerism, especially showing benevolence to poor women of her meeting. She displayed a sense of caring that cut through gender and ethnic restrictions, and she taught the women of her circle these caring skills. Her thirty-years of active leadership in the Swarthmoor women's meeting attest to these qualities of authority and benevolence. Fell's feminine interpretation of Scripture that enabled Quaker women in their public ministries and that displayed a special sensitivity to traditionally discriminated against groups also attests to her religious authority. She demonstrated her manipulative and dictatorial side in her treatment of her steward and erstwhile friend Thomas Rawlinson, whom she considered her social inferior, and of her only son George, who attempted to thwart her religious activities. Indeed, Fell could be overbearing, as seen in her attitude toward those who opposed the establishment of women's meetings. Conversely, considering the images of inferiority and weakness of women in her era, her dominant personality is further proved by her high degree of creativity in the early Quaker administration. We do Fell a service in removing the aura that surrounds her and allowing her to stand on her own. She comes down to us as a model of powerful female public ministry whose authority went far beyond that of the seventeenth-century female sectarian prophet. Leadership and religion were part of Fell's everyday experience. Her theology was not simply a doctrine but a way of life. As a religious leader Margaret Fell emerges as a shrewd, complex, fascinating, enormously energetic, and savvy woman who was moved by both high and low motives.

One final note on Fell's descriptive title "mother of Quakerism" is in order. If we infer from the word "mother," a woman who was an overseer, gentle, meek, and mild, then we have a misnomer in this term. Fell was an able overseer of Quakerism, but she was aggressive, dominant, and argumentative in her role. Her writings to her opponents reflect a style of theological polemics that often rings with invective. Although this polemical style was not unusual for religious discourse in her era, her statements to the Anglican priest of Carlisle, Allen Smallwood give witness to this invective:

> Thou hast manifested thy ignorans and unskillfulnesse in the things off God, though thou art called a Doctor and art gott upp in as high as thou coulde . . . so [thou] must neede pleade for this thy kingedome. Butt itt is the kingdome of darknesse . . . thou neuer had the true knowledge off the Father nor off the Sonne . . . whereby the simple and innocent [people] might be deceiued and beguiled, . . . [by] thee who art outwardly appeareinge to bee greate and in esteem, butt it is in that wch will vanish as the dust before the winde.[1]

Like her aristocratic friend William Penn, Fell was sometimes bellicose in her defense of Quakerism against its enemies. She, like Penn, displayed the dual personality traits of religious fervor and rebellion, and that combination was part of her attraction and success.[2] The image of "mother of Quakerism" ascribed to Fell veils much of what she really was. The metaphor would be more apt if it conveyed the sense of a "mother superior," a powerful, dominant, and somewhat aloof authority figure.

Secondly, it is evident that Fell's personal power within Quakerism was greater than traditional Quaker historiography has conceded. In her role in the early movement, Margaret was an opportunist, insofar as she used every opportunity to spread her Quaker beliefs. Fell also took every opportunity to promote George Fox's primacy *vis-à-vis* other prominent early leaders such as James Nayler, John Wilkinson, John Story, and others, including William Penn. Her efforts combined with the longevity of both Fox and Fell helped assure stability in the Society. Further, it enabled organizational continuity in a religious movement that might otherwise have gone the way of the other mid-century sectarian movements. As a longtime intimate companion to Fox, she had considerable psychological and intellectual influence on him, for Fox almost certainly made some of her ideas his own, such as his borrowing ideas from her peaceable testimony in 1660 and in the organization of the women's meetings. However, after 1681 it appears that her influence on Fox may have waned. Nonetheless, to some considerable degree she acted the role of a "gray eminence" in his life and probably influenced Fox's thought and attitudes more than any other person between their meeting in 1652 and his death in 1691. That aspect of early Quaker history has not been conveyed clearly in much of the secondary literature.

A third aspect of Fell's role in Quakerism is in need of comment here. Although evidence is incomplete, there is a compelling consistency to the evidence we do have that suggests that suppression took place in the early Quaker record. It resulted in a subsequent rewriting of the first and second generations of the Quaker experience for reasons of internal stability and survival. In recent historiography it has been suggested that Fox's charisma, longevity, and his personal *Journal* rendered him *primus inter pares* of the early Quaker leadership.[3] This is correct as far as it goes. This book has considered another factor in Fox's acknowledged leadership and has argued an additional hypothesis. Margaret Fell, through her personal affection, self-interest, social position, connections, permanence of place, practical skills, wealth, personal appeal, determination, and longevity, contributed in a significant degree to Fox's *primus inter pares* status in the sect by 1691. Along with Fell's promotion of Fox's primary position in the Society, this book has further suggested that Fell, through her old-gentry status, was an analogue to Penn's leadership in the Quaker community. Her class and wealth enabled her to become a powerful leader and pierce through traditional gender constraints of her era, and thus awe people in a way that Fox could not. Historians of the early Quakers have tended to see the "patrician influence" of Penn and Barclay as evident after 1668. It is this writer's assessment that the patrician influence was present from the beginning in the person, work, and charisma of Margaret Fell.

In her family life, affection abounded between mother, daughters, sisters, sons-in-law, and grandchildren. With the exception of George Fell and wife, all were loyal to her and to one another, and they were very devoted to George Fox as well. All of her grandchildren except one esteemed her highly and treated her reverently in her dotage. As a mother she encouraged her intelligent daughters to develop their skills and independence. She was a model for them in this. Her religious beliefs deeply influenced her family life, causing her daughters and several grandchildren to marry Quakers, and, along with most of the servants, to become lifelong active Quakers.

Margaret presided at Swarthmoor uninterrupted as a matriarch during her widowhood and second marriage, and probably during the last years of her marriage to Thomas Fell as well. At least after 1658 she was the locus of authority in the Fell clan and received the deferential treatment usually given to a landed family patriarch.

There are many blanks in our knowledge of her relationships with her two spouses. From what little we know about Thomas Fell, he appears to have been deferential to her feelings. Her relations with her second husband permits us only a partial "public" view as well. The public relationship was one of loving devotion. George Fox was the real love of her life, and it appears that her devotion to him was the greater. Fox's all-consuming quest in life was a spiritual one – he was a man too committed to his religious vocation to allow himself total personal commitment to one human relationship. The spousal relationship – as seen in Fox's account in his *Journal*, the fragments of their correspondence, and his permanent absence from Swarthmoor after 1681 – bespeaks a gradually changing attitude on the part of Fox. In the last decade of George's life, it was Margaret who felt the need to see him and who traveled to London for three extended visits. The evidence gives the overall impression that her commitment to him was the greater and that she bent her knee to no one except Fox.

Margaret Fell's social life prior to 1652 conformed to that of middling gentry woman in wealth, education, leisure, and influence. We have seen that her lifestyle, however, was significantly altered by her religious persuasion. Her social rank gave her a certain amount of influence in high places. Besides having several audiences with the later Stuart monarchs and others at court in the name of religion, she was accepted wherever she traveled among Friends as one of their most esteemed ministers. She expected that special privileged status among friends, family, and strangers even in her old age.

Her religious life after 1652 was a single-purposed, all-consuming endeavor on behalf of Quakerism. For exactly fifty years she watched it move from a persecuted, low-status radical religious movement to an organized and respectable, even sedate, church-sect after 1689. Her devotion to Quakerism denied her any rest from her labors; two years before her death her last published epistle to Friends warned them against the plain gray dress that was becoming popular among Quakers. Her words seemingly fell on deaf ears, for her advice was not followed. By the turn of the century her leadership had been superseded by younger Quaker women and men with changing Quaker values.

In view of Fell's wealth and status and her unswerving devotion to women's meetings from 1671 onward, one question persists. Why did Quakerism appeal to women, and especially to women like Margaret Fell and her daughters? To participate in seventeenth-century nonconformity – either in the Cromwellian or later Stuart periods – meant enduring persecution, social, and political ostracism, and the possible loss of property. For Margaret Askew Fell Fox, the desire to become a "despised" Quaker, to say nothing of her decision to marry a social inferior, may seem highly contradictory. But at a deeper level, neither Fell nor any seventeenth-century landed woman, could easily break through traditional English gender and class roles. Fell's relationships in her public and private life resonated an emerging gender consciousness in an age when female stereotypes precluded a public role for women in any capacity, and where birth rank defined a person's public persona and private friendships. The relationships of Fell, her six daughters, and her circle of Quaker women suggests that revolutionary Quaker theology provided a counterpoint to cultural restrictions of gender in seventeenth-century England. By cutting across both class and gender lines, Quakerism allowed women to exercise their ministries in the public sphere. This female experience undoubtedly heightened these women's awareness of their own business, speaking, and writing skills. Moreover, the main thrust of Fell's prose was to call on both women and men who were in the Spirit to be nurturers and interpreters of Scripture. It seems most probable that Quakerism appealed to Fell and other women because they had traditionally experienced personal exclusion from the church, and Quakerism gave them the opportunity to exhibit their gifts and be active outside the domestic sphere. However, Fell went beyond the prophetic authoritative stance of most Quaker women to become authoritative as a public apologist for Quaker doctrine and polity. A subtle "pecking order" existed in the Quaker community, as seen in Fell's personal status in the SWMM and LQWM. This indicates that within the Quaker group she did not lose her status and, indeed, it may have been enhanced due to her nonconventional behavior as a gentlewoman. Thus, Margaret Fell, whose religious authority rested to a significant degree on her social status, offered a new paradigm of radical Protestant womanhood. Her model of authoritative female public ministry was more enduring and effective than the model of the female prophetess, who was considered authoritative only while in a temporary and volatile state of religious enthusiasm. Mainline Protestantism failed to assimilate this alternative model until very recently. Finally, Fell's life also reminds us that patriarchy as an explanatory model for seventeenth-century female social behavior has serious limitations.

In 1700 the SWMM attributed a "thanksgiving" to Margaret, an accolade that neither the Swarthmoor men's nor women's meetings gave to George

Fox. The statement, penned by the clerk, Rachel Fell Abraham, was cast in diplomatic plural language to include all the old guard women of the Swarthmoor network. However, the content makes it evident that her daughter was writing a special tribute to her mother since most of the old guard were long dead.

> It is the Desier of ye [meeting to be] upright harted that the Anchantt Eueniety [unity] in the love of god may Continue amonge us and the Regarde and honer which is deue [due] and Doth beelonge to the Elderrs and those that hath bene as mothers in Israell and have bore the burdine and Exercyes in the morning of the Day and have washed ther garmentts in the blod of the lame [lamb]; pasing through many trabblations [tribulations]; that the younger Genration may not hurtt themselfs nor greeue [grieve] the lord; by ther Disregard and ungrattfullness: towards his worthyes whoe ar Deare and preshas [precious] in his Sightt and whoe ar knowne unto hime.[4]

Margaret Fell's interwoven public and domestic experience in early Quakerism is one of the few sustained documented records of what life was like for a gentle-born woman who was head of a household and likewise a public leader in seventeenth-century religious sectarianism.

Notes

Introduction

1. Margaret Fell (1614–1702) is best known by the name of her first husband, Judge Thomas Fell (1598–1658). I will refer to her by this name throughout this work although she did become Margaret Fell Fox in 1669.
2. Richard L. Greaves, *Deliver Us From Evil: The Radical Underground in Britain, 1660–1663* (New York: Oxford University Press, 1986) 4–5; see also J. Colin Davis, "Radicalism in a Traditional Society: The Evaluation of Radical Thought in the English Commonwealth, 1649–1660," *History of Political Thought*, No. 3 (Summer 1982) 203; Nigel Smith, *Perfection Proclaimed: Language and Literature in English Radical Religion*, Oxford: Oxford University Press, 1989, Introduction. An essential study to understand the force and ferment of radical ideas of the period is Christopher Hill's, *The World Turned Upside Down* (New York: Penguin, 1972 and 1980); also important is Hill's more recent work, *The Experience of Defeat: Milton and Some Contemporaries* (New York: Viking Penguin, Inc., 1984) Introduction.
3. Alan Macfarlane, *The Family Life of Ralph Josselin* (New York: W. W. Norton & Co., 1970); Paul S. Seaver, *Wallington's World; A Puritan Artisan in Seventeenth Century London* (Stanford, California: Stanford University Press, 1985); Keith Wrightson, *English Society 1580–1680* (New Brunswick, N.J.: Rutgers University Press, 1982); William Lamont, "Biography and the Puritan Revolution," *Journal of British Studies* 26 (July 1987): 347–53.
4. There are three older biographies of Fell: Maria Webb, *The Fells of Swarthmoor Hall and Their Friends* (London; F. Bowyer Kitto, 1867); Helen Crosfield, *Margaret Fox of Swarthmoor Hall* (London: Headley Bros., 1913); Isabel Ross, *Margaret Fell, Mother of Quakerism* (London: Longman, 1949; reprint, 1984). Webb's biography is marred by erroneous evidence and inexact transcriptions of family letters. Crosfield's biography is reliable and includes carefully transcribed letters of the Fells. Moreover, Crosfield was the first biographer of Fell to suggest the connection between Fell and Spinoza of Amsterdam in the Quaker missionary effort to Jews in Holland. The best biography of Fell, however, is the study by Isabel Ross. It is a comprehensive treatment of her life, and its chief value lies in its wide array of evidence. However, its value is diminished by an overly biased Quaker approach. Ross, claiming to be a descendent of Fell, gives a very informative account of her life, but is, unfortunately, a veritable exemplar of "ancestor worship" in support of Quaker tradition.

There is a large body of primary evidence that relates to Fell and the rise of Quakerism. The Fell family correspondence is contained in several collections of early Quaker letters and documents including, most importantly, the Swarthmoor, Abraham, and Barclay manuscript collections. The minutes of the Swarthmoor and Kendal men's and women's monthly and quarterly meetings through Fell's death in 1702 yield valuable insights into her life and work. More evidence comes from the eight-volume collection of George Fox's works. Of special significance is a collection of some of Fell's writings; see Margaret Fell's volume of her published works, *A Brief Collection of Remarkable Passages and Occurrances Relating to the Birth, Education, Life, Conversion, Travels, Services and Deep Sufferings of that ancient Eminent and Faithful Servant of the Lord, Margaret Fell, but by her Second Marriage M. Fox* (London: J. Sowle, 1712), hereafter cited as *Works*. Fell's domestic activities have been documented in Norman Penney, ed., *The Household Account Book of Sarah Fell* (Cambridge: Cambridge University Press, 1920). Two newly discovered documents on Fell include *M[argaret] fff[ell]s Answer to Allan Smallwood Dr. priest of Grastock in Cumberland* (1668); see my transcription in "An Unpublished Work of Margaret Fell," *Proceedings of the American Philosophical Society* (December 1986): 424–52; Rawlinson MS, [A book to] *goe abroad only among all Friends ith [in the] Truth [in answer] to severall papers of Margret Foxe . . . toucheing the unjust orders, papers, and illegale proceedings . . . on her behalfe . . . as toucheing my stewardship for her . . . at Forse Fordge [Force Forge] . . .* (1680). I am grateful to Dr. Craig Horle for informing me of the existence of this last document.

5. Concerning the advantages and pitfalls of biography, see Robert Skidelsky, "Exemplary Lives," *Times Literary Supplement* (13–19 November 1987), 1250; see also Bonnie Smith, "The Contributions of Women to Modern Historiography," *American Historical Review* (June 1984): 709–32.

6. Lois G. Schwoerer, *Lady Rachel Russell* (Baltimore: The John Hopkins Press, 1988), xv–xxviii.

7. See Barry Reay, *The Quakers and the English Revolution* (New York: St. Martin's Press, 1985), chap. 1. Melvin Endy, "Puritanism Spiritualism and Quakerism: An Historiographical Essay," in Richard S. Dunn and Mary Maples Dunn, eds, *The World of William Penn* (Philadelphia: University of Pennsylvania Press, 1986). Christopher Hill, *The World Turned Upside Down* (New York: Penguin Books 1972, 1980); *The Experience of Defeat* (New York: Penguin, 1984).

8. The issue of the myth of Quaker beginnings was first raised by Winthrop S. Hudson in "A Suppressed Chapter in Quaker History," *Journal of Religion* 24 (April 1944): 108–18. Hudson argued that early Quakers idealized their own past in the interest of group solidarity and prestige. In recording the origins of their movement, Quaker chroniclers

piously suppressed any evidence that did not portray George Fox as the original founder. See also the Quaker historian, Henry J. Cadbury's response, "An Obscure Chapter of Quaker History," *Journal of Religion* (1944), 201–14.
9. William C. Braithwaite, *The Beginnings of Quakerism*, 1912, 2d ed. rev. by Henry J. Cadbury (1912; Cambridge: Cambridge University Press, 1955, hereinafter cited *BQ*; *The Second Period of Quakerism* 2d ed. rev. by Henry J. Cadbury, 1919; (Cambridge: Cambridge University Press, 1961), hereinafter cited *SPQ*. For further commentary see Richard L. Greaves, "The Puritan-Nonconformist Tradition in England, 1560–1700: Historiographical Reflections," *Albion* (Winter 1985), 469–72.
10. Braithwaite, *BQ*, 134, 162, 185, 199, 541; *SPQ*, 251, 271, 274, 291–92, 401, 452, 454.
11. Luella Wright, *The Literary Life of Early Friends, 1650–1725* (New York: Columbia University Press, 1932), 74; Melvin B. Endy, *William Penn and Early Quakerism* (Princeton: Princeton University Press, 1973) *passim*; Reay, *The Quakers and the English Revolution*, chap. 5; Michael Watts, *The Dissenters* (Oxford: Clarendon Press, 1978), 200; Richard Vann, *The Social Development of English Quakerism 1655–1755* (Cambridge, Mass.: Harvard University Press, 1969; Milton D. Speizman and Jane C. Kronick, "A Seventeenth Century Women's Declaration," *Signs* (1975): 231–45; Hugh Barbour, *Margaret Fell Speaking*, 1976, 5. See also Hugh Barbour, *The Quakers in Puritan England* (New Haven: Yale University Press, 1964), *passim*; Hugh Barbour and J. William Frost, *The Quakers* (New York and Westport, Conn.: Greenwood Press, 1988).
12. Mabel Brailsford, *Quaker Women, 1660–1690* (London: Duckworth and Co., 1915); Keith Thomas, "Women and the Civil War Sects"; Trevor Aston (ed.), *Crisis in Europe*, 1975, 317–40; Maureen Bell, George Parfitt, and Simon Shepherd, eds and compilers, *A Biographical Dictionary of English Women Writers, 1580–1720* (Boston: G. K. Hall & Co., 1990; Hilda Smith, *Reason's Disciples*, 1982, 95; Barbara Kanner, ed., *The Women of England From Anglo Saxon Times to the Present*, 1979, 162; Mary Prior, ed., *Women in English Society 1500–1800*, 1985, 211–45; this contains Patricia Crawford's listing of seventeenth-century women's published writings. Antonia Fraser, *The Weaker Vessel*, 1984, chap. 18; Phyllis Mack, "Women as Prophets During the English Civil War," *Feminist Studies* (Spring 1982). Phyllis Mack, "The Prophet and Her Audience: Gender and Knowledge in the World Turned Upside Down" in Geoff Eleg and William Hunt, eds, *Reviving the English Revolution* (London and New York: Verso, 1988). Mack's study, "Visionary Women: Ecstatic Prophecy in Seventeenth-Century England," is forthcoming. A new biographical study of three seventeenth-century women is Sara Heller Mendelson's *The Mental World of Stuart Women* (Amherst: The University of Massachusetts Press, 1987). This work is a detailed portrayal of the attitudes and

mentality of three highly literate non-Quaker women, one a deeply pious Puritan, who were contemporaries of Fell. It provides a contrast and comparison to Fell and her religiously oriented mental world; note also Hilda L. Smith and Susan Cardinale, eds, *Women and the Literature of the Seventeenth Century* (New York and Westport, Conn.: Greenwood Press, 1990); Joyce Irwin (ed.), *Womanhood in Radical Protestantism*, 1979, 179–88. A new general survey of the early Quaker female experience, which mentions Fell and family, is Christine Trevett's *Women and Quakerism in the Seventeenth Century* (York: Ebor Press, 1991).

13. Lawrence Stone, *The Family, Sex and Marriage* (New York: Harper, abridged edition, 1977), 4–9 and chaps 4 and 5; see Alan Macfarlane's "Review of Stone," *History and Theory* (1979): 103–26. For a good general introduction to the family in the non-separating and non-divorcing society of early modern England, note Stone's recent study, *Road to Divorce* (Oxford: Oxford University Press, 1990) chapters 1–3; Miriam Slater, *Family Life in the Seventeenth Century* (London and Boston: Routledge & Kegan Paul, 1984); Alan Macfarlane, *Marriage and Love in England 1300–1840* (Basil Blackwell, 1986), 154, 174–75, 308, 322–25; see Lawrence Stone's review, "Illusions of a Changeless Family," *Times Literary Supplement* (16 May 1986): 525–26; Susan Dwyer Amussen, *An Ordered Society, Gender and Class in Early Modern England* (Oxford: Basil Blackwell, 1988), 41; Barbara J. Todd, "The Remarrying Widow: A Stereotype Reconsidered," in Mary Prior, ed., *Women in English Society, 1500–1800* (New York: Methuen, 1985), 54–92; Ralph Houlbrooke, *The English Family 1450–1700* (London: Longman, 1984), chap. 8. On the hypotheses concerning the formation, development, and social relations in the early modern English family, see Keith Wrightson, *English Society 1580–1680*, chaps 3 and 4; Roderick Phillips, *Untying the Knot: A Short History of Divorce* (Cambridge: Cambridge University Press, 1991), chapters 2 and 4; Gordon J. Schochet, *The Authoritarian Family and Political Attitudes in Seventeenth-Century England: Patriarchalism in Political Thought* (New Brunswick: Rutgers University Press, 1988. The Fell family did not reflect some of the more radical views of the family expressed during the Interregnum. For instance, Ranter tradition attacked the family structure. According to Christopher Durston, "Ranter theorists carried antinomian ideas to their extreme conclusion by promoting social units of people which were sexually open, to replace the traditional family"; see Christopher Durston, *The Family in the English Revolution* (Oxford: Basil Blackwell, 1988), 12; note also Margaret J. M. Ezell, *The Patriarch's Wife: Literary Evidence and the History of the Family* (Chapel Hill and London: The University of North Carolina Press, 1987) chaps 1–2.

14. Hill, *The Experience of Defeat*, 21.

15. For a discussion of religious sectarianism as the immediate legacy

of the English Revolution, see Gordon J. Schochet, "The English Revolution in the History of Political Thought," in Bonnelyn Young Kunze and Dwight D. Brautigan, eds, *Court, Country, and Culture: Essays on Early Modern British History in Honor of Perez Zagorin*, 1–20.

Chapter 1

1. "The Testimony of Margaret Fox, concerning her late husband, George Fox: together with a brief account of some of his travels, sufferings, and hardships endured for the truth's sake," in *A Journal . . . of George Fox* (London: 1694; reprint, Philadelphia: Vol. I, 1831, 50–51. It is impossible to know if Fell's self-confidence derived from her conversion experience and newfound faith or if it was natural to her because of her class. As of 1652, expressed in her autobiographical statements, she assumed a confident role toward her husband and others. In a letter to him in London in 1652 she wrote: "Be not afraid of Man . . . Greater is he that is in you than he that is in the world." To the General Council of army officers in 1659 she wrote, "You have bin the Instruments of war and the battle-axe in the hand of the Lord . . . And now are you come into their place, where you may execute that which you formerly desired [justice] and suffered for . . . now quit yourselves like men"; see *To the General Council and Officers of the Army* (London: Thomas Simmons, 1659), 2. In 1660 she wrote an epistle to Major-General Thomas Harrison in which she instructed him to read in the "Fear of the Lord," telling him that the day of his "visitation" was at hand; see Fell, *Works*, 654. Fell did not display a "weaker vessel" image or self-deprecating manner in any of her encounters with men or women. She did not write apologies for her weaker-vessel status as a woman, and she gave it only lip service in her tract, *Women's Speaking Justified* (London: 1666).

2. This chapter is mainly a conflation of the two autobiographical sources: "A relation of Margaret Fell, Her Birth, Life, Testimony and Sufferings for the Lord's Everlasting Truth in her Generation" (1690) included in her *Works*, 1–14; and "The Testimony of Margaret Fox, concerning her late husband . . . ", appended to the first edition of *A Journal . . . of George Fox* (1694), 49–58. Abbreviated selections of Fell's two autobiographical sketches are contained in Hugh Barbour's *Margaret Fell Speaking* (Pendle Hill Pamphlet, 1976), to which I am indebted partially in this portrayal. Barbour also cites a third source in the Spence MSS III, 135, which is a less polished introduction to some letters. Further references to the above autobiographies will be designated "Relation" and "Testimony." It is important to note that Fell's autobiographies fulfilled the function of stressing her conversion and life-long religious motivation. Two other early visitors to the Hall, who came on the heels of George Fox and who were known to Fox,

were James Naylor and Richard Farnsworth. Both men, who became lifelong friends of Margaret, were Seekers and then, "Children of the Light," as the first Quakers were called. See H. Larry Ingle's article, "A Letter from Richard Farnsworth, 1652," *Quaker History* v. 79 (Spring 1990), 35–38 for a portrayal of the first encounters of these four early Quaker leaders of the North country.

3. "Testimony," 51; "Relation," 2–3. For an in-depth study of the culture of hospitality in seventeenth-century Britain, see Felicity Heal, *Hospitality in Early Modern England* (Oxford: Clarendon Press, 1990), chp. 5.
4. "Testimony," 51; "Relation," 2–3.
5. "Testimony," 52–53.
6. Regrettably, I have lost this citation.
7. "Relation," 3.
8. "Testimony," 54–55.
9. "Relation," 4.
10. "Relation," 5–6.
11. "Relation," 8; Margaret Fell's trial was recorded by Quaker witnesses and published in her *Works*, 276–90. On the persecution of Quakers in the reign of Charles II, see Ronald Hutton, *Charles the Second: King of England, Scotland, and Ireland* (Oxford: Clarendon Press, 1989), 457.
12. Fell, *Works*, 328–29.
13. Fell, *Works*, 331–50. Abbreviated versions of *Women's Speaking Justified* are included in most source books of religious sectarianism. See Joyce L. Irwin, *Womanhood in Radical Protestantism, 1525–1675* (New York: Edwin Mellen Press, 1979), 179–88. Early tracts written by Quakers in defense of women preaching include Richard Farnsworth, *A Woman Forbidden to Speak* (1655); George Keith, *The Woman Preacher of Samaria* (1674); Anne Whitehead and Mary Elson, *An Epistle of True Love, Unity and Order* (1680); see also William Mather, an erstwhile Quaker who wrote against women's meetings and ministry, *A Novelty, or a Government of Women* (1694?). For some further comment, see Elaine Hobby, *Virtue of Necessity; English Women's Writings* (Ann Arbor: University of Michigan Press, 1989), 43, 45; David Latt, ed. *'Women's Speaking Justified' (Margaret Fell 1667), Epistle from the Women's Yearly Meeting at York (1688), A Warning to All Friends (Mary Waite, 1688)*, Los Angeles, 1979.
14. See chapter 7.
15. Fell, *Works*, 240, 350; *M[argaret] f[ell]s Answer to Allan Smallwood Dr. priest of Grastock in Cumberland* (1668); see my "An Unpublished Work by Margaret Fell," 424–52.
16. Bonnelyn Young Kunze, "An Unpublished Work by Margaret Fell," 424–52; see also Christopher Hill, *The World Turned Upside Down*, 103–4.
17. "Relation," 8.
18. "Testimony," 56.

19. "Testimony," 57.
20. "Relation," 10; Hugh Barbour has given a useful comparison of monetary value for the period. It cost £5 for an Atlantic crossing; a horse cost about £2, and two oxen cost about £11. Also a farm laborer could be hired for one penny per day. Barbour, *Margaret Fell Speaking*, 18; see also ABSF, *passim*.
21. See Barbour, *Margaret Fell Speaking*; "Relation," 10–12. The magistrates advised Margaret repeatedly to go home and remain there, the proper place for a married woman. In August 1669, Col. Kirkby of Lancashire was advising Fell to stay at home and stop going abroad. Gibson MSS II, 71.
22. Fell, *Works*, 531–32.
23. PWP, #282.
24. Box Meeting MSS #43; Fox's *Journal* (Ellwood edn.), 49.
25. Quoted in Ross, 37; original in Spence MSS III, 135, 124.
26. "Testimony," 49.
27. Quoted in Ross, 379–80; original in Portfolio 25 and 67, FHL.
28. Fell, *Works*, "Some of Margaret Fell's Dying Sayings," unpaginated.

Chapter 2

1. Barry Levy, *Quakers and the American Family* (New York: Oxford University Press, 1988), 3–52; Bruce G. Blackwood, *The Lancashire Gentry and the Great Rebellion 1640–1660* (Manchester: Chetham Society, 1978), 1–4; Barry Reay, *The Quakers and the English Revolution* (New York: St. Martins, 1985), 55–72; see also the VCHL for a detailed account of the economy and topography of Furness.
2. Blackwood, *The Lancashire Gentry*, 2–4.
3. Levy, *Quakers and the American Family*, 13.
4. VCHL, vol. 8, 351–54, 361–62; Ralph A. Houlbrooke, *The English Family, 1450–1700* (London: Longman, 1984), 23; Henry Barber, *Swarthmoor Hall and Its Associations* (Ulverston: 187?), 24–33; Ross, *Margaret Fell, Mother of Quakerism* (London: Longman, 1949), chap. 1. In this chapter I am indebted to Isabel Ross for her fine chronological chart giving details of the important events in Fell's life. This study includes an updated version of Ross's chart.
5. See *Dictionary of National Biography* for details of the various offices held by Judge Thomas Fell. See also Ross, *Margaret Fell, passim*; VCHL, 8:354; Barber, *Swarthmoor Hall and Its Associations*, 24–33. Judge Fell's funeral took place in October 1658 at St. Mary's, Ulverston. According to local church ritual, it was by torchlight, a tradition observed for notable persons. See C. W. Bardsley, ed. *Registers of the Ulverston Parish Church* (Ulverston; James Atkinson, 1886), 154. Between 1645 and 1646 with the expulsion of Royalists from the House of Commons, new elections were held and many Independents were elected, of which Thomas Fell was one. These MPs were known

as recruiter members. Fell remained an MP until 1647 after which time he was reported absent without cause and fined £20. Judge Fell was sympathetic to the Independents and, as lord of the manor of Ulverston, he may well have been instrumental in calling the Independent minister, William Lampitt, to the rectorship of his parish St. Mary's, Ulverston. William Lampitt, referred to as "Priest Lampitt" by George Fox in his *Journal*, encountered Fox for the first time when Fox made his somewhat disdainful "steeplehouse" entry in June 1652. Ross, 11–12; Fell's "Testimony," George Fox *Journal*, Thomas Ellwood, ed., 1891, reprinted in *Works of George Fox* (1975), I, 50.

6. The Committee of Sequestrators changed its name in 1643 to the Committee for Compounding. The committee administered royalist lands that could be restored on payment of a fine. Acting as a judicial tribunal in matters of royalist delinquency and land ownership, its powers lay "outside the ordinary legal structure" and there was no means of legal appeal on its decisions. MPs and non-MPs served on this committee. G. E. Alymer, *The State's Servants*, 13–15, 26, 61; Bruce Blackwood, *The Lancashire Gentry* . . . , 94–6, 141.

7. Harlton Hall was located in the township of Halton, south of the Sands in Londsdale Hundred, near Lancaster. A Thomas Carcus of Halton Hall was rector of Halton parish east of Lancaster from 1660–1676. It was an ancient manor with substantial manorial privileges, royalties, rents, fines, and services accompanying them. The owner of Halton Hall held the advowson for the rectory which was worth £20 p.a. In 1743 the Halton Hall estate had 140 acres of land. *Calendar of Committee of Compounding*, Part II, 1172; Part I, 481; Part III, 1846; G. E. Alymer, *The State's Servants*, 60–61.

8. Blackwood, *The Lancashire Gentry* . . . , 94, 12. This is in accord with William Hunt's findings on gentry social stratification in early seventeenth-century Essex. "Big farmers," according to Hunt's categories, were those owning or leasing more than fifty acres of land. The only stratum in Hunt's categorization that superseded big farmers were peers and gentry who included landlords, lawyers, civil servants, high ecclesiastics. Hunt maintains that the majority of justices of the peace came from the middling gentry, which corresponds to Fell's situation. William Hunt, *The Puritan Moment: The Coming of Revolution in an English County* (1983), 21–22, see Table 5. See Margaret Spufford, *Contrasting Communities* (1974), 38. A big farm by Cambridgeshire standards was one of 130 acres or more and a small farm size averaged under forty acres or a yardland. See also Perez Zagorin, *The Court and the Country*, 1971, 27; Miriam Slater, *Family Life in the Seventeenth Century: The Verneys of Claydon House*, 1984, chap. I; Lawrence Stone, *The Crisis of the Aristocracy*, 1965, chap. II.

9. Mary Prior ed., *Women in English Society 1500–1800* (1985). William Hunt, *The Puritan Moment* (Cambridge: Harvard University Press, 1983), 220–27. See essay by S. H. Mendelson, "Stuart Women's Diaries . . . ," 190 and 199; Ivy Pinchbeck, *Women Workers and*

the Industrial Revolution (1930, 1969, 1981); 1–26. Women's work on the farm was very necessary to the farm economy. For example the Quaker, William Stout, wrote that his father farmed twenty-four acres of land in the seventeenth century in the Lake District not far from Swarthmoor. Stout's mother was "not onely fully imployed in housewifry but in dressing their corn for the market, and also in the fields of hay and corn harvests along with our father and the servants," quoted in K. Wrightson, *English Society*, 93–94. See also Houlbrooke, 107–108. There was a long tradition in western Europe, of aristocratic women running estates in their husbands' absences. See Caroline Walker Bynum *et al.*, eds, *Gender and Religion*, 1986, 257; see also Michael R. Best, ed., *The English Housewife* (McGill-Queens University Press, 1986).

10. Swarthmore MS 1/166, Tr. 3/31; Swarthmore MS 4/141, Tr. 1/716.
11. Ross, *Margaret Fell*, chap. 8; Swm MS I; quoted in the Jermyn Transcripts #47, Haverford College Quaker Collection.
12. Fell, *Works*, 2, 4.
13. Fox's *Journal*, Cambridge edn., 71.
14. Preface to William Caton *Journal*; quoted in Ross, *Margaret Fell*, 117–18.
15. Abraham MSS I; quoted in Ross, *Margaret Fell*, 119–20.
16. Ibid,; cf. Houlbrooke, *The English Family*, 102–5.
17. Spence MSS 3/49; quoted in Ross, *Margaret Fell*, 96–97. Ross admitted but minimized this discord. Thomas Fell's attitude toward the Quakers is further detected in a vignette recorded in William Sewel's 2–volume chronicle, *The History of the Rise, Increase and Progress of the Christian People Called Quakers* (London: James Philips, 1795). This history was first printed in Amsterdam in 1700. I have taken this example from the 1795 edition. Sewel attributed the story to one of Judge Fell's daughters, who claimed that upon one occasion her father arrived home with his servants and found the barn full of the horses of strangers. Margaret cleared the stable to make room for her spouses's entourage and his remark to her was, "this was the way to be eaten out [of hay], and that they themselves should soon be in want." She responded "in a friendly way, that she did not believe they would have [any] the less for that." Sewel concluded that the harvest of hay that year was in greater abundance than usual and "thus the proverb was verified, that charity doth not impoverish." Sewel, *History of the Rise*, 1:142.
18. Spence MSS III, 42, 43; quoted in Ross, *Margaret Fell*, 25–26; Miriam Slater, *Family Life in the Seventeenth Century* (London: James Philips (Sewel), 1795), *passim*; John Gillis, *For Better, For Worse* (New York: Oxford University Press, 1985), 89–90. In May 1653 Margaret wrote Colonel West two letters appealing to him to heed the "Light of Jesus" that shined in his conscience, to denounce idolaters, and to be willing to suffer as one of God's true witnesses. West had requested that Margaret use her influence to restrain Friends, but

she seemingly had no intention of doing so. She chastised West for his compliance with those who criticized Quaker activity and called on him to separate from deceivers and sinners and suffer patiently with God's witnesses. Considering her persuasive approach toward West to become a Quaker, we may logically presume she placed similar or greater pressure on her own spouse. See Spence III, 3/29–30, 3/31, Horle Tr.

19. Slater, *Family Life*, 109.
20. Ibid. See Gillis, *For Better, For Worse*, 21,37; Gillis argues that in landed families parental permission and patriarchal power were not "nearly so great as many of the 'conduct books,' most of them puritan in origin, should have suggested." He does concede, however, that "patrician families" did control their children's marriage plans, which essentially substantiates Slater's findings. Another recent study of seventeenth century family relationships offers a similar view. Ralph Houlbrooke points out that financial dependence of gentry children gave parents greater control over the marriage choices of their offspring, and that their daughters enjoyed very little personal freedom of choice in marriage partners. Houlbrooke adds, however, that there was a gradual development toward greater parental readiness to take into account the wishes of their offspring in marriage choices. Alan MacFarlane has marshalled demographic, literary, diary and epistolary evidence to support the argument that the mutual consent of prospective marriage partners had a long and deeply respected tradition in England. He claims that the chief motives of the "Malthusian marriage system" in England were of social and economic nature. The stress on romantic love and mutuality in marriage undermined and indeed held in check parental coercion in matchmaking, although parental approval of marriage partners was a hoped-for ideal. Macfarlane contends that the English system of marriage presupposed equality, individual freedom of choice and consent – all features that accommodated the rise of capitalism. Macfarlane's evidence does not contradict what we do know about the Fell daughter's choices in marriage, to be addressed later in this chapter. It is apparent that the Verney daughters suffered greater hardships in their quest for marriage partners due, at least in part, to their father's untimely death and their eldest brother's insensitivity. Ralph A. Houlbrooke, *The English Family . . .* , 215, 237, 140, 142, 146; Alan Macfarlane, *Marriage and Love in England, 1300–1840*, 154, 175, 293–94, 322–25; Slater, *passim*.
21. Houlbrooke, *The English Family . . .* , 118–19.
22. Ross, Appendix 15, 241; the Spence MSS, contains approximately 200 letters of Fell and her family covering a period of family separation between 1660–61, when Fell was living in London. The Abraham MSS contain family letters written mostly between 1680 and 1702. They cover a second period of separation between mother and married daughters and sons-in-law. Unfortunately there are no letters preserved from the male side of the Fell family. A number of these letters have

been transcribed in Helen G. Crosfield, *Margaret Fox of Swarthmoor Hall* (London: Headley Bros., 1913), chaps vi and xii. See Geoffrey Nuttall, *Early Quaker Letters*, 1952, for a calendar of letters from the Swarthmoor MSS that touch on these two periods of Fell family life.
23. Ross, *Margaret Fell*, 344–45.
24. SWM, MSS I, 373; see Barry Reay, "Popular Hostility Towards Quakers in Mid-Seventeenth-Century England," *Social History* 5 (1980): 396; Braithwaite, *B. Q.*, 372.
25. Her letters of advice to the newly returned King, in which she reminded him of his promises of Breda, are contained in her *Works*, 17–19.
26. Spence III/73. quoted from Craig Horle Transcript.
27. Ross, *Margaret Fell*, chap. 10.
28. Spence III/69,70; C. Horle Tr; Crosfield, *Margaret Fox*, 72–73.
29. Ibid., 74, 78–79.
30. Ibid., 75, 85; Spence III/73.
31. Crosfield, *Margaret F*, 75.
32. Spence III/73.
33. Crosfield, 85.
34. Crosfield, 85.
35. Crosfield, 86.
36. Ross, 141.
37. Crosfield, 90–92.
38. Cf. to the Verney sisters who were left orphaned and unwed. Slater, *The Verney of Claydon House . . . , passim*.
39. Fell, Works, 5; Ross, chap. 10; N. H. Keeble, *The Literary Culture of Nonconformity in Later Seventeenth-Century England* (Athens, Georgia, 1987), chap. 1.
40. For details concerning the status, independence, and remarriage patterns of widows, see Barbara Todd, "The Remarrying Widow: A Stereotype Reconsidered," in Mary Prior, ed., *Women in English Society*, 54–92; Houlbrooke, 207–15.
41. Ross, *Margaret Fell*, 138, 76. Ross's estimates are as follows:

	Birth	Year of Marriage/Spouse	Age	Offspring
Margaret, Jr.	1663?	1662 to John Rous	29	10
Bridget	1635?	1662 to John Draper	27	–
Isabel	1637?	1664 to William Yeamans	27	3
		1689 to A. Morrice		
George	1638?	1660 to Hannah Potter	22	2
Sarah	1642	1681 to William Meade	39	1
Mary	1647	1668 to Thomas Lower	21	10
Susannah	1650?	1691 to William Ingram	41	–
Rachel	1653	1683 to Daniel Abraham	30	4

On late marriage patterns in seventeenth-century England see Keith Wrightson, *English Society, 1580–1680* (New Brunswick, N.J.: Rutgers University Press, 1982), 68; Gillis, *For Better, For Worse*, 111; see also

Alan MacFarlane, *Marriage and Love in England, 1300–1840* (Oxford: Basil Blackwell, 1986), and Miriam Slater, *Family Life passim*; see also footnote 42 below.

42. See a copy of Fell's will in Ross, *Margaret Fell*, appendix 10; Ralph Houlbrooke, ed., *English Family Life* (Oxford: Basil Blackwell, 1988), 22–27; Crosfield, *Margaret Fox*, 206. Details of George and Hannah Fell's spousal relationship are unknown. What is known is that they took an adversarial position to Margaret Fell, and after George Fell's death in 1670 litigation commenced between Hannah and Margaret Fell over the Swarthmoor estate. What little evidence we have suggests that George Fell's son and daughter never established a warm relationship with their grandmother Fell, the Fell aunts, or their first cousins. The Fell sisters did have some social contact with their sister-in-law while she lived at Marsh Grange in the 1670s. However, once Hannah Potter Fell moved away from the district, there is no evidence that this contact was maintained. Moreover, George Fell apparently did not take an interest in his sisters' marriages, and having left Swarthmoor in 1653, appears to have had only infrequent contact with mother and sisters, coming to exhibit only a nominal family allegiance.

43. According to Miriam Slater, *Family Life*, 84–90, the Verney sisters encountered demeaning hardships as adult unmarried women. It was a worry for the entire Verney family until each sister wedded and thus was removed from the family expense account. See also Houlbrooke, 41–43, 52, 70–72. For a detailed discussion on the late marriage patterns of northern Europe and England in the early modern period, see MacFarlane, *Marriage and Love*, 216, 321; Gillis, *For Better, For Worse*, 11; Vivien Brodsky Elliott, "Single Women in the London Marriage Market: Age Status and Mobility, 1598–1619," in R. B. Outhwaite, ed., *Marriage and Society: Studies in the Social History of Marriage* (New York: St. Martin's Press, 1981), 81–100; J. Hajnal, "European Marriage Patterns in Perspective," in David V. Glass and D. E. C. Eversely, eds, *Population in History: Essays in Historical Demography* (1965), 101–11; E. A. Wrigley and R. S. Schofield, *The Population History of England, 1541–1871. A Reconstruction* (London: Edward Arnold, 1982); Peter Laslett, *The World We Have Lost* (New York: Charles Scribner's Sons, 1965), chap. 4. In Hawkshead parish, Lancashire, a town not far from Swarthmoor, the average age of marriage for men in the period 1600–49 was 27.8 years, according to Keith Wrightson's study, *English Society*, 68.

44. Crosfield, *Margaret Fox*, 214.

45. Ross, *Margaret Fell*, 350–51. The loans were £50 lent to a Lancaster shopowner, £40 to someone on a parcel of land, and a £10 loan that was undesignated. Together with interest, this indicates that the Fell sisters probably had considerable financial security in their own names. The original letter is found in Abraham MSS/26.

46. Ross, *Margaret Fell*, 344–45.

47. Ibid., 179, 181–82; in June 1664 Mary wrote her mother in Lancaster:

> My duty and very dear love is freely given and remembered unto thee, as also my very dear love is to dear George Fox.
> This is chiefly to let thee understand that yesterday sister and I were at Whitehall, where we spoke with the King and told him that if he would please to signify something to the judges, before they went their circuit, to release you, otherwise it would be past, for the time drew very near of the Assizes. He said he would release you, if we would promise you would not go to meeting. Sister said we could make no such engagements, for the meeting it hath been kept many years and never hath done any harm. He said, "Cannot your mother keep within her own family, as she may have five more, that she must have such tumultuous meetings." We said, "she hath no such meetings, they are only her neighbors that come." The King said there were some Quakers in the last plot [Kaber Rigg plot]. Sister said that could not be proved. . . . I was there about a week since and told the King that now the Assizes drew very near and said if he did not do something for thee, they would run thee into a praemunire and get thy estate from thee and thy children: and I desired him to take it into consideration. He was then very loving to me and as though he pitied me . . . and he said, "They shall not have her estate from her . . . ." And I spoke to Prince Rupert and desired him to put the King in mind of it, and he said he would what he could in it. . . . Prince Rupert has always been very loving to friends and hath often spoke to the King about you. . . . "
> I am thy dutiful and obedient daughter, Mary Fell

The original letter is in Gibson MSS, V/55.

48. Ross, *Margaret Fell*, 148–50. As far as is known of all the Quaker correspondence that was preserved in this period only one letter refers to Bridget's death. That is a letter from William Caton to Bridget's mother in April 1663 from Amsterdam:

> Before my return out of England I had heard the sad and unwelcome news concerning the death and departure of dear Bridget, which thou may be assured came exceeding near me, but what shall I say? We must go after her, for here we have no continuing city. Oh that we always therefore may be found in that faith, life, power and spirit, through which the entrance is made into the eternal inheritance.

49. *JFHS* (XII, 1915), 53–58.
50. Spence MSS 3/188; Houlbrooke, 132–34; Ross, 353.
51. H. Crosfield, 203–4.
52. Crosfield, 213.
53. Ross, 363, 159.
54. Jermyn Tr. #16, Haverford College Quaker Collection.
55. Stone, 88–89; Stone's thesis has been much debated in the secondary literature on the family. See also Philippe Aries, *Centuries of Childhood* (1962). Ralph Houlbrooke gives a different explanation of the apparent

lack of grief expression over death in the early modern family. The traditional Catholic means of dealing with bereavement such as funeral rituals, wakes, intercessory prayers, funeral monuments, and mourning were abandoned during the Protestant Reformation. These ritual expressions of mourning began to reappear within Protestantism by the seventeenth century. However, the Quaker way of death continued to avoid external forms.

56. William Penn expressed the Quaker religious view of death and grieving when he wrote about the death of his friend Thomas Loe and about the separation of the body and soul at death. As Penn expressed it, Loe had "left Ye Body, and is ascended far above all visible and Created Things to ye full possession of ye pure Eternal Rest and Sabbath of ye holy God." For Penn, the body had finished its earthly course and could not share in the soul's afterlife. The Quaker custom of avoiding mourning costume and gravestones reflected the Quaker idea that men and women, as fundamentally spiritual creatures, must mourn in their souls what the body was insensible to, for the body was not the "essential part of man." Melvin Endy, *William Penn and Early Quakerism* (Princeton, N.J.: Princeton University Press, 1973), 199.

57. Crosfield, *Margaret Fox*, 214–15; see Ross, *Margaret Fell*, 314–18.

58. Crosfield, *Margaret Fox*, 228.

59. William Caton, *Journal of the Life of William Caton* (London: 1689); quoted from *JFHS* 29 (1932): 51–52. Whereas George was sent away to school, it seems that the daughters received their education at home. All seven daughters were literate. Sarah was the most gifted in written expression, while Susannah was the least articulate. See Houlbrooke, *The English Family*, 150–53.

60. Caton, *Journal*, 53; SWM MSS, Tr. I/535; *JFHS* 29 (1932): 30. The estrangement of Margaret Fell and her son has been discussed at some length. Norman Penney, "George Fell and the Story of Swarthmoor Hall," *JFHS* 29 (1932): pages 30 (1933); pages 31 (1934): pages. Ross, *Margaret Fell*, 220–29; Alfred W. Braithwaite, "The Mystery of Swarthmoor Hall," *JFHS* 51 (1965): 22–29.

61. Spence MSS 3/60; *JFHS* (29, 1932), 55–6.

62. *JFHS* (29, 1932), 56; quoted from SWM MSS 3/11.

63. Houlbrooke, chap. 9 and 229. Whether George Fell was designated as heir in some prior document rendering mention of it in Thomas Fell's will redundant, is unknown. There is no evidence in the will of a specific prior document relating to George Fell's inheritance to indicate this. However, subsequent letters and actions on the part of mother and son suggest that either the rule of primogeniture was assumed in the Fell family or George Fell attempted to prove that prior testamentary instruments, which gave him rights in property, had not been properly revoked. For definition of dower, devise, and jointure, I draw from the study by Susan Staves, *Married Women's Separate Property in England, 1660–1833* (Cambridge, Mass.: Harvard University Press, 1990), 235 and 238.

64. CSPD, 1659–60, 566; 1660–61, 50; 1664–65, 161.

65. Alfred Braithwaite, *JFHS* (1965) 25–27; cf. H. Larry Ingle and Jaan Ingle, "The Excommunication of George Fox," *JFHS* 56 (1991), 71–77.
66. Penney, *JFHS* (1933), 35; Ross, 221, 37.
67. Penney, *JFHS* (30, 1933), 33–34.
68. George Fell's last will and testament was dated 17.8.1670.
69. Thirnbeck MSS Transcript, #10. Without any evidence of prior testamentary instruments, it is difficult to know what George Fell was supposed to inherit. However, his subsequent arguments suggest that a prior will may have left him property upon which he then based his claims.
70. CSPD 1671, 171; Ext. 329.
71. Penney, *JFHS* (31, 1934), 28; quoted from Dix MSS.
72. Ross, 227; Penney, *JFHS* (1934), 33. A letter dated 15 March 1700 from George Whitehead to Margaret Fell indicated that Margaret had written him asking for information about taxes on the estate. Whitehead searched statutes under Charles II for information on fee farm rents but could not find out if the said taxes included tithes to the Crown. Fell paid her taxes without complaint, but steadily refused to pay tithes. See Gibson MS, #2, 149; Peter Brock, *The Quaker Peace Testimony 1660–1914* (York, England: William Sessions Book Trust, 1991) 184.
73. Houlbrooke, 232.
74. Miriam Slater has found that each member of the Verney family, especially the younger brothers, competed internally for the family resources and for greater favor with the eldest brother, Sir Ralph Verney. At times one member would secretly deprecate another in the hope of personal gain. There is no evidence of long-term in-fighting in the Fell family. Therefore, considering the seventeenth century English patronage system and the way heirs of landed families were known to behave, George Fell probably was not particularly unusual. See *Family Life in the Seventeenth Century: The Verney's of Claydon House.*
75. Cambridge *Journal* II, 153; according to Ross, the Leeds Journal, 308, contains a pre-nuptial agreement between Fell and Fox, that he would not involve himself with his wife's estate. See Ross, 215.
76. Cambridge *Journal* II, 154.
77. ABSF, *passim.*
78. For an analysis of Fox's Worcester imprisonment, see Craig Horle, "Changing Quaker Attitudes Toward Legal Defense: The George Fox Case, 1673–75, and the Establishment of Meeting for Sufferings," in J. William Frost and John M. Moore, eds, *Seeking the Light: Essays in Quaker History*, 1986, 17–40.
79. See *Biographical Dictionary of British Radicals in 17th Century England*, I, 274; SPQ, 262–63; Ross, 219, concerning an epistle Fox wrote on marriage which was disliked by the Second Day Morning Meeting, and was destroyed.
80. Henry J. Cadbury, "Intercepted Correspondence of William Penn," *Pennsylvania Magazine of History and Biography* 70 (1946) 354–56;

quoted from CSP Ireland (1669–70), 134–35; Ross, *Margaret Fell*, 218.
81. *JFHS* 11 (October 1914): 155; Cadbury, "Intercepted Correspondence of William Penn," 354; Fox, Cambridge *Journal*, II, 424.
82. Quoted in Cadbury, "Intercepted Correspondence of William Penn," 355. Richard Vann and Davis Eversley have interpreted the Fell-Fox marriage as a spiritual union as she was beyond the child-bearing age. It was a model of male-female love that was "spiritual and inward." Vann and Eversley maintain that once marriage was not just valued for procreation, "a crucial mental boundery" had been pierced which would ultimately lead to the "sanctioning of sex without reproduction." Richard Vann and David Eversley, *Friends in Life and Death*, Cambridge: Cambridge University Press, 162.
83. Slater, *Family Life*, 104–5; Prior, ed., *Women in English Society, 1500–1800* (London: Methuen, 1985), 74–79; Houlbrooke, *The English Family*, 209–10.
84. SWM MSS Tr I/568j.
85. SWM MSS, Jermyn Tr/63, 69, 70, 71, Haverford College Quaker collection.
86. Abraham MSS; see *JFHS* 11 (October 1914): 183–84.
87. SMMM, 13.11.1673, 81; *ABSF*, Penney's notes, 556–67. An entry appears in the Quaker Registers of births, marriages, and deaths at FHL for Edward Herd of Ulverston, who died in 1701.
88. SWMMM, *passim*.
89. Donald R. Kelley, *The Beginning of Ideology: Consciousness and Society in the French Reformation* (New York: Cambridge University Press, 1981), 93, 94.
90. Ibid., 96; Swm MS VI, 51; quoted in Barry Reay, "Popular Hostility Towards Quakers in Mid Seventeenth Century England," *Social History* 5 (1980): 396.
91. Alice Curwen, *A Relation of the Labour, Travail and Sufferings of That Faithful Servant of the Lord* (1680), 39.
92. K. Thomas, "Women and the Civil War Sects," in Trevor Aston, ed., *Crisis in Europe* (London: Routledge & Kegan Paul, 1965), 330; for an example of the Puritan female piety of Lady Joan Barrington, see William Hunt, *The Puritan Moment* (Cambridge: Harvard University Press, 1983), 221–23. This subject will receive further attention in Part IV.
93. Kelley, *The Beginning of Ideology*, 93.

Chapter 3

1. On the subject of widow's portions, note Lloyd Bonfield, "Marriage Settlements, 1660–1740: The Adoption of the Strict Settlement in Kent and Northamptonshire," in R. B. Outhwaite, ed., *Marriage and Society* (New York: St. Martin's Press, 1981), 101–16. Bonfield says that the

strict marriage settlement was a means of conveying the patrimony to the next generation, as opposed to entail, which attempted to keep intact the patrimony for generations. By the early seventeenth century, common law courts were beginning to reject entail. Bonfield contends that by the 1680s the strict settlement was the most popular form of marriage settlement for landed families, although he adds that a small proportion of the "lesser gentry" did not settle this way; see also A. W. B. Simpson, *An Introduction to the History of the Land Law* (Oxford: Oxford University Press, 1971); Ralph Houlbrooke, *The English Family, 1450–1700* (London: Longman, 1984), chap. 9; see also Barbara J. Todd, "The Remarrying Widow: A Stereotype Reconsidered," in Mary Prior, ed., *Women in English Society, 1500–1800* (London: Methuen, 1985).

2. Norman Penney, ed., *The Household Account Book of Sarah Fell* (Cambridge: Cambridge University Press, 1920), hereafter cited ABSF.

3. See Carol Shammas, "The World Women Knew," in Richard S. Dunn and Mary Maples Dunn, WWP (Philadelphia: University of Pennsylvania Press, 1986), 99–115; Shammas also used the ABSF as a source for her study. She describes the Fell household in terms of its production and sexual division of labor. Her conclusion is that occupational segregation by sex was deeply rooted in seventeenth-century northwest Lancashire as the Fell jobs were identified predominantly as masculine or feminine.

4. Kirkby distrained Margaret Fell's cattle on at least one occasion in 1684. ABSF, x–xvi, 18, 23, 27, 111, 141, 203, 221, 525, 538, 544, 554, 564; Ross, *Margaret Fell*, 325; Keith Wrightson, *English Society, 1580–1680* (New Brunswick, N.J.: Rutgers University Press, 1982), 52–57.

5. Alan Macfarlane, *The Family Life of Ralph Josselin*, 157; Wrightson, *English Society*, 45–48. Wrightson suggests that the social significance of kinship was not only dependent on propinquity but on the personal degree and depth of the individual kin relationships that varied from one person to another. Wrightson points out that wills produced by middling rank persons often showed small interest in kin outside the closest ones. This was true of the wills of Thomas and Margaret Fell; ABSF, *passim*.

6. Caton MSS, I and II, 71, 89, 104; III, 425; Swm. Ms. I, 161, 310; quoted in Ross, *Margaret Fell*, 155.

7. "State Papers of John Thurloe," VIII (4), *JFHS* (October 1911): 165; Maria Webb, *The Fells of Swarthmoor Hall* (London: 1867), 294; Shackleton Colln. letter, December 19.10.83; Hugh Barbour and A. O. Roberts, eds, *Early Quaker Letters* (Grand Rapids, Mich.: Wm. B. Eerdmans Publishing Co., 1973), 384. Abraham MSS, 12.8.1673; AR Barclay MSS, 14.ix.1657; *JFHS* 8 (October 1911): 165; Gibson MSS II, 71, letter dated 24.6.69.

8. Fell, *Works*, 534; Braithwaite, *SPQ*, 517–19; ABSF, 293, 321.

9. Swarthmoor MS I, 373; Ross, *Margaret Fell*, 24–27, 261, 316, 409.

Ross claims they traveled more than 1000 miles between May and August 1663; Wrightson, *English Society*, 41; Ann Kussmaul, *Servants in Husbandry in Early Modern England* (Cambridge: Cambridge University Press, 1981), Part 2.

10. I define "family" and "household" in the same sense that Lawrence Stone uses it in *The Family, Sex and Marriage in England, 1500–1800* (New York: Harper & Row, 1979), 28–29. Louise Tilly and Joan Scott use it similarly in *Women, Work and Family* (New York, 1978) 5, as does Miriam Slater, *Family Life in the Seventeenth Century* (London: Routledge & Kegan Paul, 1984), 26. Family means the kin group living in a single household, usually parents and their children. Kussmaul points out the closeness between servants and masters. She extends the meaning of "family" to include both kin and servants living under one master's authority, claiming that to ignore this concept of family is to fail to comprehend the "early modern 'mentalite' and the meaning of 'family' before 1700," Kussmaul, *Servants in Husbandry in Early Modern England*, 7–8. This notwithstanding, for clarity in telling the story of the Fells and their neighborly network, "family" means the kin relations at Swarthmoor and Marsh Grange and "household" designates both the Fells and their servants living under their roof.

11. VCHL, vol. 8, 354; Ivy Pinchbeck, *Women Workers and the Industrial Revolution*, (London: Virago, 1930), 1–26, especially 7–9; Prior, *Women in English Society*, 190; see also Arthur Searle, *Barrington Family Letters*, vol. 28 of Camden Fourth Series (London: RHS, 1983). There is insufficient evidence to know whether or not it was unusual for a woman to act both as the family accountant and bailiff.

12. The accounts are not precise on the number of servants at each place. Neither is it possible to determine the exact ages of those servants employed in 1676. The household servants' wages varied, suggesting an age range probably comparable to what Ann Kussmaul has found: an age span of 15 to 24 years. See also Marjorie K. McIntosh, "Servants and the Household Unit in an Elizabethan English Community," *Journal of Family History* 9 (1984): 3–23; Miriam Slater, *Family Life*, chap. 5.

13. Kussmaul believes that servants and laborers were in competition with one another. Servants constituted about 60 percent of the unmarried population ages 15–24. Day laborers were adults hired on a day-to-day basis and were not resident on the owner's estate. Kussmaul's study indicates that wages in the sixteenth and seventeenth centuries generally did not exceed the assessed wages granted by the justices of the peace in county Quarter Sessions. County courts attempted to regulate maximum wages at the local level, as provided by the Statute of Artificers in 1563. Workloads, age, and sex caused wages to vary significantly. Very young servants were sometimes given only room and board for one year without wages. Women were paid less than men, which is evident in the ABSF.

14. ABSF, 261, 263.

15. Alice Clark, *The Working Life of Women* (London:), 204–9; Carole Shammas also refers to Clark's study in reference to mobile female petty retailers. Shammas estimates that nearly 90 percent of the Furness district women were involved in this form of labor; Dunn, *WWP*, 103–4, 107. See also William Hunt, *The Puritan Moment* (Cambridge, Mass.: Harvard University Press, 1983), 25; Joan Thirsk, "Seventeenth Century Agriculture and Social Change," in Paul Seaver, ed., *Seventeenth Century England* (New York: New Viewpoints, 1976). Thirsk analyzes the century-long trends of changing grain and wool prices and the long-term effects it had on the lower social orders; see also Pinchbeck, *Women Workers*, 1–26.
16. ABSF, 98, 120, 171, 83, 537–38, 84, 545, 547, 69; SWMMM, 27–28.
17. ABSF, xxx, 108, 114, 120, 166, 200, 274, 533, 177.
18. ABSF, 18, 30, 56, 72, 518, 169, 174, 93, 107, 111, 113, 287, 137, 149, 211, 217, 220–21, 254, 223, 71.
19. ABSF, 23, 26, 29, 221, 531, 189, 2.
20. ABSF, 331.
21. ABSF, 75, 85, 93, 123, 129, xxxi.
22. The absence of rental entries in the account book in the 1670s could indicate that this ledger was an incomplete account of the transactions of the Fell farm and that the rental accounts were kept separate. However, there is no indication that another account book existed alongside this one. For the year 1676 and for other years in the accounts there are a few entries for "mothers acc't" that indicate the receiving as well as paying of rents for various meadows. Parcels of land referred to as Angsley, Petty Croft, Round Leva Heads, Leva Heads, "dodgson wife close," as of 1674, were either rented, cultivated, or left fallow for "beast grasse." In October 1676 Sarah paid £1.10s for a close of grasse at Rownheade, from "mothers account," and another pound two days later marked "pd. in full." In the same month she paid £1.8s6d for "oxen grasse of Mothrs at Sandscale." Also in the same month, Sarah received two payments from John Cowell totaling £12, "yet hee pays Mother pr sistr ffells order." This referred to rent p.a. of Marsh Grange property, and Hannah Fell, widow of George Fell, was acting as the intermediary. Assuming that the ABSF records the total p.a. income of Swarthmoor for the 1670s, there is another possible explanation for the absence of rental income. It may have been that the renters or tenants of Swarthmoor gave farm service and/or provisions in lieu of rents. Considering the Fell's charitable behavior and Quaker network business practices to be discussed later, it would not have been out of character for Margaret Fell and her daughters to accept farm service from their Quaker neighbors and poor cottagers. See John Fell, *Some Illustrations of Home Life in Lonsdale North of the Sands* (Ulverston: 1904); ABSF, 512.
23. Brian G. Awty, "Force Forge in the Seventeenth Century," *Transactions of the Cumberland and Westmorland Antiquarian and Archeological Society* (TCWAAS) 77 (1977): 97–112; ABSF, 536. Among

her commercial investments, Sarah and family members and some neighbors shipped grain via Liverpool to Bristol and Cornwall. Several entries appear in the ABSF referring to "our voyage with corne to Bristoll" between January 1673/4 and January 1674/5. Partners in this shipping venture included Margaret Fell, Thomas Lower, Sarah, Rachel, and Susannah Fell, James Lancaster of Walney Island, and Joseph Sharpe, the bailiff of Marsh Grange, among others. See Ross, *Margaret Fell*, 272–73; ABSF, 536–37.

24. Hunt, *The Puritan Moment*, 28; see Alan Macfarlane, *The Family Life of Ralph Josselin* (New York: W. W. Norton, 1970); Slater, *Family Life in the Seventeenth Century*.
25. Gregory King's statistics of 1688 indicate that one acre produced twenty bushels of wheat. Wheat production on the Fell farm yielded at least one or two acres of wheat in excess of grain needed for home consumption, which was sold to neighbors or at the local market. Margaret Fell had sufficient wheat production to rent barn storage. Earlier in July 1676, she paid Kendall's son for helping "gett hay at Gleaston" and had rented in April 1676 one end of a barn for the purpose of holding hay from James Kendall for 5s. ABSF, 550–51.
26. Joan Thirsk, ed., *Agricultural History of England*, IV, 128–42; Pinchbeck, *Women Workers*, chap. 1, 136–37.
27. Alice Clark, *Working Life of Women*, 67.
28. Thirsk, *Agricultural History*, 128–42; see also Eric Kerridge, *The Agricultural Revolution* (London, 1967). William Hunt, *The Puritan Moment*, 9, 16–17, 24–25.
29. Macfarlane, *Josselin*, 36, 39, 46; ABSF, 253.
30. Awty, "Force Forge," *TCWAAS*, 97–112. Awty derived his information from the Thomas Rawlinson Force Forge Accounts, 1658–63, located at the Preston PRO, Lancashire; DDHJ.
31. A letter from A. Pearson of 9 May 1653, sent from Ramshaw to a Quaker, made reference to his visit to Margaret at Swarthmoor: "Oh how gracious was the Lord to me in carrying me to Judge Fell's, to see the wonders of His power and wisdom, – a family walking in the fear of the Lord, conversing daily with him, crucified to the world, and living only to God. I was so confounded, all my knowledge and wisdom became folly; my mouth was stopped, my conscience convinced. . . . I have seen at Judge Fell's and have been informed from that precious soul his consort in some measure what those things mean, which before I counted the overflowings of giddy brains." Pearson remained a Quaker until the Restoration when he returned to the Anglican church. See Swarthmoor MSS I, 87 and III, 33.
32. John Bradshaw (1602–1659) practiced as a provincial barrister and held a considerable estate and had much influence in Stockport. Judge Fell, as circuit Judge of Assize for Chester and North Wales, had frequent opportunity to repair to his home in Cheshire; see *Calendar of the Committee for Compounding*, Pt. III, 1846; D. C. Coleman, *Sir John Banks* (Oxford: Clarendon Press, 1963), 67.

33. Richard Vann, *The Social Development of English Quakerism, 1655–1755* (Cambridge: Cambridge University Press, 1967), 175; ABSF, 445, 527. To each of her daughter's children (numbering eight), she bequeathed two guineas; to her son's two children she gave one guinea each. There is a legal document in the Abraham MSS, xxiii, written by Sarah Fell, dated March 1678 and signed by Hannah Fell as mother and guardian of Charles Fell, heir. It stipulates that a certain parcel of land of the estate in "The Mannor of Lordppc of Osmunderley, alias Osmotherly" be released by the Fell sisters to the said Charles Fell. This may have been done to appease Hannah Fell. Margaret Fell's will is reprinted in *JFHS* 2 (July 1905): 104–6.
34. Wrightson, *English Society*, 52–57.
35. Macfarlane, *Josselin*, Part III; Wrightson, *English Society*, 52–57; Slater, *Family Life*, chaps 1, 2 and 8. Lawrence Stone, *Family, Sex and Marriage*, passim; Houlbrooke, *English Family*, passim. George Fell did not mention his mother in his will in 1670. He did bequeath £4 to his uncle Richardson of Dalton and requested his uncle's favorable attention to his wife and children after his death. The ABSF indicates that Matthew Richardson died in 1677. Aunt Richardson is seldom mentioned except after her husband's death and concerning his estate, she was "not in a capacity to intermeddle there with . . . not hath been for many years," according to Sarah. There may have been a cousin, Jane Richardson, who is mentioned once in the ABSF. It appears that the families were on friendly terms, but the Richardsons were not Quakers and the minimal contacts between the two families, despite their proximity to one another, was probably due to their different religious persuasions. ABSF, 108, 213, 543–44.
36. Frederick B. Tolles, *Quakers and the Atlantic Culture* (New York: Octagon Books, 1960), 88; see also Braithwaite, *BQ*, 498.
37. In the beginning of Margaret Fell's *Works* (1710), a testimony from her children who compiled her writings into one volume after her death, begins; "we find ourselves Conscientiously Concerned to give forth this Testimony (which we her Daughters, are true Witnesses of), of her Holy Life, and Pious Conversation, who was a dear tender, and loving Mother, in all respects; and did in true Wisdom, educate and instruct her Children in the Nurture and Fear of the Lord; and constantly exhorted us to keep Humble, that the Blessing of the Lord might be our Portion for ever; and her Memory is very Precious to us."
38. The term is used by Melvin Endy, *William Penn and Early Quakerism* (Princeton, N.J.: Princeton University Press, 1973), to denote a Quaker minister who was a recognized leader and well-known spokesperson for the Quaker cause.

Chapter 4

1. A portion of this chapter was published as an article, "'Poore and in Necessity': Margaret Fell and Female Philanthropy in Northwest England in the Late Seventeenth Century," in *Albion* 21 (Winter 1989): 559–580.
2. Thomas Lawson, *An Appeal to the Parliament Concerning the Poor, That There may not be a Beggar in England* (London: Robert Wilson, 1660), 1, 4; Richard T. Vann, *The Social Development of English Quakerism* (Cambridge, Mass.: Harvard University Press, 1969), 50, 73. For a general overview of early Quaker poor relief, see Arnold Lloyd, *Quaker Social History 1669–1738* (London: Longman, Green and Co., 1950), chap. 3.
3. Milton D. Speizman and Jane C. Kronick, "A Seventeenth-Century Quaker Women's Declaration," *Signs* 1:1 (1975): 243.
4. A. L. Beier, *Masterless Men: The Vagrancy Problem in England* London: Metheun, 1985), xxi-xxii, chap. 1, 52–54, 118–19, chap. 9; A. L Beier, *The Problem of the Poor in Tudor and Early Stuart England* (London: Lancaster Pamphlets, London, 1983), 2–7, 13, 29–35; Tom Arkell, "The Incidence of Poverty in England in the Later Seventeenth Century," *Social History* (January 1987): 23–47. Two older histories that give an overview of poverty and poor relief in the period include E. M. Leonard, *The Early History of English Poor Relief* (New York: Barnes and Noble, 1900); Ephraim Lipson, *The Economic History of England*, 3 vols. (London: A & C Black, 1915–31). A detailed new study of Tudor and Stuart poor relief is by Paul Slack, *Poverty and Policy in Tudor and Stuart England* (London: Longman, 1988); see also Diane Willen, "Women in the Public Sphere in Early Modern England: The Case of the Urban Working Poor," *Sixteenth Century Journal* 19 (1988): 559–75; Tim Wales, "Poverty, poor relief and the life-cycle: some evidence from seventeenth-century Norfolk," in Richard M. Smith, ed., *Land, Kinship and Life-Cycle* (Cambridge: Cambridge University Press, 1984), 351–404.
5. Slack, *Poverty and Policy*, 43–44; see also Keith Wrightson, *English Society, 1580–1680* (New Brunswick, N.J.: Rutgers University Press, 1982), chap. 5; Beier, *Masterless Men*, chap. 2; Arkell, "The Incidence of Poverty," 41 *passim*; Margaret James, *Social Problems and Policy During the Puritan Revolution* (London: G. Routledge, 1930), chap. 6; Joan Thirsk, "Agricultural Conditions in England, Circa 1680," Dunn, *WWP*, 87–97. See also R. H. Tawney and Eileen Power, eds, *Tudor Economic Documents* (London: Longman, 1924), 346–62; Susan Dwyer Amussen, *An Ordered Society* (Oxford: Basil Blackwell, 1988), 7–8.
6. Slack, *Poverty and Policy*, 200, 205; see C. G. A. Clay, *Economic Expansion and Social Change; England 1500–1700* 2 vols. (Cambridge: Cambridge University Press, 1984), vol. I, chap. 7; Clay likewise claims that the "hospitality" of the rich in the late six-

teenth century gave way to compulsory poor rates by the end of the seventeenth century and the emergence of workhouses for the poor, especially in larger towns.

7. Beier, *Masterless Men*, 52–54, 188–19, chap. 4; For a detailed discussion of female welfare services at the parish level, and through the state public relief system see D. Willen, "Women in the Public Sphere in Early Modern England: The Case of the Urban Working Poor," ibid., 559–75. See also A. L. Beier's articles, "Poverty and Progress in early modern England," in A. L. Beier, David Cannadine and James M. Rosenheim, eds, *The First Modern Society* (Cambridge: Cambridge University Press, 1989), 209–39, esp. 215; and Beier's article, "The social problems of an Elizabethan country town: Warwick, 1580–90," in Peter Clark, ed., *Country Towns in Pre-Industrial England* (New York: St. Martin's Press, 1981), 46–85; esp. 60–61. Beier points out that gender and geography were variable factors which the statistics from the Tudor-Stuart period gloss over. He claims that in general in English towns poor females outnumbered poor males by a small majority. In the town of Warwick, however, poor women outnumbered men by two to one in the year 1587 and represented a majority as heads of households. See also Slack, 75–77; 80; Tim Wales, "Poverty, poor-relief and the life-cycle," 360, 366; and B. A. Holderness, "Widows in pre-industrial society: an essay upon the economic functions," both in R. M. Smith, ed., *Land, Kinship and Life-Cycle*, 428.

8. James, *Social Problems*, 244; *Victoria County History, Lancashire*, II, 289–99, Thirsk, "Agricultural Conditions," in *WWP*, 89; W. K. Jordan, *The Social Institutions of Lancashire* (Manchester: Chetham Society, 1962), 5–8.

9. W. K. Jordan's two studies, *Philanthropy in England* (N.Y.: Allen & Unwin, 1959) and *The Social Institutions of Lancashire* cover the period preceding and overlapping the rise of Quakerism. Jordan dismissed the subject of sectarian charity with the comment, "The sectaries, who were lending fanatical . . . attention to religious questions during these years, were so engrossed with their spiritual vision, of the Kingdom of God which seemed to be at hand, that they gave but scant attention to the more pedestrian problem of poverty." *Philanthropy in England*, 205. More recently Paul Slack, *Poverty and Policy*, has alluded to Quaker close involvement in local poor relief, as has B. A. Holderness. Without going into detail, Holderness claims that "Quakers and Baptists . . . lent money and offered sales credit as promiscuously and extensively as conformists among the countrymen who acted as money lenders." See "Widows in pre-industrial society," in Smith, ed., *Land, Kinship and Life-Cycle*, 441. See Lloyd, *Quaker Social History*, 34, on Quaker independence from parish poor relief; note also Michael Mullett, *Radical Religious Movements in Early Modern Europe* (London: George Allen & Unwin, 1980), 41–46, 65–68.

10. *The Journal of George Fox*, 2 vols. (Cambridge: Cambridge University

11. Bonnelyn Young Kunze, "An Unpublished Work of Margaret Fell," *Proceedings of the American Philosophical Society* (December 1986): 451.
12. Margaret Fell, *Works*, 97.
13. Mullett, *Radical Religious Movements*, 41–46, 65–68; see Michael Mullett, "The Assembly of the People of God," 19–20, in Michael Mullett, ed., *Early Lancaster Friends* (Lancaster: Northwest Regional Studies, 1978); Braithwaite, 560, 567; Lloyd, *Quaker Social History*, chap. 3; Lloyd rightly commented that Quakers classified their poor into three categories: "those poor by the hand of Providence . . . those poor by 'sloth and carelessness,' . . . those poor by oppression of persecutors," 33. See also Hugh Barbour, *The Quakers in Puritan England* (New Haven: Yale University Press, 1964), 175.
14. Swarthmoor Men's Monthly Meeting Minutes (SMMMM); PRO, Preston and Lancashire 1668–74, 43, 93. Fox's contemporary and friend Robert Barclay similarly expressed Quaker concern for poor relief. See Robert Barclay, *Anarchy of the Ranters* 1674), 84–85.
15. In his younger years, William Penn expressed an idealized notion of social obligation, only to alter it somewhat in his later years. In an early edition of *No Cross, No Crown*, he called for an alleviation of the misery of "oppressed tenants" with their "pale faces" due to "pinched bellies" and "naked backs," whose "sweat and tedious labour" was given for the benefit of the rich. However, after his European mission in 1677, Penn "urged the poor to be silent and patient and to trust in the Lord," and in the 1682 edition of his *No Cross, No Crown*, he suggested to the "civil magistrate":

    > That if the Money which is expended in every Parish in such vain Fashions, as wearing of Laces, Jewels, Unnecessary Ribbons, . . . Costly Furniture . . . with what is commonly consumed in Taverns, Feasts, Gaming etc. could be collected into a Public Stock . . . there might be Reparation to the broken Tenants, Work-Houses for the Able and Almshouses for the Aged and Impotent.

    George Fox had made a similar suggestion to Parliament in 1659. See Barbour, *Quakers in Puritan England*, 170, 250; Melvin Endy, *William Penn and Early Quakerism* (Princeton, N.J.: Princeton University Press, 1973), 346–47. For an example of a successful Quaker "workhouse" in Bristol in the 1690s see Lloyd, *Quaker Social History*, 40–41.
16. Fox, *Works* 7, Epistle 200, "The line of righteousness and justice stretched forth over all merchants," 194–95; see also J. William Frost, *The Quaker Family in Colonial America* (New York: St.

Martins's, 1973), 56, 197, 201; Lloyd, *Quaker Social History*, 36–37. For an interpretation of Puritan attitudes toward the deserving and undeserving poor see Daniel A. Bough, "Poverty, Protestantism and Political Economy: English Attitudes Toward the Poor, 1660–1800," in Stephen B. Baxter, ed., *England's Rise to Greatness* (Berkeley: University of California Press, 1983), 63–107.

17. Along with the quantitative analysis of the year 1676, I draw examples from the other years between 1673 and 1678. Carole Shammas cites briefly Sarah Fell's banking and charitable activities. She simply comments that aid to the poor women recorded in the ABSF was the result of the personal interest of the Fell women, who as leaders of the SWMM, simply focused more heavily on indigent females than "might otherwise be the case." Shammas, *WWP*, 102.

18. Because no banks existed outside London before the eighteenth century, local financial services provided in towns were very important. Characteristically, attorneys frequently fulfilled the function of local banker, investing the surplus capital of wealthy clients in loans to local borrowers, secured by mortgage or bond with interest. Extensive use of credit was also commonplace in rural areas where cash was scarce. By extending credit through "bills of exchange," cash need not be transported. Clay, *Economic Expansion and Social Change*, 1: 179–87; Note Shammas, in *WWP*, 102–3. One atypical feature of the banking business carried on at Swarthmoor Hall was that it was conducted by a single woman.

19. Ross, *Margaret Fell*, 62, 64, 266.

20. Clay, *Economic Expansion and Social Change*, 176–91; cf. Wrightson, *English Society*, 52–54. Wrightson claims that the loaning and borrowing of small sums of money without interest was part of traditional duty and goodwill toward one's neighbors. Such cooperation promoted harmonious social relations between non-kin and helped overcome isolation in a fluid society where kinfolk were too widely dispersed to be of help. Note also Holderness, "Widows in pre-industrial society," 428, 435–36, 440.

21. Fox, *Works*, 8, 191, 195; LBC, 965–66.

22. ABSF, 267, 277, 281, 309, 313, 327, 383, 369, 338, 387; *passim*. See Chapter 3 for other examples of Sarah's loans to female petty retailers.

23. SWMMM; PRO, Preston, Lancs., 38–39; ABSF, 250, 257.

24. SWMMM, 102, *passim*; ABSF, 306, 309, 433.

25. SWMMM, 34–51, 167; ABSF, 297; *passim*.

26. SWMMM, 28, 32, 39–41, 50, 58–59, 86, 88, 107, 138, 142–43; ABSF, *passim*. See Slack, *Poverty and Policy*, for a detailed discussion of "shallow" and "deep" poverty, 7, 39, 52–54, 190, *passim*.

27. SWMMM, 76; ABSF, 17, 176–77, 179, 525.

28. SWMMM, 67.

29. SWMMM, 61; ABSF, 265.

30. SWMMM, 105–7, 113, 168–69.

31. Mullett, ed., *Early Lancaster Friends*, 19–20; Arkell, "The Incidence of Poverty," 38–39.
32. SWMMM, 86.
33. SWMMM, 83–139.
34. Lloyd, *Quaker Social History*, 34–5; London Six Weeks Meeting Minute Book, Vol. I, 20 (April 1680): 120–21; see Diane Willen, "Women in the Public Sphere," 569.
35. SMMMM, 13, 23, 27, 54, 77, 79, 81, 85–86, 92–93. I have compared the activities of the SWMM with those of the Swarthmoor Men's Monthly Meeting (SMMM), since they worked in close cooperation with one another and the SMMM seldom nullified the activities of the SWMM. The minutes of the SMMM are extant for the years 1668 to 1674, with a gap in the record between 1674 and 1691, while the minutes of the SWMM are continuous from 1671 onward. The SWMM and the SMMM regularly collected funds called the "stock" from their members to finance their Quaker religious, political, and charitable activities.
36. Penn's philanthropy over the same period as the Fell accounts consisted mostly of poor taxes paid in this period. Richard Dunn and Mary Maples Dunn, eds, *The Papers of William Penn* (Philadelphia: University of Pennsylvania Press, 1981), 1:600–15. The Daniel Fleming Accounts were kept by his steward, John Bankes. See Fleming Papers, ED/RY, Kendal Public Record Office, Lancashire.
37. The account of the Countesse of Pembroke of Appleby Castle in Westmorland covers only August and October 1673. The entries for her household staff indicate that she owned several residences and personally oversaw the management of her estate. See Joseph Whiteside, "Some Accounts of Anne, Countesse of Pembroke, *Transactions of the Cumberland and Westmoreland Archeological and Antiquarian Society* 5, New Series (1905): 188–201; cf. Ann Clifford, *The Lives of Ann Clifford . . . and her Parents* (London: 1916), 25 *passim.*
38. Alan Macfarlane, *The Family Life of Ralph Josselin* (New York: W. W. Norton, 1970) 51–52. Beier has pointed out that the "Protestant ethic" did not bring about any new "critique of poverty." This may well be true on the whole. However, it is necessary to modify this statement when speaking of the Quakers, who not only issued a renewed "critique on poverty" but also implemented their ideas in a coherent fashion; Beier, *Problem of the Poor*, 14–15; *Masterless Men*, 5.
39. It is possible that charity was also given in kind. This may help explain the small portions of money given. Note Lloyd, *Quaker Social History*, 34.
40. Lloyd, *Quaker Social History*, 34, 36, 42; Endy, *William Penn and Early Quakerism*, 356; cf. Vann, *Social Development*, 143–47.
41. Slack, *Poverty and Policy*, 200, 205–8; Mullett, *Early Lancaster Friends*, 19–20; Vann, *Social Development*, 148; see Lotte Mulligan and Judith Richards, "A 'Radical' Problem: The Poor and the English Reformers in the Mid-Seventeenth Century," *Journal of British Studies* 29 (April 1990): 188–46, and especially 121.

42. K. Wrightson and D. Levine, 183–84; Vann, *Social Development*, 148; cf. Baugh, "Poverty, Protestantism and Political Economy," in Baxter, *England's Rise to Greatness*, 63–107; cf. also the individual philanthropy of puritan, Mary Rich, Countess of Warwick in Sara Heller Mendelson, *The Mental World of Stuart Women*, chap. 2.

Chapter 5

1. Barbara J. Todd, "The Remarrying Widow: A Stereotype Reconsidered," in Mary Prior, ed., *Women in English Society 1500–1800* (London: Methuen, 1985), 55. See also Susan Staves, *Married Women's Separate Property in England, 1660–1833* (Cambridge: Harvard University Press, 1990).
2. Rawlinson MS, Friends House Library, London, 3, 35, 90. This is a two-volume document (rebound from one) entitled, "[A book to] goe abroad onely among all Friends i[n] th[e] Truth [in answer] to seuerall papers of Margret Foxe . . . formerly called Margret Fell . . . toucheing the unjust orders, papers and illegale proceedings . . . on her behalfe . . . as toucheing my Stewardship for her . . . at Forse [Force] Fordge [Forge] . . . ." (1680). In 1851 the manuscript was left in the custody of the then recording clerk at FHL, J. Crosfield. I am grateful to Craig Horle for informing me of this valuable document only recently discovered among the rare early Quaker documents in the FHL. Mr. Horle has completed a partial transcript of approximately the first 34 pages of the 455-page document. As far as is known, this manuscript is a unique copy. The handwriting in the document may not be that of Thomas Rawlinson. As internal evidence indicates, several copies were made, and this extant copy may have been written by an amanuensis for Rawlinson. The MS consists of Rawlinson's two letters to the Swarthmoor men's meeting and two letters to Margaret Fell in February 1663 and May 1667. Following these are a letter to the quarterly meeting at Lancaster in May 1670 and a second letter to the same meeting and Margaret Fell in June 1672. These are followed by ten queries to Margaret Fell and her children and dated the same month, June 1672. The longest epistle, to the men's monthly and quarterly meetings, containing four parts, was sent in June 1673. Finally, a brief "SALME of praise," written in April 1680 concludes the document.
3. Braithwaite, *BQ*, 456; Ross, *Margaret Fell*, 267–68; Hugh Barbour and A. O. Roberts, *Early Quaker Writings* (Grand Rapids, Mich.: Wm. B. Eerdmans Publishing Co., 1974), 607; Bruce Blackwood, "The Lancashire Gentry," *Transactions of the Lancashire and Cheshire Historical Society*, 15; ABSF, 538, 558–59. There was a Curwen Rawlinson of Cartmel in Furness whose father, Justice Robert Rawlinson, sent Fox to Lancaster Castle in 1663. One of Rawlinson's daughters, Lydia, married James Lancaster and became a prominent Quaker preacher

in the early eighteenth century. A puzzling note is contained in the *Calendar of the Committee for Compounding*, under Marsh Grange, Margaret Fell's family home. It indicates that Marsh Grange was compounded in 1648 as the estate of the late Robert Rawlinson who died a papist in arms. This does not agree with the accepted record that Marsh Grange was in the John Askew family. If there is a connection between the Rawlinson family and Margaret (Askew) Fell's ancestral home, it is as yet unknown. The question needs further investigation.

4. Swm. Tr. 3/167. Spence III, 3/62.
5. Helen Crosfield, *Margaret F of Swarthmoor Hall* (London: Headley Bros., 1913), 87–89.
6. Rawlinson MS 4; see also C. Horle Tr. 1. I have retained the first person pronouns in quoting Rawlinson to preserve the original sense and pungency of his statements despite the fact that it renders the flow of the text somewhat bumpy.
7. Ross, *Margaret Fell*, 267–68. For more information on the history of iron production in Lancashire, see A. Fell, *The Early Iron Industry of Furness and District* (Ulverston: 1908, reprinted, London: 1966). A. Fell cites the 1681 deed of transfer for Force Forge, 193–94; see also H. R. Schubert, *History of the British Iron and Steel Industry* (London: 1957); C. B. Phillips, "Iron-Mining in Restoration Furness: The Case of Thomas Preston," *Recusant History* 14 (1977–78): Isabel Ross has recorded the heretofore known details of the feud. Ross believed that Rawlinson's stewardship practices became suspect in 1668. She was apparently unaware of the fact that his work at Force Forge had terminated five years earlier.
8. Fox's *Works* VI, 179; VIII, 173. Robert Barclay, *Anarchy of the Ranters* (1676), 40–41.
9. Ross, *Margaret Fell*, 267–68. For a more recent study of Force Forge, see Brian G. Awty, "Force Forge in the Seventeenth Century," *Transactions of the Cumberland and Westmorland Antiquarian and Archeological Society*, 77 (1977): 97–112. Ross and Awty agree that the Forge appears to have been operating at a loss between 1658 and 1663 and that that may have been the underlying cause for the disagreement.
10. Ross, *Margaret Fell*, 268, quoted in ibid., 272. Original in the Dix MSS, 2/3, 10 July 1684; Awty, "Force Forge," 97–98, italics are mine.
11. The accounts are located in the PRO Preston, Lancashire, and have been transcribed by Brian Awty. According to Awty's findings, the accounts have survived in two versions. One covers the years to 1660 and is not in Rawlinson's hand but gives more details on the repairs of the forge. The second is in his handwriting and is more complete in information concerning customers and locations. I have compared Awty's notes to the original accounts and to the Rawlinson MS. Both versions used Quaker-style dating. Awty, "Force Forge," 97–107.
12. Awty, "Force Forge," 98; DD#J, MSS 278/2; see Inventory 2 of

original accounts. Reginald Walker's stewardship at Force extended from 1663 until 1676, at which time William Wilson succeeded him. Both names appear frequently in the Rawlinson MS and the ABSF.

13. Awty, "Force Forge," 103, 104–6, 110. Iron ore was mined in the district and purchased at the price of 4s.6d. per quarter during at least part of the period of the accounts. Some iron was mined, for example, on the property of Matthew Richardson of Elliscales, who was a brother-in-law of Margaret Fell and an investor in Force Forge. Adgarley was another mining area and its iron ore was known for its high quality. Horseshoes made from Adgarley iron ore were known to last far longer than other iron shoes; Ross, *Margaret Fell*, 269. Charcoal was also carted to the forge by pack horse at the average of ten loads per ton of ore smelted.
14. Awty, "Force Forge," 105–7; there were six forges operating in the Furness-Cartmel area and two more forges in operation south of the River Kent and Morecambe Bay. See map, ibid., 105.
15. DDHJ/MSS 278/2, 17.
16. Awty, "Force Forge," 107.
17. Ibid.
18. Rawlinson MS, 167, 175, 210–11, 248, 299.
19. Ibid., 206, 304. M:ff – initials of Margaret Fell.
20. Ibid., 102–3.
21. Braithwaite, *BQ*, 236–37; Mabel Brailsford, *Quaker Women* (London: Gerald Duckworth & Co., 1915), 150–57.
22. Spence MSS III, 35; Cambridge *Journal* I, p. 291; see also Norman Penney's notes in the Cambridge *Journal*, 450–51. Penney states that Fox's original *Journal* recorded that there was a large general meeting in 1657 at the home of Christopher Fell in Cumberland. The Ellwood edition deleted Christopher Fell's name and substituted "Langlands," the home of John Fell, probable father of Christopher Fell. According to the *Dictionary of Quaker Biography*, Christopher Fell died in 1706 and was buried at Swarthmoor, Lancashire. This raises the possibility that he may have been related to Margaret Fell. However, the Fell name was very common in the Furness district. It is also a possibility that he was reinstated in the Quaker church and only temporarily "fell away from Truth."
23. Spence MSS III, 3/32, 1653; Braithwaite, *SPQ*, 303, 242. Fox occasionally expressed his exasperation against those whose "openings" thwarted his own.
24. Rawlinson MS, 114; 109, 111, 74.
25. Ibid., 209.
26. Ibid., 294.
27. Ibid., 417–18.
28. Ibid., 281 and 287.
29. Fox's *Works* VI, 179; VIII, 173. Within the meeting it was the custom for "weighty" or important members to arbitrate final settlement of differences if the contenders would not agree to the meeting's

consensual decision. One instance of this is seen in the SMMMM in September 1673 when the meeting turned to Margaret Fell to arbitrate a dispute between Thomas Fisher and Will Hawthornthwaite. The minute reads: "And whereas the diferences betwixt Thomas ffisher and Will Hathornethwaite being not desided, it is with the consent of thos parties refered to [the] heareing and determination of margrett ffox" (SMMMM, 12.6.1673, 73). See Craig W. Horle, *The Quakers and the English Legal System 1660–1688* (Philadelphia: University of Pennsylvania Press, 1988), chaps 5 and 6. Note also Lawrence Stone, *Road to Divorce*, 24, who maintains that most litigants in the early modern era, went to law as a "tactical manoeuvre" which was "part blackmail and part bluff," with the object being to gain an out-of-court settlement with dispatch and thus reestablish order.

30. Rawlinson MS, 120–21.
31. Ibid., 134. Rawlinson's reference to "father" was probably his father-in-law, Thomas Hutton. The text is unclear.
32. Ibid., 136.
33. Ibid., 203–5; 213.
34. Ibid., 211 and 212.
35. Ibid., 211–12.
36. Ibid., 213, 214–16, 243. With Friends as witnesses, the disputants sealed bonds to obey the order and end result of arbitration within the Quaker fellowship, according to Rawlinson.
37. Ibid., 212–13; the cover page of the original account book of Rawlinson was dated March 1667 and March 1668. It was docketed as follows: "showed to the witnesses at the time of their exa[m]i-[n]acon before us," and signed by Matthew Richardson and William West. There are two other names below these that appear unrelated to the original examination and signatures, and may have been added later.
38. Rawlinson MS, 214, 216–17; there is no mention in the SMMMM that Fell took her case to civil lawyers.
39. Ibid., 224–25.
40. Ibid., 232, 240, 245, 271, 273, 276, 298, 246–48.
41. Rawlinson MS, 287–88. As early as 1668 the SMMM was exasperated with Rawlinson's behavior. A minute in the SMMMM of 9.12.68/69 refers to an order given at the last LQM "that a paper should be given forth against Tho:Rawlinson and left to the last monthly meeting at Cartmel to publish itt wch they then Referred to ye considerasion of this meeting." The quarterly meeting duly considered it and judged that a paper should be read at the meetings at Hawkshead, Cartmel, and Swarthmoor after friends at the monthly meeting at Lancaster had seen it and subscribed [to] it "that then itt may be further published" (SMMMM 9.12.1668/9). See Rawlinson MS, 195–96.
42. Ibid., 258; see Kunze, "An Unpublished Work of Margaret Fell," for Fell's hostile opinion of the Anglican prelates.
43. Rawlinson MS, 197. I have not found a minute to this effect either in the SWMMM or the LWQMM.

44. Ibid. 158.
45. Ibid., 303, 309, 313, 321.
46. Ibid., 321; see SMMMM 11.6.1668 and SMMMM 8.7.1668. A minute of the SMMMM of 8.10.1668, 13, at William Satterthwaite's home indicated that Rawlinson had been notified by friends from the last quarterly meeting to come to the next quarterly meeting at Lancaster. The minute reads that he had promised to give his answer in six weeks.
47. Rawlinson MS, 326–27, 329. The minutes of the SMMM give a different slant to the proceedings between Fell and Rawlinson. There is a minute of 11 August 1668 at Swarthmoor that states that Robert Penington was sent to tell Thomas Rawlinson to come to the next monthly meeting to be held at Robert Penington's house. Rawlinson had missed the two previous meetings at Cartmel and Swarthmoor. A minute of the men's monthly meeting 8 October 1668 referred to the previous meeting at Penington's home, which Rawlinson had also failed to attend. According to the minute, Rawlinson had "promise[d] [but] hee hath not performed. It is therefore now ordered, that James Tayler and Charles Sill goe to him, And lay it upon him, to be at the next quarterly meetinge at Lancaster, And to take his answer And returne it to ffriends; if he Refuse to come, accordinge to the desire of ffriends at the Meetinge, who are stil disastisfied, that an End Accordinge to truth, and Equity is not yett putt to ye Aforesaide differance; And that they alsoe show him, A Letter from G[eorge] ff[ox] to ye same purpose, with the names of severall good ffriends in it, whose Assistance may bee desired to deside the said differance, and put An End to it" (SMMMM, 5, 7). Nothing is mentioned of Rawlinson's illness.
48. Rawlinson MS, 331, 336–40.
49. Ibid., 340–45. Ross recorded the story of the lame horse, which she drew from the Spriggs MSS; Ross, *Margaret Fell*, 268.
50. Rawlinson MS, 157–62.
51. Ibid., 172–73.
52. Ibid, 177.
53. Ibid., 191.
54. Ibid., 354. The minutes of the SMMM of 8.1.1669, convey a *deja vu* attitude toward Rawlinson. They state: "Tho. R[awlinson] we have all this time borne a great share of thy burthen in this longe and tedious busines of thyne and ye time is com yt yu wilt beare lest to thyselfe in it for frends now must of nesecitie: publiclly deny ye seeing ye thinge is soe publicke to ye world: unlese yu goe to ye next quarterly meetinge and give satisfaction to frends: for yu hast not done it by thy pap[ers]: but hast guien in thy paper if yu did give satisfaction to Willm Wilson and Robert Penington accordinge to an order mad[e] at Swarthmore ye 8.1.1669 wch we find yu has not done." The meeting requested that he own his own condemnation in writing prior to any further communication.

55. Rawlinson MS, 367. The men's monthly meeting of 12.11.1669 states, "A paper against Tho. Rawlinson as A resister of the order of truth and out of it" was left by the quarterly meeting for the consideration of the monthly meeting to publish such an order.
56. Rawlinson MS, 369, 380. An example of his temporary suspension is found in the Swarthmoor men's minutes. A minute of the men's meeting on 10.6.1669 states that Thomas Rawlinson had issued a complaint for the meeting to settle a disagreement between him and two men, William Grave and Reginald Holmes. Apparently Rawlinson had bargained with these two men and wanted a bond on the bargain. The minute reads, "wee Judge it meet thatt they speake to Tho. Rawlinson and give him such security as is suficient. If hee have it not or if they cannot agree of it themselves that both he and they be att the next monthly meeting at Satherwhaite there to have the Busines Desided." Four years later Rawlinson registered another complaint to the meeting in December 1673, against Reginald Holme and William Grave. He claimed that they had not paid him the sum of £10, overdue since March 1672/3. The meeting then ordered William Wilson and William Satterthwaite to "speake to them and desire them . . . to sattisfie Thomas Rawlinson concerning the sd money or give us their Reason to the Contrary the next monthly meeting." Approximately six months later the meeting reported that "wee fine noe End put to Thomas Rawlinson['s] complaint." This time three men were picked to speak to Holme and Grave to "know whether ye money be payed as friends desired and to give Acount what is done to the next m.m." (12 March 1674). The men paid Rawlinson the money in arrears and the account was closed. Such litigation by the Friends meeting on Rawlinson's behalf indicates that he was only temporarily suspended from the fellowship until his settlement of the dispute with Fell (see SMMMM, 27, 79, 84, 86).
57. Rawlinson MS, 409.
58. Ibid., 385.
59. Ibid., 385–91; 408.
60. Ibid., 267.
61. Ibid., 435–36. There is an extant letter of self-condemnation by Reginald Walker that alludes to the fact that there was something amiss between Rawlinson and Walker. Unfortunately the letter is undated, but judging from Rawlinson's account of Walker's suspicious behavior at Force, this letter of confession may have been written sometime between 1673 and 1680. Apparently the SMMM did not continue to meet regularly at Walker's home due to some misbehavior on his part. In Walker's words, "I have not desired ye Meettinge to be att my house since . . . [due to] ye iniquitty yt I have mett wth amonge ffriends, ffirst of all how you proceeded agt me att ye Quarterly Meetinge att Lancashire concerninge Tho. Rawlinson." Apparently his confession was sufficient to place him in good standing with Friends again, for a second minute reads: "Soe hee haveinge Judged

ye afor[e]s[ai]d paper, ffriendes may meet att his house as before" (SMMMM, 102–3).
62. Rawlinson MS, 250.
63. Ibid., 286–87, querie 10; 250.
64. Ibid., 252–53.
65. Ibid., 271.
66. Ibid., 140–41.
67. Ibid., 419–20.
68. Ross, *Margaret Fell*, 272.
69. This question of suppression of evidence and the rewriting of early Quaker history was posed by Winthrop S. Hudson in connection with other missing documentary evidence in the early record of Quakerism. For instance, the first pages of Fox's *Journal*, which would have given valuable evidence of the earliest years of Fox's ministry, are missing. See Winthrop S. Hudson, "A Suppressed Chapter in Quaker History," *Journal of Religion* 24 (April 1944): 108–18; cf. Henry J. Cadbury, "An Obscure Chapter in Quaker History," *Journal of Religion* 24 (April 1944): 201–14.
70. Rawlinson MS, 451–52. When Rawlinson died in 1689, he was buried in the Quaker burial ground at Colthouse, indicating he was reconciled to his meeting.
71. See Craig Horle, *The Quakers and the English Legal System*.
72. Rawlinson MS, 158, 385, 447.
73. Rawlinson MS, Swarthmoor men's meeting, 149.
74. William Beck and T. Frederick Ball, *The London Friends' Meetings* (London: F. B. Kitto, 1869), 50.
75. Leonard Trinterud, "AD 1689: The End of the Clerical World" (Paper delivered at the William Andrews Clark Library Seminar on Theology in Sixteenth and Seventeenth Century Englan at the University of California at Los Angeles, 1971), 48.
76. Fox became notorious in the eyes of orthodox Christians when, in 1650, he was accused and tried for blasphemy. Fox claimed to his judges that, because he was sanctified in Christ, he was without sin. Fox meant that Christ, who dwelt in him, was perfect and had taken away his sin. See Cambridge *Journal* I, 2–3. Later Robert Barclay approached justification and sanctification more conditionally, thus modifying the earlier Foxian concept. Barclay, like Baxter and other mainline Puritan thinkers, claimed that good works within one were done by the grace of Christ, and these good works were a necessary fruit of justification. Therefore, Barclay defined perfectionism in a way that was less open to the attack of Pelagianism. An individual may not be entirely perfect, according to Barclay, but this does not inhibit the "good and perfect works from being brought forth in him by the spirit of Christ." Thus God's Spirit in the believer accomplishes the good works that bring justification for "Christ's works in his children are pure and perfect." See Dean Freiday, ed., *Barclay's Apology* (1967), 153–54. The early Quaker who felt him/herself a regenerate person believed that the truly

converted heart and soul brought with it a gradual elimination of sin, as the indwelling Christ made the believer a perfected creature. This state was achieved only gradually. It was not an imitation of righteous living, rather a complete surrender to the power of Christ in one's life. In so doing, the new man and new woman became "free from conscious sin" or self-emptied (of self-will) and Spirit-filled. See Melvin Endy, *William Penn and Early Quakerism* (Princeton, NJ: Princeton University Press, 1973), 50, 64–68, 168, 181–82; LBC, xii–xv; Trinterud, "AD 1689," *passim*, 48–49. For a discussion of the continental origins of post-Reformation spiritualism and perfectionism, see Nigel Smith, *Perfection Proclaimed: Language and Literature in English Radical Religion 1640–1660* (Oxford: Clarendon Press, 1989) 8–10, 17–18, 23, 107–9, *passim*.
77. Paul S. Seaver, *Wallington's World* (Stanford: Stanford University Press, 1986).
78. Ibid., 14–15, 119–20.

Chapter 6

1. Hilda L. Smith and Susan Cardinale, eds, *Women and the Literature of the Seventeenth Century* (Westport, Conn.: Greenwood Press, 1990), 116–19, 109–11. See also Patricia Crawford, "Women's Published Writings 1600–1700" in *Women in English Society*, Mary Prior, ed. (London and New York: Metheun, 1985), 211–82; Elaine Hobby, *Virtue of Necessity: English Women's Writing 1649–88* (Ann Arbor: University of Michigan Press, 1989).
2. Smith and Cardinale, *Women and the Literature of the Seventeenth Century*, xii–xiii, 55–59. Included in Margaret Fell's *Works*, published posthumously by her children, are numerous epistles written to Quakers and non-Quakers that are not included in this bibliography.
3. Fell, *Works*, 15–17; Ross, *Margaret Fell*, 39.
4. Fox, *Journal*, Nickalls edn., 264–65.
5. Quoted in Frederick B. Tolles, *Quakers and the Atlantic Culture* (New York: Octagon Books, 19), 38; Braithwaite, *SPQ*, 94. Note also Ross, *Margaret Fell*, 39, 52, 55, 107–8, 112–13.
6. Barry Reay, *The Quakers and the English Revolution* (New York: St. Martin's Press, 1985), chap. 5, 81, 87 gives an analysis of the widespread hostility toward Quakers in 1659. Julius Hutchinson, ed., *Memoirs of the Life of Colonel Hutchinson* (London: George Routledge, 1906), 316–17, 320; see Perez Zagorin, *Rebels and Rulers* (Cambridge: Cambridge University Press, 1982), vol. 2, chap. 12, for an in-depth analysis of the political events leading to the Restoration. For a discussion of the religious issues of the Civil War period, see William Lamont, "The Religious Origins of the English Civil War," in Gordon J. Schochet, ed., *Religious Resistance and the Civil War: Proceedings of the Folger Institute* (Washington, D.C.: The

Folger Shakespeare Library, 1990), 1–12. See also Austin Woolrych, *Commonwealth to Protectorate* (Oxford: Clarendon Press, 1982). The opposition to Quakers in 1659–60 was expressed in two Quaker letters sent to Fell. A letter by Henry Fell was dated February 1659; the other by Richard Hubberthorne was dated March 1660. As they describe the political context, these letters are worth quoting.

> London, the 7 of the 12 mo 59 [Feb 1659/60].
> M.F. my dearly beloved in the Lord, my soul greets thee, and honours the. . . . General Monk's soldiers begin to be rude concerning Friends' Meetings. John Scafe is come to town, and went yesterday to the meeting in the Palace-yard at Westminster; but soon after he began to speak, they began to pull Friends out of the house violently, and beat them very sore, and would not suffer any of them to stay in the house; yea, they beat and abused Friends exceedingly in the streets. I came there when they had hailed almost all Friends out, and scattered them; and they pulled me out, and beat me much, and knocked me down in the street, and tore all my coat. Edward Billing and his wife were much abused, he especially. I hear he went presently and wrote to the Parliament . . . and acquainted some of them with their usage, and that he would endeavour to lay it before General Monk and the rest. . . . Great distractions and disaffections there are in people, as things now stand.
> Thy brother,
> Henry Fell

> Dear Sister,
> Our meetings at present are peaceable and quiet, though we have had rudeness by some soldiers and disturbances. I was moved to write something to Monk about it, upon which he gave out a few words as an order to the officers and soldiers, which did stop them for the present from their rage. I intend to stay in the city about two weeks. . . . F. H. [Francis Howgill], Samuel Fisher and John Stubbs, are in the city.
> Thy dear Brother,
> Richard Hubberthorne.

Monk's order read as follows:
> St. James, 9th March.
> I do require all officers and soldiers to forbear to disturb the peaceable meetings of the quakers, they doing nothing prejudicial to the Parliament or Commonwealth of England.
> George Monk.

7. Reay, *The Quakers and the English Revolution*, chap. 5. Reay fails to mention Fell's political role 1659–61 *vis-à-vis* other Quaker leaders in connection to military activity or the tithe petition to Parliament; note also Alan Cole, "The Quakers and Politics, 1652–1660" (Ph.D. thesis, Cambridge University, 1955); Stephen A. Kent, "Seven Thousand

Handmaids and Daughters of the Lord: English Quaker Women's Political Protest in 1659," unpubl. paper, University of Alberta, Edmonton, Canada.

8. M. Fell, *To the Generall Council and Officers of the English Army . . .* (London: Thomas Simmons, 1659).

9. M. Fell, *The Citie of London Reproved* (London: Robert Wilson, 1660).

10. Fell, *This was given to Major Generall Harrison and the Rest* (London: Thomas Simmons, 1660). Fell also wrote a letter from London dated 24.July.1660 to A. P. and W. G. saying: "I have Laboured this five or six weeks hereabout those great ones yt [that] are in power to Informe them and have made known unto them ye sufferings of all ffriends in the Nation. And the King has promised much often [to] me severall times . . . and that we should not be abussed nor suffer for our religion. . . . " SWM, vol. 378.

11. Quoted in Mabel Brailsford, *Quaker Women 1650–1690* (London: Gerald Duckworth Co., 1915), 30–31.

12. Quoted in Ross, *Margaret Fell*, 129–30. Years later, Fell recalled her visits to Charles II. She claimed, "I spake often with the King, and writ many letters unto him, and many books were given by our Friends to the Parliament." and they were fully "informed of our peaceable Principles and Practices," *Works*, "A Relation," 6.

13. Quoted in Ross, *Margaret Fell*, 127; original letter in Swm MSS III, 146; letter from Howgill to Fell, 29, vii [166] also quoted in *JFHS* (1952): 27–28.

14. Fell, *Works*, 202–10.

15. Fox's *Journal*, Nickalls edn., 398.

16. Ibid. Although noting that Fell "drafted a paper" on pacifist principles in June 1660, Christopher Hill has stated: "In January 1661, the 'peace principle' was announced, henceforward characteristic of Quakerism." See Hill, *Experience of Defeat*, 161; cf. Reay, *The Quakers and the English Revolution*, 42. Reay fails to mention Fell in his discussion of Quaker pacifism.

17. Parts of *A Declaration from the People Called Quakers* (30.December.1659), *To the whole English Army*, and *To the Parliament of the Common-Wealth of England* were expunged from Burrough's *Works*. See Hill, *Experience of Defeat*, 165 and 167. The issue of early Quaker suppression of evidence was first raised by Winthrop S. Hudson in 1944. See "A Suppressed Chapter in Quaker History," *Journal of Religion* 24 (April 1944): 108–18.

18. Barry Coward, *The Stuart Age* (London: Longman, 1980), chap. 8; for the nonconformist response in literary form after 1660, see N. H. Keeble, *The Literary Culture of Nonconformity* (Athens, Ga.: University of Georgia Press, 1987), chap. 2.

19. Ibid.; Joseph Besse, *A Collection of the Sufferings of the People Called Quakers*, 3 vols. (London: J. Soule, 1733), 1: xlii. The Five Mile Act of 1665 applied to unlicensed preachers. All nonconformist preachers

were forbidden from traveling within a five-mile radius of the place where they had formerly held an incumbency. The above acts are as follows in the *Statutes of the Realm*: 13 Car. II, st. 2, C.I (Corp. Act, 1661); 14 Car. II, C. I (Conventicle Act 1670), see John Kenyon, *Revolution Principles* (New York: Cambridge University Press, 1977), chap. 10; see also Keeble, *The Literary Culture*, chap. 2.

20. The term "she-soldiers" comes from a book review written by Melvin Maddocks on Antonia Fraser's work, *The Weaker Vessel*. *Time* (September 17, 1984), 86. I have used this expression because it aptly conveys the energy and zeal of Fell and other Quaker women who would not back down in their rebukes of political figures who persecuted Quakers.

Chapter 7

1. Journal of George Fox, Norman Penney, ed., I, 266–67; 343. Men's meetings were set up possibly as early as 1653 and certainly between 1656 and 1660 after conferences at Skipton and Balby in Yorkshire, and elsewhere in the north country.
2. Braithwaite, *BQ*, chap. xiii; especially 307; *SPQ*, chap. x. See also William Beck and T. Frederick Ball, *The London Friends' Meetings* (London: F. B. Kittos, 1869), 47–51.
3. Braithwaite, *BQ*, chap. xiii; *SPQ*, 273–74; Fox's *Works*, 7, epistle #248; 8, #291, #296.
4. Arnold Lloyd, *Quaker Social History*, (London: Longman, Green and Co., 1950), chap. 8; Ross, *Margaret Fell*, chap. 19; Richard Vann, *The Social Development of English Quakerism* (Cambridge: Cambridge University Press, 1967), chap. 3 and 4; Michael Watts, *The Dissenters* (Oxford: Clarendon Press, 1978), 200; see also Hugh Barbour and Arthur Roberts, eds, *Early Quaker Writings* (Grand Rapids, Mich.: W. B. Eerdmans Publishing Co., 1973), 491, *passim*; Beck and Ball, *London Friends' Meetings*, chap. xxii; Antonia Fraser, *The Weaker Vessel* (New York: Alfred A. Knopf, 1984), chap. 18; cf. Christopher Hill, *The Experience of Defeat* (New York: Viking 1984), 21. See also Hugh Barbour, "Quaker Prophetesses and Mothers in Israel," in J. William Frost and John M. Moore, eds, *Seeking the Light* (Wallingford, Pa: Pendle Hill Publications, 1986), 41–60. See also Phyllis Mack's forthcoming book on Quaker women.
5. The chief sources for the early women's meetings that I have investigated are various documents and letters to Fell and Fox, preserved in the Swarthmore MSS, Spence MSS, and Abraham MSS. Other sources include George Fox's *Journal*, Margaret Fell's *Works*, and the minute books of the earliest women's monthly and quarterly meetings at Swarthmoor (1671–1700), Kendal, and Lancaster (1672–1700) in Lancashire. The contemporary sources on the Wilkinson-Story schism also yield evidence. See notes below.

6. Beck and Ball, *The London Friends' Meetings*, 344.
7. Anne Whitehead and Mary Elson, *An Epistle for True Love, Unity and Order* (London: Andrew Soule, 1680?), 8–12.
8. Beck and Ball, *London Friends' Meetings*, 345. The name of the Box Meeting was taken from the box in which charitable donations were placed.
9. Beck and Ball, *London Friends' Meetings*, 347.
10. Ibid., 348–49.
11. Ibid., 349–51; Whitehead and Elson, *An Epistle for True Love, Unity and Order*, 7, 11. Rebecca Travers was a London woman of considerable stature. In the 1670s she was one of the few women who sat on the important Second Day Morning Meeting, the chief editorial board for early Quaker published writings. In 1790 the Women's Two-Week Meeting and the Box Meeting amalgamated and agreed to have monthly meetings.
12. Fox's *Works*, 8, #317, 79–80. Braithwaite, *SPQ*, chap. xi, gives an overview of the dispute over the adjustment from individual spiritual authority to group authority. In the process of Quaker organization after the Restoration, the Quakers vacillated between the practice of the rule of the spiritual elders and the egalitarian practice that espoused the Reformation principle of a lay apostolate based on the inner light experience that could occur in the least of them, including, therefore, women. See Melvin Endy, *William Penn and Early Quakerism* (Princeton: Princeton University Press, 1973), chap. 7; H. Larry Ingle, *Quakers in Conflict* (Knoxville: University of Tennessee Press, 1986), chap. 1. Wilkinson and Story drew their support from various northern meetings from Westmorland to Yorkshire as well as from meetings in Bristol, Hertford, and Wiltshire. See W. Mucklow, *The Spirit of the Hat* (London: 1673), concerning Fox's alleged haughty demeanor toward Quaker ministers who disagreed with him.
13. The important primary sources of the Wilkinson-Story dispute are William Rogers, *The Christian-Quaker Distinguished from the Apostate and Innovator* (London: 1680); and John Blaykling *et al.*, *Anti-Christian Treachery Discovered and Its Way Blocked Up* (1683); both documents reveal, in some detail, Fell's instrumental role and the resentment engendered by her intrusion. Note also the Kendal Women's Monthly Meeting Minutes, 1671–1719; Swarthmore MSS, Tr. 5, 9–14. The Wilkinson-Story controversy developed over a number of issues that go beyond the focus of this study. Besides the cases made against Fox's assumed primary leadership and Fell and her women's meetings, other issues included the Wilkinson-Story resistance to the growth of new church laws and the church funding of itinerant ministers. For further discussion of the other specific issues causing the schism, see Braithwaite, *SPQ*, chaps 11–13, especially 296–97.
14. Braithwaite, *SPQ*, 280, 292–92; Fox's *Works*, 8, #317, 79–83. Fox directed that this epistle be read in the men's and women's meetings wherever they assembled, but the SDMM squelched it. George Fox

wrote a letter to the London women Friends, in which he expressed his anger that the SDMM which Fox had originally organized to censor and edit Quaker books, had rejected the printing of one of his own letters. He claimed with some petulance,

> I was not moved to set up that meeting [SDMM] to make orders against the reading of my papers; but to gather up bad books that was scandalous to Friends . . . and not for them to have an authority over the M. & Q. and other meetings or for them to stop things to the nation which I was moved of the Lord to give forth to them.

Since several of Fox's closest colleagues sat on the SDMM, the rejection of one of his epistles was an extraordinary move. In 1675–76 the SDMM was attempting to smooth over internal disputes to present a united front to the world, and they apparently found Fox's letter too acerbic. Braithwaite, *SPQ*, 280; Fox's *Works*, 8, 79–80. See also Jeffrey E. Crosby, "Friends See It Not Safe to Print: The Historical Development of Censorship Among the Quakers in the Seventeenth Century" (M.A. thesis, Brigham Young University, 1983), chaps 3, 4.

15. Rogers, *Christian-Quaker*, Part IV, 7–14, 37–40.
16. Ibid., Part IV, 10–13, 14–15.
17. Blaykling, *Anti-Christian Treachery*, 88. According to Blaykling's account, when John Story was asked if he would go to see Fell for the sake of unity, his response was, "That he would not go over the street to meet her." This grieved Margaret, in Blaykling's words, "having a tender desire of the Man's good, and of his being preserved in the ancient Love and Power, in which he once was an instrument of Good in the church of God." Blaykling also added:

> Several Accusations he [John Story] cast out against her afterwards, to blemish her withal, both publick and private, to the Disparagement of Truth, and our Fellowship therein . . . and a Paper was writ against her . . . urged by him . . . to be sent from the Meeting to her (but that it was stopt by some that shewed a Dislike thereof).

18. Ibid., 88–89,; italics mine.
19. Ibid., 29.
20. Story died in 1681. Wilkinson died c. 1683. *Journal of George Fox*, Nichols ed., 691–92; italics mine.
21. Folio ALS/case 12/Box 24, Pennsylvania Historical Society; also quoted in *JFHS* 50 (1962): 15.
22. Richard Dunn and Mary Maples Dunn, eds, *Papers of William Penn* (Philadelphia: University of Pennsylvania Press, 1981), vol. 1, 360. *The Household Account Book of Sarah Fell* reveals that Penn visited Swarthmoor in April 1676 enroute to the meeting called at Draw-well. Margaret Fell, William Penn, Sarah Fell, and Thomas Lower attended

the meeting together to settle this matter. Curiously, Fox chose to remain at Swarthmoor. Ross, *Margaret Fell*, 288.
23. Braithwaite, *SPQ*, chap. xix; Vann, *Social Development of English Quakerism*, 103–4.
24. Quoted in Braithwaite, *SPQ*, 296–97. Original letter in Spence MSS, III, 165. John Story was not a man to be talked down to so easily, for Lower went on to admit in the letter that upon leaving Fox to return to Westmorland, Story was well mounted, well dressed, and probably outfitted by his "great friends," the Curtises. He sported an "extraordinary broad-brimmed beaver hat and his periwig and broad belt with silver buckles and great hose etc. hath great obeisance rendered to him in the country where he comes by those that know him not." One detects not only a little pique in this description but possibly some fear of Fox's potential rival for leadership. See Howard R. Smith, "The Wilkinson-Story Controversy," *JFHS*, 1; Thomas Curtis and Benjamin Coale, *Reasons why the Meeting House Doors were shut up at Reading* (1686). The Reading separatists did not rejoin the mainline Quakers until 1716, thirty-six years after the controversy began.
25. Fox's *Works*, 7, #35, 43; see LBC, 418–22, 565–66.
26. See Braithwaite, *BQ*, 322, Cambridge *Journal*, I 266, 355.
27. SWM, MSS vii, #168, quoted in Braithwaite, *SPQ*, 218; Fox's *Works*, 7, 14–15; LBC, 418.
28. Fox's *Works*, 7, 284, 347, 341; LBC, 418–22; the 1666 Epistle is reprinted in Barbour and Roberts, *Early Quaker Writings*, 491.
29. Kendal WMMMB, 25; Fox's *Works*, 8, #291, 39–41; quoted in Braithwaite, *SPQ*, 273. In a testimony written in 1676, Fox recalled that the monthly men's meetings were set up in 1667 and 1668, "and afterwards the women's meetings throughout the nation," see *Works*, 7, 14–15.
30. SWMMM, 11; see also Fox's *Works*, 8, #308, 61; xxxiii; Quakers of the primitive period held some beliefs on the restoration of Christ's true church that were closely akin to ideas for which the radicals of the continental Reformation a century earlier lost their lives. However, no evidence exists that proves a direct link between the two similar bodies of ideas.

Restoration ideas were a part of a whole range of sectarian thought current in England in the commonwealth era. One instance of this is seen in Gerrard Winstanley, a contemporary of Fox, who envisioned a secularized version of a regenerate society. Winstanley's ideal was in the form of a protocommunist community where all would share the earth in equality restored to this perfected condition through the use of reason and knowledge of nature. See George Williams, *The Radical Reformation* (1962), 236, 271–73, 375–78; Ian Horst, *The Radical Brethren* (1972); Perez Zagorin, *History of Political Thought in the English Revolution* (London: Routledge & Kegan Paul, 1954), 43–58; Winthrop S. Hudson, "Gerrard Winstanley and the Early Quakers," *Church History* (1943), 177–94.

31. Fell, *Works*, 97–99; see also 56–59.
32. Fell, *Works*, 331–51, *Women's Speaking Justified, Proved, and Allowed of by the Scriptures, all such as speak by the Spirit and Power of the Lord Jesus*. Tracts written in defense of women preachers by Quaker men and women include: Richard Farnsworth, *A Woman Forbidden to Speak* (1655); George Keith, *The Woman Preacher of Samaria* (1674); Anne Whitehead and Mary Elson, *An Epistle for True Love, Unity and Order* (1680). William Mather, an erstwhile Quaker wrote against women in meetings and ministry, *A Novelty, or a Government of Women* (1694?). See also Phyllis Mack, "Women as Prophets During the English Civil War," *Feminist Studies* (Spring 1982); Keith Thomas, "Women and the Civil War Sects," in *Crisis in Europe*, ed. Trevor H. Aston (1975), 317–40; Lloyd, *Quaker Social History*, 108, 119. John Bunyan gave a different exegesis of the same Genesis story of the Fall, namely, that women "are not the image and glory of God, as men are." For an example of the "weaker vessel" motif, as expressed by a contemporary non-Quaker, see Lucy Hutchinson's *Memoirs of the Life of Colonel Hutchinson*, C. H. Firth rev. ed. (London: George Routledge, 1906), xix.
33. Ross, *Margaret Fell*, 409–10.
34. Braithwaite, *SPQ*, 273; Lloyd, *Quaker Social History*, 111; Braithwaite and Lloyd claim that a women's meeting had been established at Bristol prior to 1669. See Rogers, *Christian-Quaker*, pt. i, 64. Ross claims it was the WTWM of London that formed the model for women's meetings in 1671; see Ross, *Margaret Fell*, 285. Fox sent out his circular letter in 1671 with specific definition of woman's place in the restorationist church of Christ:

> Keep your womens meetings in the power of God, which the devil is out of; and take your possession of that which you are heirs of, and keep the gospel order. For man and woman were helps-meet in the image of God, and in righteousness and holiness in the dominion, before they fell; but after the fall in the transgression, the man was to rule over his wife; but in the restoration by Christ, into the image of God, and his righteousness again, in that they are helps-meet a man and a woman, as they were before the fall.

Fox, *Works*, vol. 8, 39–41; this epistle (#291) is included under the year 1672; see Braithwaite, *SPQ*, 273.
35. Ross, *Margaret Fell*, 237–38; Braithwaite, *SPQ*, 273.
36. SWMMM, 27. Fell was in a position to be a competent organizer of the women for she had had twenty years of experience in overseeing and coordinating the infant movement. Gradually, with more and more business done through the London meetings, London by 1670 replaced Swarthmoor as pivotal center for aiding prisoners and traveling ministers. It was at this juncture that Fell, freed from other administrative duties, commenced organization of the local women into a business

meeting of their own. See Fell, *Works*, 97; Braithwaite, *BQ*, 318; *JFHS* 6 (1909).
37. *JFHS* 11, no. 3 (July 1914); Swarthmore MSS I, 365; Ross, *Margaret Fell*, 286.
38. Caton MSS III, 460–2; quoted in Ross, *Margaret Fell*, 287; italics mine.
39. Braithwaite, *SPQ*, 287–88. A yearly meeting of women was established in Ireland in 1679 and one also in the American colonies very early. However, in England, women (with few exceptions) did not sit on important policy-making committees such as the Committee for Sufferings and the SDMM.
40. Ibid., 287–88. The date of these two minutes is 1701.
41. Lloyd, *Quaker Social History*, 49–50; Barbour, 177; C. H. Firth and R. S. Raid, eds, *Acts and Ordinances of the Interregnum* (London: Stationery Office, 1911) 2:715–18.
42. Lloyd, *Quaker Social History*, 50, 62, #15; Fox's *Works*, vii, 336.
43. The LWQM drew together several women's monthly meetings around Lancashire. They met semi-annually in Lancaster. Ross discovered the womens epistle in Nottingham FMH, but she devoted only one paragraph and a footnote to it in *Margaret Fell*. More recently, another copy has come to light in the Arch Street FMH in Philadelphia. It has been transcribed with a commentary. See Milton D. Speizman and Jane C. Kronick, "A Seventeenth-Century Quaker Woman's Declaration," *Signs* 1, no. 1 (1975): 231–45, hereafter abbreviated Speizman-Kronick epistle; Ross, *Margaret Fell*, 298. According to Speizman and Kronick, a third copy of a portion of this epistle has been located in the Haverford College Quaker collection. It probably belonged to the Newport meeting; see Speizman-Kronick epistle, 234–35. Another informative description of the ordering of women's meetings is found in a letter by Sarah Fell to her mother and sisters, written after her marriage to William Meade and her removal to London. As former clerk of the monthly and quarterly women's meetings, she was asked for advice from her successor Rachel, who became clerk after Sarah moved away. See Ross, *Margaret Fell*, 298–99 and *JFHS* 9 : 135–37. The original letter is in Thirnbeck MSS, 15.
44. Speizman-Kronick epistle, 235–41; Fell, *Works*, 231–40.
45. Speizman-Kronick epistle, 235.
46. SWMMM, 136, 140, 196.
47. SWMMM, 85, 189.
48. SWMM, 94–95.
49. SWMMM, 1683, 125. For the tithe controversy, see William W. Spurrier, "The Persecution of the Quakers in England, 1650–1714" (Ph.D. diss., University of North Carolina at Chapel Hill, 1976).
50. SWMMM, 191.
51. Speizman-Kronick epistle, 242.
52. SWMMM, 27–28, 48–49, 93.
53. SWMMM, 92.

54. Speizman-Kronick epistle, 245.
55. SWMMM, 178; see Barry Levy, *Quakers and the American Family* (New York: Oxford University Press, 1988), chaps 1–3.
56. SWMMM, 232B.
57. Anonymous, *Piety Promoted* (London: 1701), pt. iii, 208. This ideal of female deference strongly suggests the predominant social values of a patriarchal society permeated early Quakerism. It also clarifies why an outspoken woman like Fell would encounter hostility among Friends.
58. Mucklow, *Spirit of the Hat*, 28–29; Braithwaite, *SPQ*, 286.
59. Winthrop Hudson, "A Suppressed Chapter in Quaker History," *Journal of Religion* 24 (April 1944): 108–18; see Henry J. Cadbury's response to Hudson's article, *Journal of Religion* 24 (1944): 201–13. Cadbury agreed with Hudson's thesis in part, accepting the idea that Fox was not the first nor completely original leader. More recently, Christopher Hill and Barry Reay have argued a conspiracy theory of suppression. They have also pointed out that there were other contemporary leaders of equal or almost equal status with Fox, such as Edward Burrough and James Nayler, but by 1691 the Quaker movement was definitely "Fox's Movement." Christopher Hill, *World Turned Upside Down* (New York: Penguin Books, 1972), 231, 243; Hill, *The Experience of Defeat*, chap. 5. Barry Reay, *The Quakers and the English Revolution* (New York: St. Martin's Press, 1985), 8, 109. See Mucklow, *The Spirit of the Hat*, 28, for an accusation made against George Fox for his deliberate suppression of the writings of his opponents, written by one of these opponents.
60. Quoted in Braithwaite, *BQ*, 249–50.

Chapter 8

1. A portion of this chapter was published in an article, "Religious Authority and Social Status in Seventeenth-Century England: The Friendship of Margaret Fell, George Fox, and William Penn," *Church History* 57 (1988): 170–186.
2. Of the numerous biographies of William Penn, only a few mention the Fell-Penn nexus: Edward C. O. Beatty, *William Penn as Social Philosopher* (New York: Columbia University Press, 1939); Bonamy Dobree, *William Penn, Quaker and Pioneer* (London: Constable & Co., 1932); William Hull, *William Penn, A Topical Biography* (London: Oxford University Press, 1937); William Comfort, *William Penn, 1644–1718* (Philadelphia: University of Pennsylvania Press, 1944); Isabel Ross, *Margaret Fell*, described the friendship of these early leaders in brief and included some excerpts of their correspondence. See also Richard S. Dunn and Mary Maples Dunn, eds, *The Papers of William Penn* (*TPWP*) 2 vols. (Philadelphia: University of Pennsylvania Press, 1981 and 1982). These volumes include some heretofore unpublished letters of Fell, Fox, and Penn. Other studies of Penn that have a

bearing on this chapter include Melvin B. Endy, *William Penn and Early Quakerism* (Princeton, New Jersey: Princeton University Press, 1973), and Caroline Robbins, "William Penn, 1689–1702: Eclipse, Frustration, and Achievement," in Dunn and Dunn, eds, *The World of William Penn (WWP)* (Philadelphia: University of Pennsylvania Press, 1986). Hugh Barbour, "William Penn, Model of Protestant Liberalism," *Church History* 48 (1979): 156–73; Linda Ford, "William Penn's Views on Women," *Quaker History* 72 (Fall 1983). None of the above characterize the significance of this friendship; cf. Edwin B. Bronner, "George Fox and William Penn, unlikely yoke fellows and friends," *JFHS* 56 (1991): 78–95.

3. Julius Hutchinson, ed. *Memoirs of the Life of Colonel Hutchinson*, C. H. Firth rev. ed. (London: George Routledge, 1906), 2, 9–14; see also the *Dictionary of National Biography*.

4. Keith Wrightson, *English Society, 1580–1680* (New Brunswick, N.J.: Rutgers University Press, 1982), 17; Richard Vann, *The Social Development of English Quakerism* (Cambridge: Cambridge University Press, 1969), 85; Barry Reay and J. F. McGregor, eds, *Radical Religion in the English Revolution* (Oxford: Oxford University Press, 1984), 151, 162; Alan B. Anderson, "A Study in the Sociology of Religious Persecution: The First Quakers," *Journal of Religious History* 9 (1976–77): 255.

5. Most of the twenty-six letters are preserved in the *Papers of William Penn* (Pennsylvania Historical Society, Philadelphia, text fiche), hereafter cited *PWP*. I have compared these to the letters printed in Dunn and Dunn, *TPWP*, and Ross *Margaret Fell*.

6. Although Quakers were not involved in the northern Kaber Rigg plot of 1663 or the earlier Fifth Monarchy uprising in London in 1661, they were accused of conspiring with these radical groups. See Richard Greaves, *Deliver Us from Evil* (New York: Oxford University Press, 1986); Bernard S. Capp, *The Fifth Monarchy Men* (Totowa, NJ: Rowman and Littlefield, 1972).

7. The political oath of supremacy and allegiance to the king harked back to Elizabeth I and James I and was intended originally to curtail subversive Catholic activity in the country. The oath was reinstated and used against Quakers and nonconformists during the Commonwealth period but with greater intensity after the Restoration in the Quaker Act of 1662. For a discussion of the Quaker battle for an affirmation to replace the oath, see J. William Frost, "The Affirmation Controversy and Religious Liberty," in Dunn and Dunn, *WWP*, 303–22.

8. Fell, *Works*, 276–90.

9. Richard Bauman, *Let Your Words Be Few* (Cambridge: Cambridge University Press, 1983), chap. 7.

10. Ibid. *Praemunire* meant confiscation of the defendant's estate by the Crown. Moreover, the person was no longer under the king's protection and remained a prisoner "at the king's pleasure." One further comment concerning Fell's prominence and her leadership in other Quaker

concerns: her influence is seen in a letter sent to Fell from the political democrat, John Lilburne. Lilburne's letter of May 1657 was written after the demise of the Leveller movement and while he was a prisoner at Dover Castle. Lilburne petitioned Fell to help him receive a just settlement in a dispute over some land in County Durham that had been bequeathed to him and subsequently confiscated in the Commonwealth period by Sir Arthur Haselrig. He asked Fell's aid because of her legal and social connections as Judge Fell's wife, in the hope of restoring his title to his formerly-owned land. Lilburne added a postscript in his letter to salute George Fox if he were at Swarthmoor. George Fox, not of gentry status, lacked these political and social connections. The original letter is in Thirnbeck MSS. It is printed in *JFHS* 9 (January 1912): 53–58. Pauline Gregg, *Free-born John* (George G. Harrays & Co., 1961), 339–45; Perez Zagorin, *A History of Political Thought in the English Revolution* (London: Routledge & Kegan Paul, 1954), chap. 2.

11. Braithwaite, *BQ*, 200, 205; Braithwaite, *SPQ*, 282, 311, 454; Bauman, *Let Your Words Be Few*, chap. 7.
12. Braithwaite, *SPQ*, 69–74; Hull, *William Penn*, 185–92; Dunn and Dunn, *TPWP*, 1: 171–80. William Meade (1628–1713) was the London merchant tailor of some means who married Sarah Fell in 1681. Penn's own account of the trial is contained in *The People's Ancient and Just Liberties Asserted, in The Tryal of William Penn, and William Mead at the sessions held at the Old-Baily in London . . .* (London: W. Butler, 1682). See also Bauman, *Let Your Words Be Few, passim*. For a careful study of Quaker legal strategy in the face of persecutions in this period, see Craig Horle, *The Quakers and the English Legal System 1660–1688* (Philadelphia: University of Pennsylvania Press, 1988).
13. Ross, *Margaret Fell*, 216, 246–47. Fox returned to England in June 1673. His wife was also present to welcome him home. On the Fell-Fox wedding note, Edwin B. Bronner, "George Fox and William Penn . . . ," *JFHS* 56 (1991), 81.
14. Abraham MSS; quoted in *JFHS* (1914), 157–58; see also *George Fox's Journal*, 2 (Philadelphia: Gould, 1831 ed. reprint), 149.
15. *PWP*, #605; see Craig Horle, "Changing Quaker Attitudes Toward Legal Defense: The George Fox Case, 1673–5, and the Establishment of the Meeting for Sufferings," in J. William Frost and John M. Moore, eds, *Seeking the Light: Essays in Quaker History* (Wallingford: Pendle Hill Publications, 1986) 17–39.
16. Dunn and Dunn, *TPWP*, 1: 287–89; Hull, *William Penn*, 110–12; Ross, *Margaret Fell*, 352–53.
17. Ross, *Margaret Fell*, 254; Horle, "Changing Quaker Attitudes," 17–39.
18. *PWP*, #610; see Hugh Barbour, "The Young Controversialist," in Dunn and Dunn, *WWP*, 25.
19. Dunn and Dunn, *TPWP*, 1: 334–37.
20. *PWP*, #611.
21. *PWP*, #2691; see Dunn and Dunn, *TPWP*, 1: 359–61.
22. *PWP*, #2691; #2577.

23. *Friends Miscellany*, v. 5, 228; *PWP*, #2577. Printing of books was mentioned in six of twenty-six letters. Early Quaker pamphlet literature was extensive; see Joseph Smith, *A Descriptive Catalogue of Friends' Books*, 2 vols. (London: Joseph Smith, 1867).
24. Endy, *William Penn and Early Quakerism*, 101–2.
25. Dunn and Dunn, *TPWP*, 1: 376–77.
26. Lawrence Stone, *Family, Sex and Marriage in England, 1500–1800* (New York: Harper and Row, 1979), 79–80; see also Irene Q. Brown, "Domesticity, Feminism and Friendship: Female Aristocratic Culture and Marriage in England, 1660–1760," *Journal of Family History* (Winter 1982): 406–24.
27. Dunn and Dunn, "Changing Quaker Attitudes Toward Legal Defense . . . " *TPWP*, 1: 292, 295. *The Works of George Fox* I (1831; reprint, 56). See also Horle, 30–32; Mary Maples Dunn, "The Personality of William Penn," in Dunn and Dunn, *WWP*, 33–14.
28. Fox, *Works*, xxxii–vii.
29. Endy, *William Penn and Early Quakerism*, 136, 178, 335; Mary Maples Dunn, "The Personality of William Penn," in Dunn and Dunn, *WWP*, 3–14.
30. Dunn and Dunn, *TPWP*, 2: 277–78.
31. Norman Penney, ed., *The Household Account Book of Sarah Fell at Swarthmoor Hall* (Cambridge: Cambridge University Press, 1920), 275, 287; Dunn and Dunn, *TPWP*, 2: 460–61.
32. Dunn and Dunn, *TPWP*, 2: 460–61.
33. Fell, *Works*, 10; Ross, *Margaret Fell*, 325; Dunn and Dunn, *TPWP*, 2: 597–98.
34. Ibid; *TPWP*; Miriam Slater, *Family Life in the Seventeenth Century* (London: Routledge & Kegan Paul, 1984), *passim*; see also Ford, "William Penn's Views on Women," 75–102.
35. *PWP*, #1705; see Dunn and Dunn, *TPWP*, 1: 518–19.
36. Hull, *William Penn*, 266–68, 273–76.
37. *PWP*, #1652.
38. Hull, *William Penn*, 178–79, 276; Endy, *William Penn and Early Quakerism*, 135–36, 326; Ross, *Margaret Fell*, 373–75. Ross argues against this interpretation, saying that the evidence is too flimsy. See also Braithwaite, *SPQ*, *passim*. A reading of Penn's *Preface* conveys a sense of Penn's self-assumed leadership. The first twenty-nine pages contain a statement of Quaker principles, followed by eight pages eulogizing Fox, and ending with twelve pages devoted to advice for the continuance of the Society. Penn's *Preface* was subsequently printed under the title *A Brief Account of the Rise and Progress of the People Called Quakers* (London: T. Sowle, 1695). Hugh Barbour has posed the question, Why did Penn not become the theological leader of Quakerism? See Hugh Barbour, "The Young Controversialist," in Dunn and Dunn, *WWP*, 29; see also Caroline Robbins, "William Penn, 1689–1702: Eclipse, Frustration and Achievement," in Dunn and Dunn, *WWP*, 82, on Penn's later life and leadership.

39. *PWP*, #282. The new editions by Dunn and Dunn, *Papers of William Penn*, vols. 3 and 4, contain no material to support the contention that Fell ceased to be friends with Penn over the William Meade affair in 1693–94.
40. Endy, *William Penn and Early Quakerism*, 310–22.
41. Robert Barclay, *The Anarchy of the Ranters* . . . ([London]: 1676), 10–11, 34, 84.
42. Endy, *William Penn and Early Quakerism*, 310–22; Barclay, *The Anarchy of the Ranters*, 60.
43. Endy, *William Penn and Early Quakerism*, 310–22.
44. Ibid., 317–18; Braithwaite, *SPQ*, 228–50; Bonnelyn Young Kunze, "An Unpublished Work of Margaret Fell," *Proceedings of the American Philosophical Society* (December 1986): 424–520.
45. *PWP*, #282; Robbins, "William Penn, 1689–1702: Eclipse, Frustrations and Achievement," in Dunn and Dunn, *WWP*, 81.

Chapter 9

1. I draw on Alan Macfarlane's fine analysis of Ralph Josselin's mental world in this chapter. See Macfarlane, *The Family of Ralph Josselin* (New York: W. W. Norton, 1970), Part IV.
2. Patricia Crawford, "Women's Published Writing 1600–1700," in Mary Prior, ed., *Women in English Society 1500–1800* (London: Methuen, 1985), 211–82; see also Hilda L. Smith and Susan Cardinale, eds, *Women and The Literature of the Seventeenth Century* (Westport, Conn.: Greenwood Press, 1990). See chapter 6 above.
3. Transcript of a letter from Margaret Fell to John Rous, 1 December 1664, Haverford Quaker Collection. #16.
4. Macfarlane, *Ralph Josselin*, 164–65.
5. Quoted from a reprint of the Abraham and Barclay MSS in *JFHS* 11 (1914): 169; *JFHS* 44 (1952): 37; Ross, *Margaret Fell*, 186.
6. Michael Mullet, *Radical Religion*, 122; cf. Ralph A. Houlbrooke, *The English Family, 1450–1700* (London: Longman, 1948), chap. 8; J. William Frost, *The Quaker Family in Colonial America* (New York: St. Martin's Press, 1973), 43–44.
7. Fell, *Works*, A2 and A10.
8. Ibid., 23.
9. Ibid., 45, 29, 27, 299, 300, 53, 45.
10. Margaret Fell, *A Testimonie of the Touchstone . . . Some of the Ranters Principles Answered* (London: Thomas Simmons, 1656), 276–81. Fell rose to a crescendo of curses against the Ranters with whom the Quakers were frequently connected in the popular mind. Christopher Hill has pointed out that because Quakers and Ranters both preached universal salvation and the guidance of the light within and because both groups demoted Scriptural authority in favor of God's Spirit within and condoned female ministers, they were frequently seen as

one and the same group. John Bunyan linked Ranter and Quaker doctrines, making one disparaging contrast: Whereas Ranters "made [these doctrines] threadbare at an ale-house . . . the Quakers have set a new gloss upon them by an outward legal holiness." See Hill, *The World Turned Upside Down* (New York: Penguin Books, 1972, 1980), chap. 10. George Fox expressed equal antipathy toward the Ranters. In one letter to the Second Day Morning Meeting, Fox alluded to their alleged sexual predisposition and license when he wrote: "And who are the same that say the soul of man is a woman's under-petticoat . . . and whether such an Expression is not of Ranterism . . . " Portfolio 10, no. 13, FHL. Letter of 19.11.1683/4. I am indebted to Larry Ingle for this citation.
11. Macfarlane, *Ralph Josselin*, 170–76, 193.
12. Fell, *Works*, 271, 46, 70, 497; Macfarlane, *Ralph Josselin*, 193.
13. Fell, *Works*, 220, 307–10, 496–503; 125–26, 95–97, 197, 297, 530, 319, 98, 487, 82, 76, 80, 128, 90–91, *passim*; Macfarlane, *Ralph Josselin*, 168. See also Bonnelyn Young Kunze, "An Unpublished Work of Margaret Fell," *Proceedings of the American Philosophical Society* (December 1986): 424–52.
14. Fell, *Works*, 2, 53, 70, 273, 254, 288, 533, 48.
15. Katharine R. Firth, *The Apocalyptic Tradition in Reformation Britain, 1530–1645* (Oxford: Oxford University Press, 1979); Paul Christianson, *Reformers and Babylon* (Toronto: University of Toronto Press, 1978); John Collins, *The Apocalyptic Imagination* (New York: Crossroad, 1984); Fell, *Works*, 98, 307, 310, 341–42, 351–56; Macfarlane, *Ralph Josselin*, 163, 184–86.
16. Collins, *Apocalyptic Imagination*, 214–15.
17. Fell, *Works*, 220, 71–91.
18. Quoted in Ross, *Margaret Fell*, 27. see Cambridge *Journal* I, 231. The original is in Swm MSS VII, 24.
19. Macfarlane, *Ralph Josselin*, 189–93.
20. Fell, *Works*, 269, 298, 303, 496, 231, 238–39, 290, 293, 299, 76, 243–44, 395, 257; Kunze, "An Unpublished Work," 434–51.

Chapter 10

1. Perry Miller, *The New England Mind: The Seventeenth Century* (New York: Macmillan Co., 1939) and *Orthodoxy in Massachusetts, 1630–1650* (Cambridge, Mass.: Harvard University Press, 1933); Alan Simpson, *Puritanism in Old and New England* (Chicago: University of Chicago Press, 1955); see Melvin Endy, "Puritanism, Spiritualism, Quakerism," in Richard S. Dunn and Mary Maples Dunn, eds, *The World of William Penn* (Philadelphia: University of Pennsylvania Press, 1986), 281–301, especially 284. Endy's essay delineates the nuances of Puritan and Quaker theological similarities and differences in the historiography covering the past generation of scholars of English

religious sectarianism. The essay raises anew the question: were Puritans and Quakers one movement with a "continuity of experience" or was there a definite "rupture" in the thought and practice of these two groups who hated one another? Endy argues cogently that most Quakers were not Puritans but "Spiritualists." By this he means that, 1) Quakers believed the spirit could operate "apart from Scripture"; 2) the operation of the spirit did not derive from the reason or conscience but from a "special spiritual sense"; 3) the spiritual experience provided infallible certainty of full sanctification and at times; 4) "objective divine knowledge and directives." See also Michael Mullett, *Radical Religious Movements in Early Modern Europe* (1980), chp. 4. Finally, Quaker spiritualism meant reliance on the leadings of the spirit in worship. Endy points out a number of historians, notably Geoffry Nuttall and Hugh Barbour, who have observed that there were deep differences between these antagonistic groups but who see their doctrines as more similar than different in the important matters. Others did not emphasize the similarities at the expense of the differences such as Henry J. Cadbury and Howard H. Brinton. I will draw from Endy's careful study in this chapter.

2. Endy, "Puritanism, Spiritualism, Quakerism," *passim*. Spiritualism has had a variety of meanings which can lead to confusion. For more definition of the Reformation and post-reformation concept of spiritualism, see George Williams, *The Radical Reformation*, Philadelphia: Westminster Press, 1962; Joyce L. Irwin, *Womanhood in Radical Protestantism, 1525–1675*, New York: Mellen Press, 1979, xv-xvii. For a delineation of Puritan-Quaker theological differences, see Allison Coudert, "Henry More, Kabbalah, and Quakers" in Richard Kroll *et al.*, eds, *Philosophy, Science, and Religion*, 32–33.

3. The Baxter-Penn debate and the Fell-Smallwood debate are examples of the ongoing theological disagreements of Quakers and Puritans. See Mary Maples Dunn and Richard S. Dunn, eds, *The Papers of William Penn* (Philadelphia: University of Pennsylvania Press, 1986), 1: 337–52 for an example of the Baptist-Quaker christological debate; John Bunyan, *Some Gospel Truths Opened . . . or, The Divine and Human Nature of Christ Jesus* (London: 1656); Bonnelyn Young Kunze, "An Unpublished Work of Margaret Fell, *Proceedings of the American Philosophical Society* December 1986): 424–52; Endy, "Puritanism, Spiritualism, Quakerism," 285–89, 299; Maurice Creasey, "Early Quaker Christology with Special Reference to the Teaching and Significance of Isaac Penington, 1616–1679" (Ph.D. diss., University of Leeds, 1956).

4. Fox, *Works*, vol. 5, 217. Endy, "Puritanism, Spiritualism, Quakerism," 287, 291–92, 300, n. 43; Nigel Smith, *Perfection Proclaimed . . .*, 9.

5. Fell, *Works*, 242–43; Endy, "Puritanism, Spiritualism, Quakerism," 292.

6. Endy, "Puritanism, Spiritualism, Quakerism," 283; see Dean Freiday,

ed., *Barclay's Apology* . . . (Manasquan, N.J.: Religious Society of Friends, 1967), proposition IV, 78–79.
7. Fell, *Works*, 29.
8. Margaret Fell and George Fox, *A Paper Concerning Such as are made ministers* (London: 1659).
9. Ibid., 294; Kunze, "An Unpublished Paper of Margaret Fell," 438.
10. Ibid., 436; Fell, *Works*, 201, 296–97.
11. Adoptionism was an early church heresy and subsequently an eighth-century Spanish heresy that revived again in the seventeenth century. While orthodox Christianity claims Jesus from conception was God and human, adoptionism repudiated any distinction between the second person of the Trinity and the historic Christ. Christ was seen as the adopted son of God who, through the Word, adopted human nature. Adoptionist ideas were expressed in medieval thought by mystic Duns Scotus, who claimed that Jesus the historical man was the adopted son of God. These ideas were consistently rejected by the church as weakening orthodox christology. See the *Oxford Dictionary of the Christian Church*; Fell, *Works*, 199, 295.
12. Fell, *Works*, 257; Kunze, "An Unpublished Paper of Margaret Fell," 451; Endy, "Puritanism, Spiritualism, Quakerism," 284.
13. Fell, *Works*, *passim*. On Quaker Christology and the concept of the inward light or Light of Christ, see Hugh Barbour, *The Quakers in Puritan England*, 94, 98, 102, 108, 110, 146 *passim*.
14. Kunze, "An Unpublished Paper of Margaret Fell," 434–35, 452; Leonard Trinterud, "A.D. 1689: The End of the Clerical World" (Paper presented at seminar, Theology in Sixteenth and Seventeenth Century England, University of California at Los Angeles, 1971), 29–37. Note also Richard Greaves, *Deliver Us From Evil* (1985), chapters 1–2.
15. Kunze, "An Unpublished Paper of Margaret Fell," 442; Endy, "Puritanism, Spiritualism, Quakerism," 295–96.
16. Kunze, "An Unpublished Paper of Margaret Fell," 443–44.
17. Ibid., 446; Fell, *Works*, 294, 243–44. Pauline reference is I. Corinthians 11: 27–29.
18. Kunze, "An Unpublished Paper of Margaret Fell," 431; Fell, *Works*, 271.
19. Kunze, "An Unpublished Paper of Margaret Fell," 438–40; 447–48.
20. Fell, *Works*, 331; 337; 340.
21. Bernard S. Capp found millenarian ideas in 70 percent of the works published by Puritans between 1640 and 1653. Bernard S. Capp, *The Fifth Monarchy Men* (London: 1972) 39; see also William M. Lamont, *Godly Rule: Politics and Religion, 1603–1660*, London: Macmillan, 1969; Allison P. Coudert, "Henry More, the Kabbalah, and the Quakers," in Richard Kross *et al.*, eds, *Philosophy, Science, and Religion in England 1640–1700* (Cambridge: Cambridge University Press, 1992), 33; Katharine R. Firth, *The Apocalyptic Tradition in Reformation Britain, 1530–1645* (Oxford: Oxford University Press,

1979), 1–2. For a detailed discussion of millennial ideas current in Britain that had pre-Reformation roots, see chaps 1, 6, and 7. See also Perez Zagorin, ed., *Culture and Politics from Puritanism to the Enlightenment* (Berkeley and Los Angeles: University of California Press, 1980), especially the essay by Richard H. Popkin, "Jewish Messianism and Christian Millenarianism," 67–90; Christopher Hill, *The Century of Revolution* (New York: W. W. Norton, 1961, 1980), 143–44. Hill points out that millenarian beliefs, long held by radical proponents of the lower orders, became respectable when espoused by scholars like Thomas Brightman, Joseph Mede, and John Milton. Milton held that Christ was the "shortly-expected King" whose Second Coming would put down all earthly tyrants. See also C. Hill, *The World Turned Upside Down* (New York: Penguin Books, 1972, 1980), 91–98.
22. David S. Katz, *Philo-Semitism and the Readmission of the Jews to England, 1603–1655* (Oxford: Clarendon Press, 1982), 232–44, *passim*. See also Cecil Roth, *A History of the Jews in England* (Oxford: Clarendon Press, 3d ed., 1964), and Cecil Roth, *A Life of Menasseh ben Israel* (Philadelphia: 1934, 1945); Leslie Hall Higgins, "Radical Puritans and Jews in England, 1648–1672" (Ph.D. diss., Yale University, 1979).
23. It has been argued that the framework of Fox's writings was apocalyptic. See Douglas Phillip Gwyn, *Apocalypse of the Word: The Life and Message of George Fox 1624–1691* (Richmond, Ind.: Friends United Press, 1984).
24. Zagorin, *Culture and Politics*, 69–71; there is a substantial body of secondary literature on Puritanism and millenarianism; see, for example, Peter Toom, *Puritans, The Millennium and the Future of Israel, Puritan Eschatology, 1600 to 1660* (Cambridge: James Clarke & Co., 1970); Ernest Tuveson, *Millennium and Utopia* (Berkeley and Los Angeles: University of California Press, 1972); Paul Christianson, *Reformers and Babylon: English Apocalyptic Visions From the Reformation to the Eve of the Civil War* (Toronto: University of Toronto Press, 1978); William M. Lamont, *Godly Rule: Politics and Religion 1603–60* (London: Macmillan, 1969); Christopher Hill, *Antichrist in Seventeenth-Century England* (London: 1971); B. S. Capp, *The Fifth Monarchy Men* (1972); Robert M. Healey, "The Jew in Seventeenth-Century Protestant Thought," *Church History* 46 (March 1977): 63–79; John J. Collins, *The Apocalyptic Imagination* (New York: Crossroad, 1984).

Chapter 11

1. Richard S. Popkin, "Jewish Messianism and Christian Millenarianism" in Perez Zagorin, ed., *Culture and Politics from Puritanism to the Enlightenment* (Berkeley and Los Angeles: University of California Press, 1980), 67–90; David S. Lovejoy, *Religious Enthusiasm in the*

New World (Cambridge, Mass.: Harvard University Press, 1985), 22; David Katz, *Philo-Semitism and the Readmission of the Jews to England, 1603–1655* (Oxford: Oxford University Press, 1982), 156, 232–244; see also Robert M. Healey, "The Jew in Seventeenth-Century Protestant Thought," *Church History* 46 (March 1977): 63–79. See Gerhard Scholem Gershom, *Sabbatai Sevi: The Mystical Messiah* (Princeton, N.J.: Princeton University Press, 1973) for the Sabbatian movement. The Fifth Monarchists also calculated that Christ would return in the 1660s. The Whitehall Conference adjourned at the end of the 1656 without issuing an official invitation to Jews to return to England. Menasseth and English millenarians were mutually disappointed.

2. Margaret Fell, *Works*. The Thomason collection gives the date of *For Menasseth-ben-Israel* as 20 February 1656 and for *A Loving Salutation* . . . as 31 October 1657.
3. Abraham MSS III, case 17.
4. The original Stubbs letter is in Swm. Mss I, 92. There is some uncertainty about the date of publication of *A Loving Salutation* and her *Certain Queries*. Another extant undated letter from Fell to Stubbs indicates that her book *A Loving Salutation* was first printed in 1656. That earlier edition included *Certain Queries* for publication. The Joseph Smith catalogue dates *A Loving Salutation* in 1656, while her *Works* dates it and the *Certain Queries* in 1657. See Leslie Higgins, "Radical Puritans and Jews in England, 1648–1672 (Ph.D. diss., Yale University, 1979), 102–6 for more details on the intricacies of Fell's early publications and translations for the Dutch Jews. Isabel Ross, in *Margaret Fell*, ignored the *Queries* as a separate publication and included two other tracts instead: *A Call to the Universal Seed of God* (1664) and *The Daughter of Zion Awakened* (1677). These treatises addressed the Jews and a wider audience as well.
5. Zagorin, *Culture and Politics*, 74–77.
6. Quoted in Richard H. Popkin, "Spinoza's Relations with the Quakers in Amsterdam," *Quaker History* (Spring 1984): 15, 14–28. I am relying on Popkin's excellent article for the chronology of the Dutch Jewish mission.
7. Ibid., 15–16.
8. Ibid., 15–17.
9. Ibid., 18–19; Ross, *Margaret Fell*, 92; Fell, *Works*, 99–100.
10. Popkin, "Spinoza's Relations with the Quakers in Amsterdam," 19. For a study of Ames' theology see Keith L. Springer, *The Learned Doctor William Ames* (Urbana, Ill.: University of Illinois Press, 1972).
11. Ross, *Margaret Fell*, 91–92; see Caton MSS III, 13, 23–25, 35.
12. Popkin, "Spinoza's Relations with the Quakers in Amsterdam," 20.
13. Ibid., 20–21.
14. Ibid., 19–21. Fox wrote several tracts to the Jews. See Fox's *Works* 4: 53–57; *A Looking Glass*, in 5: 61–84; *A Declaration*, in 4: 290–97; Fox's other treatises to the Jews include *An Answer to the Arguments*

of the Iewes (1661), *An Epistle to All Christians, Jews, and Gentiles* (1682), and *An Epistle to the Household of the Seed of Abraham* (1682).

15. Popkin, "Spinoza's Relations with the Quakers in Amsterdam," 22–23.
16. Swm MSS I, Tr., 468; see also Swm MSS IV, 28; Caton MSS II, 48, 50, 507.
17. Hugh Barbour, *The Quakers in Puritan England* (New Haven, Conn.: Yale University Press, 1964), 181–83.
18. Fell, *Works*, 109, 112; Barbour, *The Quakers in Puritan England*, 185–87.
19. Fell, *Works*, 473, 478; Barbour, *The Quakers in Puritan England*, 189; Healey, "The Jew in Seventeenth-Century Protestant Thought," *passim*. For further analysis of the influences of Jewish messianic thought on Christian millennial thought see Gershom Scholem, *Sabbatai Sevi*, part v, 93–102.
20. Fell, *Works*, 191, 470, 152, 112, 109; Fox, *Works*, 4: 58.
21. Fell, *Works*, 118 and 475; Fox's *Journal* I, liv. A study of the Quaker-Jewish debate over theological issues such as christology, election, and sanctification is found in Higgins, "Radical Puritans and Jews in England, 1648–1672," chaps 4 and 5. Fell's and Fox's use of the term "Jew" was marked by their conception of the "Jew Inward" and "Jew Outward," which became a fundamental motif of their theological ideas. According to L. Higgins, by "using typologies such as 'Jew Inward' and 'Jew Outward,' leading Quaker thinkers worked out essential tenets of election, sanctification, and the doctrine of the divine light within."
22. Fell, *Works*, 468.
23. Ibid., 108–11, 119–23, 155, 171–73, 476, 478, 479; Healey, "The Jew in Seventeenth-Century Protestant Thought," 67–68.
24. Fox, *Works*, 4, 53; 296–97; 357.
25. Fell, *Works*, 472; Matthew 5:17–18; Deuteronomy 6:6.
26. Healey, "The Jew in Seventeenth-Century Protestant Thought," 72–73.
27. Fell, *Works*, 153–54.
28. Ibid., 182–83.
29. Healey, "The Jews in Seventeenth-Century Protestant Thought," 73.
30. Fox, *Works*, 5: 66–68. Higgins, "Radical Puritans and Jews in England, 1648–1672," sees Fox's later writings as open to the Jewish question and he, Fell, Pennington, and others sustained frequent criticisms from other Quakers as "Judaizers"; see chap. 5.
31. Healey, "The Jew in Seventeenth-Century Protestant Thought," 73–74.
32. Fell, *Works*, 189.
33. Ibid., 124, 165.
34. Ibid., 124, 189, 324, 105.
35. Fox, *Works*, page citation lost; 70.
36. Higgins, "Radical Puritans and Jews in England, 1648–1672," chap. 4; Fell, *Works*, 471–79; Braithwaite, *SPQ*, 385, *passim*; Maurice Creasey, "Early Quaker Christology with Special Reference to the Teaching and

Significance of Isaac Pennington, 1616–1679" (Ph.D. diss., University of Leeds, 1956), section II.
37. Barbour, *The Quakers in Puritan England*, 188.
38. Fox, *Journal* II, 464; see LBC 787; Fox, *An Answer to the Arguments of Jews* (1661); see LBC, 787.
39. Fox, *Journal* II, 278, 311; LBC 787–88.
40. Fell, *Works*, 514, 525.
41. Fell, *Works*, 525–26; Fell's quote is a blend of quotations from Galatians and Colossians; Fell's tract, *Women's Speaking Justified* in her *Works*, 331–51, is her most clear-cut statement supporting women in public ministries.
42. The radical nature of the Quaker message on female ministry cannot be overemphasized. The Levellers of the 1650s were one of the most radical political sects of the Interregnum, and they did not advocate total male suffrage, nor suffrage for women. Quakerism offered a unique opportunity to women to exercise their gifts in publicly recognized roles. That the Levellers were suppressed says much about English societal attitudes of the mid-century, which early Quakers, especially women like Fell, confronted head-on. See Christopher Hill, *The Century of Revolution* (New York: W. W. Norton, 19), 151.
43. Douglas Gwyn, *Apocalypse of the Word* (Richmond, Ind.: Friends United Press, 1984), 137–38; 442–44, *passim*. See also J. Van den Berg, "Quaker and Chiliast: The 'contrary thoughts' of William Ames and Patrus Serranius," in R. B. Knox, ed., *Reformation, Conformity and Dissent* (London: Epworth Press, 1977), 195.
44. Gwyn, *Apocalypse of the Word*, 301, 431–33; 1 Thess. 1:8.
45. Ibid., 421–22, 429, 301.
46. Hill, *The Century of Revolution*, 143–46.
47. Fox, *Journal* I, 341–43, 384; II, 165. Thomas Ellwood found the passages written by Fox on his depressions material "fit to be left out" of the 1694 *Journal*. See N. Penney's introduction in the Cambridge Journal I, xviii. See A. N. Brayshaw, *The Personality of George Fox* (1919 and 1933), 38, 42–44, 48, 79–80.
48. Fell, *Works*, 520, 526.
49. Hill, *The Century of Revolution*, 146.
50. Fox, *Journal* II, 169; Brayshaw, *The Personality of George Fox*, 79–80.
51. Fox, *Works* 4: 303; quoted in Brayshaw, *The Personality of George Fox*, 80.
52. Scholem, *Sabbatai Sevi*, 94.

Chapter 12

1. See Bonnelyn Young Kunze, "An Unpublished Work of Margaret Fell," *Proceedings of the American Philosophical Society* (December 1986): 424–452.

2. Mary Maples Dunn, "The Personality of William Penn," in Mary Maples Dunn and Richard S. Dunn, *The World of William Penn* (Philadelphia: University of Pennsylvania Press, 19), 5–7.
3. Barry Reay, *The Quakers and the English Revolution* (New York: St. Martin's Press, 1985), 7–9; Christopher Hill, *The Experience*, 166.
4. SWMMM, 5 November 1700.

Appendix

Table 1 Fell Farm Income from 1673 to 1678

Years	Total Recorded P. A. Income	Incomplete Years & Dates
1673	£215	Mar. 21–Sept. 25
1674	£589	Complete
1675	£263	June 3–Oct. 28; Jan. 5–Feb. 19
1676	£408	Complete
1677	£765	Complete
1678	£245	Aug. 1–Mar. 21

Table 2 Swarthmoor Hall and Marsh Grange: Total Number of Servants and Laborers Employed by Fells in 1676

	Male	Female
Servants	6	6
Day-Wage Laborers	26	18
TOTAL LABOR FORCE	32	24

Percent of total labor force that was female = 43%

Table 3 Comparison of Charity Between Fells of Swarthmoor Hall, 1676 Sir Daniel Fleming of Rydal Hall, 1676 And Anne, Ctsse of Pembroke in Cumberland, Part of 1673

	Males	Female	% of Total Who Were Female
Swarthmoor Hall Charity Recipients	2	6	75%
Daniel Fleming Charity Recipients	11	2	15%
Anne, Ctsse of Pembroke	1	1	50%

Table 4 *How the Fells Spent Their Income in 1676*

Cost of Running the Farm and Household		Additional Expenses – Family, Business, and Religious	
1. House and farm expenses	£42.12.19	1. Charity – individual and corporate giving (3% of total annual income)	£13.14.06
2. Wages of servants and laborers	41.04.04	2. Loans – interest free to relatives and friends (16% of total annual income)	64.10.00
3. Food for family and laborers	18.10.07	3. Travel expenses	10.17.01
4. Clothing for Fells and Lowers (11 persons)	18.14.03	4. Correspondence (including Fox's)	4.11.00
Clothing for servants	1.09.03	5. George Fox's expenses while in residence	14.06.00
Total Cost	£122.11.06	Total Cost	£107.18.07
Savings for 1676: £179			
Charitable Activities for 1676 (real charity plus loans) (19% of total annual income)	£78.04.06		

Bibliography

Manuscripts

Anonymous. "The Burning Lamp Set Up Before the Throne," MS. ADD 62932.
Abraham MSS. A collection of letters, deeds, etc., originally collected at Swarthmoor Hall and subsequently presented by the Abraham descendents of Margaret Fell. Located at Friends' House Library, London.
Abram Rawlinson Barclay MSS. "Letters of Early Friends." A collection of about 250 original letters of early Quakers from 1659–1688, edited by A. R. Barclay. Friends' House Library, London.
Box Meeting MSS. Contains numerous letters of women's meetings in America and England, sent to and preserved by the London Women's Yearly Meeting, seventeenth and eighteenth centuries. Friends' House Library, London.
Brownsword, William. *The Quaker-Jesuite, or Popery in Quakerism*. Fleming MSS, 1660.
Cash MSS. Letters from seventeenth-century Quakers. Friends' House Library, London.
Caton MSS. A folio volume of 158 pages containing copies of letters of early Friends. Most are in the handwriting of William Caton. Friends' House Library, London.
Etting Papers. Contains letters of Fox to the Women's Meetings. 1673, 1674, and 1677 and a 1672 letter from Rebecca Travers to Margaret Fell. Located at Historical Society of Pennsylvania, Philadelphia.
Fell, Margaret. "M[argaret] ff[ell]s Answer to Alan Smallwood, priest of Gravestock in Cumberland, (1668)." MS. ADD 62931/352I. This is a vigorous rebuttal to Smallwood's criticism of some of Fell's writings. Located in British Library, London. See Bonnelyn Young Kunze, "An Unpublished Work of Margaret Fell," below.
Fell, Margaret, James Taylor, and Thomas Atkinson. "The standard of the Lord lifted up against the Beast which is pushing with his iron horns against the family of the Lamb" (1653). MS. ADD 62930/352I. A unique MS that is an answer to a book entitled *A Looking Glass for George Fox*. Located at British Library, London.
Fleming Mss of S. H. le Fleming, Esq. of Rydal Hall, 1890, LPRO, Kendal.
Force Forge Account Book, MSS. DD #J, 278.
Gibson Mss, FHL.
Kendal Women's Monthly Meeting Minute Book 1672–1702, LPRO, Kendal.
Lancashire Men's and Women's Quarterly Meeting Minutes, 1671–1702, LPRO, Preston.

London Yearly Meeting MS. Manuscript records of the yearly meeting complete from 1672 to present in 31 vols. Located at Friends' House Library, London.

MSS Portfolio 10/53. A manuscript fragment of the Fell-Fox marriage.

Minute Book of the Reading Monthly Meeting 1668–1716 (Curtis Party). Transcribed by Nina Saxon Snell, n.d. in the FHL.

Papers of William Penn (PWP). Albert Cook Myers Collection. Located at Historical Society of Pennsylvania, Philadelphia.

Portfolio no. 10, FHL contains miscellaneous letters of early Quakers.

Rawlinson MS. "[A book to] goe abroad onely among all Friends ith [in the] Truth [in answer] to seuerall papers of Margret Foxe . . . formerly called Margret Fel . . . toucheing the unjust orders, papers and illegale proceedings . . . on her behalfe . . . as toucheing my stewardship for her . . . at Forse Fordge [Force Forge]" (1680). A 455-page, handwritten document giving Rawlinson's account of his nineteen-year-long business dispute with Margaret Fell. Located at Friends' House Library, London.

Rawlinson Ms. 278/2 Account Book of Thomas Rawlinson 1658–63. Located at LPRO, Preston. Transcribed by Brian Awty, 1976.

Spence MSS. A Collection of Seventeenth-century MSS. belonging to Robert Spence of North Shields and London. 3 vols. The first and second volumes comprise the MS. of the "Journal of George Fox" and the third volume contains numerous letters to and from the Fell family. Located at Friends' House Library, London.

Swarthmore MSS. Early Quaker Letters from the Swarthmore MSS. to 1660. A collection of about 1400 original seventeenth-century letters, papers, etc. These and many other manuscript records of early Quakerism were preserved at Swarthmoor Hall until 1759, when they were dispersed on the sale of the estate. They were gradually collected and preserved at Friends' House Library, London.

Swarthmore MSS. Transcripts by Emily Jermyn at Friends' House Library, London. There is one small volume of Emily Jermyn transcripts in the Quaker Collection at Haverford College, Haverford, Pennsylvania.

Swarthmoor Men's Monthly Meeting Minutes 1668–1674; 1691–1715. Located at Friends' House Library, London.

Swarthmoor Women's Monthly Meeting Minutes, 1671–1700, LPRO, Preston, Lancashire.

Thirnbeck MSS. A lesser-sized collection of letters originally kept by Fell at Swarthmoor Hall. Located at Friends' House Library, London.

Printed Works

Anonymous, ed. *Journal of the Life of William Caton*. London: 1689.
———. *The Lamb's War*. 1658.
———, ed. *The Life of Margaret Fox, Wife of George Fox. Compiled From Her Own Narrative and Other Sources*; with a selection from her epistles, etc. Philadelphia: Book Association of Friends, 1885.

———. *The Memory of John Story Revived*. 1683.

———. *These several papers were sent to the Parliament . . . being above 7,000 of the Names of the Handmaids and Daughters of the Lord*. London: 1659.

———. *A Friendly Dialogue . . . Concerning Women's Preaching . . . "* Swarthmore MSS. Friends' House Library, London. Port. 31, 21.

———. *Piety Promoted in a Collection of Dying Sayings of Many of the People Called Quakers by Various Editors*. London: 1st ed., 1701, 2d ed. 1829.

Barclay, Robert. *An Apology for the true Christian Divinity, as the same is held forth and preached by the people called in scorn, Quakers: being a full explanation and vindication of their principles*. John Forbes, 1678.

———. *Anarchy of the Ranters and other Libertines; the hierarchy of the Romanists, and other pretended churches, equally refused and refuted, in a twofold apology for the church and people of God called in derision Quakers, wherein they are vindicated*. (1674).

Bardsley, C. W., and L. R. Ayre. *Registers of the Parish of Ulverston*. Ulverston: 1886.

Baxter, Richard. *The Quakers catechism, or the Quakers questioned, their questions answered, and both published, for the sake of those of them that have not yet sinned unto death; and of those ungrounded*. Printed by A. M. for Tho. 1655.

———. *One Sheet Against the Quakers*. London: Printed by Robert White for Nevil Simmons, 1657.

Besse, Joseph. *A Collection of the Sufferings of the People Called Quakers, from 1650–1689. Taken from Original Records and Other Authentick Accounts*. 3 vols. London: J. Soule, 1733.

Birch, Thomas, ed. *A Collection of the State Papers of John Thurloe. Containing authentic memorials of the English affairs from the year 1638, to the Restoration of King Charles II*. London: F. Gyles, 1742.

Bishop, George. *Vindication of the Principles and Practices of the People Called Quakers from the False Aspersions of being Monstrous in their Opinions as to Religion*. 1665. London.

———. *New England Judged by the Spririt of the Lord*. London: T. Sowle, 1703.

Blaugdone, Barbara. *Account of the Travels of Barbara Blaugdone*. London: 1691.

Blaykling, John et al. *Anti Christian Treachery Discovered and its way block'd up, in a clear distinction betwixt the Christian apostolical spirit*. (London? 1686?)

Bunyan, John. *Some Gospel Truths Opened according to the scriptures, or, The Divine and Human Nature of Christ Jesus, his coming into the world; his righteousness, death, resurrection, ascension, intercession*. London: 1656.

Burrough, Edward. *A Declaration from the People Called Quakers to the present distracted nation of England with mourning and lamentation over it because of its breaches and the cause there-of laid down, with advice *and council how peace, union, and happiness may be restored, and all the present troubles removed.* London: Printed for Th. Simmons, 1659.

———. *A Trumpet of the Lord Sounded Forth in Sion: which sounds forth the controversie of the Lord of Hosts.* London: Giles Calvert, 1656.

———. *Works.* 1672.

Calamy, Edmund. *The Nonconformists Memorial being an Account of the Lives, Sufferings and Printed Words of the Two Thousand Ministers ejected from the Church of England chiefly by the Act of Uniformity.* London: W. Harris, 1775.

Curwen, Alice. *A Relation of the Labour, Travail and Sufferings of That Faithful Servant of the Lord.* 1680.

Dewsbury, William. *The Faithful Testimony of that ancient Servant of the Lord, and Minister of the Everlasting Gospel William Dewsbery his Books, Epistles and Writings, collected and printed.* London: A. Sowle, 1689.

Ellwood, Thomas, ed. *The Journal of George Fox.* 1694.

Elson, Mary. *Piety Promoted by Faithfulness . . . Testimonies re. Anne Whitehead.* London: 1686.

Farnsworth, Richard. *A Woman Forbidden to Speak in the Church.* London: 1655.

Fell, Margaret. *A Brief Collection of Remarkable Passages and Occurrences Relating to the Birth, Education, Life, Conversion, Travels, Services and deep Sufferings of that ancient Eminent and Faithful Servant of the Lord, Margaret Fell, but by her second Marriage M. Fox.* London: J. Sowle, 1712. (Cited as *Works.*)

———. *Aen Manasseth ben Israel, den roep der Joden nyt Babylonien.* Amsterdam: 1657.

———. *A call to the Universall seed of God.* London: 1665.

———. *A call to the Seed of Israel.* Robert Wilson, 1668.

———. *The Citie of London Reproved.* London: Robert Wilson, 1660.

———. *The Daughter of Sion awakened.* London: 1677.

———. *False Prophets Antichrists, Deceivers Which Are in the World . . .* London: Giles Calvert, 1655.

———. *A loving salutation to the seed of Abraham.* London: Th. Simmons, 1656.

———. *A Paper concerning such as are made Ministers.* H. W., 1659.

———. *The Standard of the Lord revealed.* London: 1667.

———. *This is to the clergy.* London: Robert Wilson, 1660.

———. *This is to Major Generall Harrison.* London: Th. Simmons, 1660.

———. *A Testimonie of the Touchstone for all Professions, and all Farms and Gathered Churches . . . and A Tryal by the Scriptures, who the False Prophets are . . . also Some of the Ranters Principles Answered.* London: Thomas Simmons, 1656.

———. *This is an Answer to John Wiggon's Book.* London: 1655.

———. *To the General Council of Officers of the English Army.* London: Thomas Simmons, 1659.

———. *To the general councel.* London: Thomas Simmons, 1659.

———. *A Touch-stone, or a perfect tryal.* London: 1667.
———. *A true testimony from the people of God.* London: Robert WIlson, 1660.
———. *Two general epistles to the flock of God.* London: 1664.
———. *Women's Speaking Justified.* London: 1666; 2nd edn., 1667.
Fell, Margaret, and George Fox. *A Paper Concerning such as are made ministers by the Will of Man and an exhortation to all sober minded people to come out from among them.* London: MW, 1659.
Fell, Margaret, and James Parke. *Two General Epistles to the Flock of God (at the severall assizes held at Lancaster 1664. Tracts,* 101/26.
Fox, George. *Journal of George Fox.* Edited by Norman Penney. Cambridge: Cambridge University Press, 1911.
———. *The Works of George Fox.* 8 vols. 1831. Reprint, New York: AMS Press, 1975.
Fox, Margaret Fell. "The Testimony of Margaret Fox, concerning her late husband, George Fox; together with a brief account of some of his Travels, Sufferings, and Hardships endured for the Truth's Sake."
Friends' Registers, 1650 to Present. London: Friends' House Library.
Gough, John. *History of the People Called Quakers.* 4 vols. Dublin: R. Jackson, 1789–90.
Green, Mary Ann Everett, ed. *Calendar of the Proceedings of the Committee for Compounding with Delinquents etc., 1643–1660.* London: H. M. Stationery Office, 1990; reprint, 1967.
Calendar of State Papers, Domestic of the reign of Charles II, 1660–1685. 28 vols. London, H. M. Stationery Office, 1860–1968.
Calendar of State Papers, Domestic of the reign of James II, 1685–1688. 3 vols. London, H. M. Stationery Office, 1960–1972.
Calendar of State Papers, Domestic of the reign of William and Mary, 1689–1693. 5 vols. London, H. M. Stationery Office, 1895–1906.
Higginson, Francis. *A Brief Relation of the Irreligion of the Northern Quakers, Wherein their horrid principles and practices, doctrines and manners are plainly exposed.* London: T. R., no date.
Howgill, Francis. *The Works of darkness brought to Light.* London: Th. Simmons, 1659.
Keith, George. *The Woman-preacher of Samaria.* London: 1674.
Latham, Robert, and William Matthews, eds. *The Diary of Samuel Pepys.* Berkeley and Los Angeles: University of California Press, 1970–83.
Loder, Robert. *Robert Loder's Farm Accounts 1610–1620.* Edited by G. E. Fussell. Series LIII. London: Offices of the Society, Russell Square, 1936.
London Yearly Meeting Minutes. Washington, D.C.: Folger-Shakespeare Library, 19.
Marshall, J. D., ed. *The Autobiography of William Stout of Lancaster, 1665–1752.* Manchester: The Chetham Society, 1967.
Mather, William. *A Novelty: On a Government of Women Distinct From Men Erected Amongst Some of the People Called Quakers Detected.* c. 1694.

Matthews, A. G., ed. *Calamy Revised, being a Revision of Edmund Calamy's Account of Ministers and Others Ejected and Silenced, 1660–2.* Oxford: The Clarendon Press, 1988.

Minute Book of the Men's Meeting of the Society of Friends in Bristol 1667–1686. Vol. XXVI. Bristol: Record Society Publications, 1971.

Mucklow, W. *The Spirit of the Hat, or, The government of the Quakers among themselves as it hath been exercised of late years by George Fox and other leading-men in their Monday or second-dayes meet.* London: F. Smith, 1673.

Nuttall, Geoffrey, ed. *Early Quaker Letters from the Swarthmore MSS to 1660.* Unpublished MS. London: Friends' House Library, 1952.

Pagitt, Ephraim. *Heresiography: or a description of the hereticks and sectaries sprang up in these latter times.* 5th edition. London: William Lee, 1654.

Penn, William. *The People's Ancient and Just Liberties Asserted [sic] in the tryal of William Penn, and William Mead, at the sessions held at the Old Baily in London, the first, third, fourth and fifth of Sept. 70 against the most arbitrary procedure of that court.* London, 1670.

———. *A Brief Account of The Rise and Progress of the People Called Quakers.* London: T. Sowle, 1695.

———. *No Cross, No Crown; a discourse, showing the nature and discipline of the holy cross of Christ.* York, England: William Sessions, reprint, 1981.

Pennington, Mary. *Experiences in the Life of Mary Pennington.* London: Headley Brothers; Philadelphia: The Biddle Press, 1911.

Penney, Norman, ed. *Extracts from State Papers Relating to Friends.* London: Headley Brothers; Philadelphia: H. Newman, 1910–1913.

———. *The Household Account Book of Sarah Fell of Swarthmoor Hall.* Cambridge: Cambridge University Press, 1920.

———, ed. *Journal of George Fox.* 2 vols. Cambridge: Cambridge University Press, 1911.

Pyot, Edward, and William Salt. *West Answering to the North in the Fierce and Cruel Persecution of George Fox, Ed. Pyot, and William Salt at Lancaster in Cornwall.* London: 1657.

Quaker Women's Diaries. London: World Microfilm Publications, 1978.

Quaker Scrap Book. Philadelphia: Historical Society of Pennsylvania.

Rogers, William. *The Christian Quaker Distinguished from the Apostate and Innovator.* London: 1680.

Sewel, William. *History of the Rise, Increase and Progress of the Christian People Called Quakers.* London: J. Sowle, 1722.

Smallwood, Alan, D.D. *A Sermon Preached at Carlisle, August 17, 1664.* York: Stephen Bulkley, 1665.

———. *A Reply to a Pamphlet Called Oaths no Gospel-Ordinance . . .* York: Stephen Bulkley, 1667.

Smith, Joseph. *A Descriptive Catalogue of Friends' Books* 2 vols. London: Joseph Smith, 1867.

———. *Bibliotheca Anti-Quakeriama.* London: Joseph Smith, 1873.

Spencer, E. *A Brief Epistle to the Learned Manasseh Ben Israel.* London: 1650.
Stillingfleet, John. *Seasonable Advice Concerning Quakerism.* 1702.
Stirredge, Elizabeth. *Strength in Weakness Manifest in the Life of . . . E. Stirredge, Philadelphia.* London: T. Sowle, 1711, 1810.
Tomlinson, W. *A Bosome opening to the Jewes: Holding forth to others some Reasons for our receiving of them into our Nation.* London: Giles Calvert, 1656.
Turner, George Lyon, ed. *Original Records of Early Nonconformity under Persecution and Indulgence.* London: T. F. Unwin, 1911.
Vokins, Joan. *God's Mighty Power Magnified: As Manifested . . . in his Faithful Handmaid, Joan Vokins.* London: Th. Northcott, 1691.
Webb, Maria. *The Fells of Swarthmoor Hall and their Friends.* London: 1865.
Whitehead, Anne, and Mary Elson. *An Epistle for True Love, Unity and Order in the Church of Christ Against the Spirit of Discord, Disorder and Confusion.* London: Andrew Sowle, 1680.
Whiting, John. *A Catalogue of Friends' Books Written by many of the people, called Quakers, from the beginning or first appearance of the said people.* London: J. Sowle, 1708.
Whitrow, Joan. *The Work of God in a Dying Maid.* London: 1677.
Wilkinson, John et al. *The Memory of the Servant of God John Story, Received,* . . . . John Gain, 1683.

Selected Secondary Works

Anonymous. *The Life of Margaret Fox.* Philadelphia: Book Association of Friends, 1885.
Abbott, Dilworth. *Quaker Annals of Preston and the Fylde, 1653–1900.* London: Headley Brothers, 1931.
Ackroyd, Peter, ed. *Cambridge History of the Bible.* Cambridge: Cambridge University Press.
Alymer, G. E. *The State's Servants.* London: Routledge & Kegan Paul, 1973.
Amussen, Susan Dwyer. *An Ordered Society.* Oxford: Basil Blackwell, 1988.
Anderson, Alan B. "Lancashire Quakers and Persecution." M.A. diss., University of Lancaster, 1971.
Armytage, W. H. G. *Heavens Below: Utopian Experiments in England.* London: 1961.
Arnold, H. G. *Early Meeting Houses.* London: Friends' House Library, unpublished typescript.
Aston, Trevor, ed. *Crisis in Europe.* London: Routledge & Kegan Paul, 1965, 1975.
Bagley, J. J. *A History of Lancashire with Maps and Pictures.* London: Darwen Finlayson, 1956.

Baines, Edward. *History of the County Palatine and Duchy of Lancaster.* Revised and edited by John Harland. 4 vols. London: George Routledge and Sons, 1863, 1970.

Baker, D. *Sanctity and Secularity: The Church and the World.* Oxford: 1973.

Barber, Henry. *Notes on Furness and Cartmel; or Jottings of Topographical, Ecclesiastical and Popular Antiquities.* Ulverston: James Atkinson, 1894.

———. *Swarthmoor Hall and Its Associations.* London: F. B. Kitto, [1872].

Barbour, Hugh. *Margaret Fell Speaking.* Wallingford, Pa.: Pendle Hill Publishers, 1976.

———. *The Quakers in Puritan England.* New Haven, Connecticut: Yale University Press, 1964.

Barbour, Hugh, and Arthur O. Roberts, eds. *Early Quaker Writings 1650–1700.* Grand Rapids, Michigan: Wm. B. Eerdman's Publishing Co., 1973.

Barclay, Robert. *Inner Life of the Religious Societies of the Commonwealth:, considered principally with reference to the influence of church organization and the spread of Christianity.* London: Hodder and Stoughton, 1879.

Barclay, A. R., ed. *Letters etc. of early Friends illustrative of the history of the society from nearly its origin to about the period of George Fox's decease with documents respecting its early discipline, also epistles of counsel and exhortation.* London: Harvey and Darton, 1841.

Bauman, Richard. *Let Your Words Be Few.* Cambridge: Cambridge University Press, 1983.

Baxter, Stephen B., ed. *England's Rise to Greatness.* Berkeley and Los Angeles: University of California Press, 1983.

Beatty, Edward C. *William Penn as Social Philosopher.* New York: Columbia University Press, 1939.

Bebb, Evelyn D. *Nonconformity and Social and Economic Life 1660–1800.* Philadelphia: Porcupine Press, 1980.

Beck, William, and T. Frederick Ball. *The London Friends' Meetings.* London: F. B. Kitto, 1869.

Beier, A. L. *Masterless Men: The Vagrancy Problem in England.* London: Methuen, 1985.

———. *The Problem of the Poor in Tudor and Early Stuart England.* London: Lancaster Pamphlets, 1983.

Beier, A. L., David Cannadine, and James M. Rosenheim. *The First Modern Society.* Cambridge: Cambridge University Press, 1989.

Bell, Maureen, George Parfitt, and Simon Shepherd. *A Biographical Dictionary of English Women Writers 1580–1720.* Boston: G. K. Hall and Co., 1990.

Benson, Lewis. *Notes on George Fox.* London: Friends' House Library, unpub. typescript, 1981.

———. *What did George Fox Teach about Christ?* Gloucester, England: George Fox Fund, 1976.

Best, Michael, ed. *The English Housewife*. Montreal: McGill-Queens University Press, 1986.
Bickley, A. C. *George Fox and the Early Quakers*. London: Hodder and Stoughton, 1884.
Blackwood, Bruce G. *The Lancashire Gentry and the Great Rebellion 1640–1660*. Manchester: The Chetham Society, 1978.
Blunt, John H. *Dictionary of Sects, Heresies, Ecclesiastical Parties and Schools of Religious Thought*. London: Longman, Green & Co., 1903.
Bowden, James. *The History of the Society of Friends in America*. 2 vols. London: 1850–54.
Brailsford, Mabel Richmond. *Quaker Women, 1650–1690*. London: Gerald Duckworth & Co., 1915.
Braithwaite, William C. *The Beginnings of Quakerism*. 1912. 2d ed. rev. by Henry J. Cadbury. Cambridge: Cambridge University Press, 1955.
———. *The Second Period of Quakerism*. 2d ed. rev. by Henry J. Cadbury. Cambridge: Cambridge University Press, 1961. London: Macmillan and Company, 1919.
Brayshaw, Alfred Neave. *The Quakers: Their Story and Message*. New York: ALlen & Unwin, 1938.
———. *The Personality of George Fox*. London: Headley Bros., 1919; Allensons and Company: 1933.
Brock, Peter. *The Quaker Peace Testimony, 1660–1914*. York, England: William Sessions Book Trust, 1991.
Bronner, Edwin B. *William Penn's "Holy Experiment."* Philadelphia: Temple University Publications, 1962.
Brown, Elizabeth Potts, and Susan Stuard, eds. *Witness for Change: Quaker Women 1650–1987*. New Brunswick, New Jersey: Rutgers University Press, 1989.
Broxap, E. *The Great Civil War in Lancashire, 1642–51*. 2d ed. Manchester: 1973.
Budge, Frances Anne. *Annals of the Early Friends*. London: Samuel Harris, 1877.
Buranelli, Vincent. *The King and the Quaker*. Philadelphia: University of Pennsylvania Press, 1962.
Butterworth, Edwin. *A Statistical Sketch of the County Palatine of Lancaster*. Manchester: Lancaster and Cheshire Antiquarian Society, 1841.
Bynum, Caroline Walker. *Gender and Religion*. Boston: Beacon Press, 1986.
Cadbury, Henry J. *Annual Catalogue of George Fox's Papers Compiled in 1694–7*. Ann Arbor, Michigan: Edwards Bros., 1939.
———. *Narrative Papers of George Fox*. Richmond, Indiana: Friends United Press, 1972.
———. *George Fox's Book of Miracles*. New York: Octagon Books, 1973.
Campbell, Mildred. *The English Yeoman under Elizabeth and the Early Stuarts*. New Haven, Connecticut: Yale University Press, 1942.
Capp, Bernard S. *The Fifth Monarchy Men: A Study in Seventeenth Century English Millenarianism*. London: 1972.

Chandaman, C. D. *The English Public Revenue 1660–1688.* Oxford: Clarendon Press, 1975.
Childs, John. *The Army, James II, and the Glorious Revolution.* Manchester: Manchester University Press, 1980.
Christianson, Paul. *Reformers and Babylon: English Apocalyptic Visions from the Reformation to the Eve of the Civil War.* Toronto: University of Toronto Press, 1978.
Clark, Alice. *The Working Life of Women.* Reprint, London: Cass, 1968.
Clark, George. *The Later Stuarts 1660–1714.* Oxford: Clarendon Press, 1955.
Clay, C. G. A. *Economic Expansion and Social Change: England 1500–1700.* 2 vols. Cambridge: Cambridge University Press, 1984.
Cliffe, John T. *The Puritan Gentry: The Great Puritan Families of Early Stuart England.* London: Routledge & Kegan Paul, 1984.
Cole, Alan. "Quakers and Politics, 1652–1660." Ph.D. thesis, Cambridge University, 1955.
Cole, C. Robert, and Michael E. Moody. *The Dissenting Tradition.* Athens, Ohio: Ohio University Press, 1975.
Coleman, D. C. *Sir John Banks: Baronet and Businessman.* Oxford: Clarendon Press, 1963.
Collins, John C. *The Apocalyptic Imagination.* New York: Crossroad, 1984.
Comfort, William W. *William Penn, 1644–1718.* Philadelphia: University of Pennsylvania Press, 1944.
Comly, John, and Isaac Comly, eds. *Friends Miscellany.* Vol. 5. Philadelphia: 1834.
Coward, Barry. *The Stuart Age.* London: Longman, 1980.
Cragg, Gerald R. *Puritanism in the Period of the Great Persecution.* Cambridge: Cambridge University Press, 1957.
Creasey, Maurice. "Early Quaker Christology of Isaac Pennington, 1616–1679." Ph.D. diss., University of Leeds, 1956.
Crosby, Jeffrey E. "Friends See it not Safe to Print: The Historical Development of Censorship Among the Quakers in the Seventeenth Century." M.A. thesis, Brigham Young University, 1983.
Crosfield, Helen G. *Margaret Fox of Swarthmoor Hall.* London: Headley Bros., 1913.
Danielou, J. *The Theology of Jewish Christianity.* London: Darton, Longman and Todd, 1964.
deHartog, Jan. *The Peaceable Kingdom.* 1971.
Dictionary of Quaker Biography. London: Friends' House Library transcript.
Dobree, Bonamy. *William Penn, Quaker and Pioneer.* London: Constable & Co., 1932.
Doncaster, Leonard Hugh. *Quaker Organization and Business Meetings.* 1958.
Dunn, Mary Maples. *William Penn: Politics and Conscience.* Princeton: Princeton University Press, 1967.
Dunn, Mary Maples, and Richard S. Dunn, eds. *The Papers of William*

Penn. Vol. I, 1644–1679. Vol. II, 1680–1684. Philadelphia: University of Pennsylvania Press, 1981.

———. *The World of William Penn.* Philadelphia: University of Pennsylvania Press, 1986.

Durston, Christopher. *The Family in the English Revolution.* Oxford: Basil Blackwell, 1989.

Earle, Alice Morse. *Colonial Dames and Goodwives.* New York: 1895.

Eleg, Geoffrey, and William Hung, eds. *Reviving the English Revolution.* London: Verso, 1988.

Endy, Melvin B. *William Penn and Early Quakerism.* Princeton, New Jersey: Princeton University Press, 1973.

Emmott, E. B. *A Short History of Quakerism.* 1923.

Everitt, Alan. *Change in the Provinces: The Seventeenth Century.* Leicester: Leicester University Press, 1972.

Ezell, Margaret J. M. *The Patriarch's Wife: Literary Evidence and the History of the Family.* Chapel Hill and London: The University of North Carolina Press, 1987.

Fantel, Hans. *William Penn: Apostle of Dissent.* New York: William Morse Co., 1974.

Fell, A. *The Early Iron Industry in Furness and District.* Ulverston, 1908. Reprint, London: 1966.

Fell, John. *Some Illustrations of Home Life in Lonsdale North of the Sands.* Ulverston: 1904.

Fell, William. *The History and Antiquities of Furness.* Ulverston: 1887.

Ferguson, Margaret, M. Quilligan, and N. Vickers, eds. *Rewriting the Renaissance.* Chicago: University of Chicago Press, 1986.

Fildes, Valerie A. *Women as Mothers in Pre-industrial England: Essays in Memory of Dorothy McLaren.* London: Routledge, 1990.

Firth, C. H., and R. S. Raid, eds. *Acts and Ordinances of the Interregnum.* 2 vols. London: Stationery Office, 1911.

Firth, Katharine B. *The Apocalyptic Tradition in Reformation Britain, 1530–1645.* Oxford: Oxford University Press, 1979.

Fletcher, Anthony, and John Stevenson, eds. *Order and Disorder in Early Modern England.* Cambridge: Cambridge University Press, 1985.

Fraser, Antonia. *The Weaker Vessel.* New York: Alfred A. Knopf, 1984.

Freiday, Dean. *The Bible: Its Criticism, Interpretation and Use in 16th and 17th Century England.* Manasquan, NJ: Catholic and Quaker Studies, 1979.

Frost, J. William. *The Quaker Family in Colonial America, A Portrait of the Society of Friends.* New York: St. Martin's Press, 1973.

Frost, J. William, and John M. Moore, eds. *Seeking the Light: Essay in Quaker History.* Wallingford, Pennsylvania: Pendle Hill Publications and Friends' Historical Association, 1986.

Gadt, Jeanette Carter. "Women and Protestant Culture: The Quaker Dissent from Puritanism." Ph.D. diss., University of California at Los Angeles, 1974.

Gaythorpe, Harper. *Swarthmoor Meeting House, Ulverston, a Quaker Stronghold.* Kendal: Titus Wilson, 1910.

Gershom, Gerhard Scholem. *Sabbatai Sevi: The Mystical Messiah.* Princeton, New Jersey: Princeton University Press, 1973.

Gillis John. *For Better, For Worse: British Marriages 1600 to the Present.* Oxford: Oxford University Press, 1985.

Gittings, Clare. *Death, Burial and the Individual in Early Modern England.* London: Croom Helm, 1984.

Glass, David V. and D. E. C. Eversley, eds. *Population in History: Essays in Historical Demography.* London: E. Arnold, 1965.

Gloel, Elizabeth. *Die Frau bei den Quakern des 17. Jarhunderts in England.* Halle: R. Mayr, 1939.

Gooch, G. P. *English Democratic Ideas in the 17th Century.* Cambridge: Cambridge University Press, 1927, 1954.

Goody, Jack. *The Development of the Family and Marriage in Europe.* Cambridge: Cambridge University Press, 1976, 1983.

Gosling, R. *The Penal Laws Against Papists and Popish Recusants, Nonconformists and Nonjurors: with the statutes relating to the succession of the Crown, forfeited estates, tumults and riots, imprisonment of suspected persons, and the late Acts for obliging Papists and Nonjurors to register their estates.* E & R. Natt and R. Gosling, 1723.

Greaves, Richard L. *Deliver Us from Evil.* New York: Oxford University Press, 1986.

——. *Enemies under his Feet: Radicals and Nonconformists in Britain, 1664–1677.* Stanford, CA: Stanford University Press, 1991.

Greaves, Richard, and Robert Zaller. *Biographical Dictionary of British Radicals in the Seventeenth Century.* 3 vols. Brighton, Sussex: Harvester Press, 1982–84.

Gregg, Pauline. *Free-born John. A Biography of John Lilburne.* London: George G. Harrays & Co., 1961.

Green, I. M. *The Re-Establishment of the Church of England, 1660–1663.* New York: Oxford University Press, 1978.

Green, Mary Ann Everett, ed. *Calendar of Committee for Compounding, 1643–1660.* London: PRO, Kraus Reprint, 1967.

Gummere, Amelia Mott. *The Quaker: A Study in Costume.* Philadelphia: Ferris and Leach, 1901.

Gwyn, Douglas Philip. *Apocalypse of the Word: The Life and Message of George Fox, 1624–1691.* Richmond, Indiana: Friends United Press, 1984.

Haller, William, ed. *Foxe's Book of Martyrs and the Elect Nation.* 1963.

Heal, Felicity. *Hospitality in Early Modern England.* Oxford: Clarendon Press, 1990.

Higgins, Leslie Hal. "Radical Puritans and Jews in England, 1648–1672." Ph.D. diss., Yale University, 1979.

Hill, Christopher. *Antichrist in Seventeenth Century England.* London: 1971.

——. *The Century of Revolution, 1603–1714.* New York: Norton, 1982.

———. *The Experience of Defeat*. New York: Viking Press, 1984.
———. *A Nation of Change and Novelty: Radical politics, religion, and Literature in Seventeenth-century England*. London: Routledge, 1991.
———. *Society and Puritanism in pre-revolutionary England*. Reprint, Harmondsworth, Penguin, 1986.
———. *The World Turned Upside Down*. New York: Penguin Books, 1972, 1980.
Hobby, Elaine. *Virtue of Necessity: English Women's Writings, 1649–88*. Ann Arbor, Michigan: University of Michigan Press, 1989.
Hodgkin, L. V. *A Quaker Saint of Cornwall: Loveday Hambly and Her Guests*. London: Longman, Green & Co., 1927.
Holder, William Frederick. *The Quakers in Great Britain and America*. New York: Nainer, 1913.
Horle, Craig W. *The Quakers and the English Legal System 1660–1688*. Philadelphia: University of Pennsylvania Press, 1988.
Horst, Ian. *The Radical Brethren*. 1972.
Houlbrooke, Ralph A. *The English Family, 1450–1700*. London: Longman, 1984.
———, ed. *English Family Life*. Oxford: Basil Blackwell, 1988.
Hudson, Winthrop S., and Leonard J. Trinterud. *Theology in Sixteenth and Seventeenth Century England*. William Andrews, William Andrews Clark Memorial Library: UCLA, 1971.
Hull, William I. *The Rise of Quakerism in Amsterdam 1655–1665*. Swarthmore, Pennsylvania: Swarthmore College, 1938.
———. *Benjamin Furly and Quakerism in Rotterdam*. 1941.
———. *William Sewel of Amsterdam, 1653–1720: The First Quaker Historian of Quakerism*. 1933.
———. *William Penn: A Topical Biography*. London: Oxford University Press, 1937.
Hunt, William. *The Puritan Moment: The Coming of Revolution in an English County*. Cambridge, Massachusetts: Harvard University Press, 1983.
Hutchinson, Julius, ed. *Memoirs of the Life of Colonel Hutchinson, Governor Nottingham by his Widow Lucy*. Rev. ed., C. H. Firth. London: George Routledge, 1906.
Hutton, Ronald. *Charles the Second: King of England, Scotland, and Ireland*. Oxford: Clarendon Press, 1989.
———. *The Restoration*. Oxford: Clarendon Press, 1985.
Ingle, H. Larry. *Quakers in Conflict*. Knoxville: The Universtiy of Tennessee Press, 1986.
Irwin, Joyce L. *Womanhood in Radical Protestantism, 1525–1675*. New York: Edwin Mellen Press, 1979.
James, Margaret. *Social Problems and Policy During the Puritan Revolution*. London: George Routledge, 1930.
Janney, Samuel M. *Life of William Penn with selections from his correspondence and autobiography*. Philadelphia: Lippincott, Grambo, 1852.
Johnson, James. *A Society Ordained by God: English Puritan Marriage*

Doctrine in the First Half of the 17th Century. Nashville: Abingdon Press, 1970.
Jones, J. R. *Country and Court: England, 1658–1714.* Cambridge, Massachusetts: Harvard University Press, 1978.
Jones, Rufus M. *Spiritual Reformers of the 16th and 17th Centuries.* London: Macmillan, 1914.
——. *Studies in Mystical Religion.* London: Macmillan, 1909.
Jordan, W. K. *Philanthropy in England, 1480–1660: A study of the changing pattern of English social aspirations.* New York: Allen & Unwin, 1959.
——. *The Social Institutions of Lancashire.* Manchester: The Chetham Society, 1962.
——. *The Development of Religious Toleration in England.* London: George Allen & Unwin, 1938.
Kanner, Barbara, ed. *The Women of England from Anglo-Saxon Times to the Present.* Hamden, Connecticut: Archen Books, 1979.
Katz, David S. *Philo-Semitism and the Readmission of Jews into England, 1603–1655.* Oxford: Clarendon Press; Oxford University Press, 1982.
——. *Sabbath and Sectarianism in Seventeenth-Century England.* New York: E. J. Brill, 1988.
Keeble, N. H. *The Literary Culture of Nonconformity in Later Seventeenth-Century England.* Athens, Georgia: University of Georgia Press, 1987.
Kelley, Donald R. *The Beginning of Ideology: Consciousness and Society in the French Reformation.* New York: Cambridge University Press, 1987.
Kennedy, William. *English Taxation, 1640–1799.* London: G. Bell and Sons, 1913.
Kenyon, John P. *Revolution Principles: The Politics of Party, 1689–1720.* New York: Cambridge University Press, 1977.
Kerridge, Eric. *The Agricultural Revolution.* London: Allen and Unwin, 1969.
King, Gregory. *Seventeenth Century Economic Documents – Statistics.* Oxford: Clarendon Press, 1972.
King, R. H. *George Fox and the Light Within.* Philadelphia: 1940.
Knight, Rachel. *The Founder of Quakerism: A Psychological Study of the Mysticism of George Fox.* London: 1922.
Knox, R. B., ed. *Reformation, Conformity and Dissent.* London: Epworth Press, 1977.
Kroll, Richard, Richard Ashcroft, and Perez Zagorin, eds. *Philosophy, Science, and Religion in England 1640–1700.* Cambridge: Cambridge University Press, 1992.
Kunze, Bonnelyn Young and Dwight D. Brautigam. *Court, Country, and Culture: Essays on Early Modern British History in Honor of Perez Zagorin.* Rochester: University of Rochester Press, 1992.
Kussmaul, Ann. *Servants in Husbandry in Early Modern England.* Cambridge: Cambridge University Press, 1981.
Lamont, William M. *Godly rule: Politics and Religion: 1603–60.* London: Macmillan, 1969.

———. *Richard Baxter and the Millennium*. London: Helm, 1979.
Lampson, Mrs. Godfrey Locker. *A Quaker Postbag*. London: Longman, 1910.
Laslett, Peter. *The World We Have Lost*. New York: Charles Scribners Sons, 1965.
Latt, David J., and Suzanne W. Hull. *Works Published by Women in England, 1475–1700*. Huntington Library.
Laycock, Bettina. *A Quaker Missions to Europe and the Near East, 1655–1665*. 1950.
Leach, Robert J. *Quaker Women Ministers*. Wallingsford, Pennsylvania: Pendle Hill, 1979.
Leonard, E. M. *The Early History of the English Poor Relief*. New York: Barnes and Noble, 1900.
Leverenz, David. *The Language of Puritan Feeling – An Exploration in Literature, Psychology and Social History*. New Brunswick, New Jersey: Rutgers University Press, 1980.
Levy, Barry. *Quakers and the American Family*. New York: Oxford University Press, 1988.
Lipson, Ephraim. *The Economic History of England*. 3 vols. London: A. & C. Black, 1915–31.
Llewellyn, Nigel. *The Art of Death: Visual Culture in the English Death Ritual c. 1500 – c. 1800*. London: Reakton/Victoria and Albert Museum, 1991.
Lloyd, Arnold. *Quaker Social History 1669–1738*. London: Longman, Green & Co., 1950.
Lovejoy, David S. *Religious Enthusiasm in the New World*. Cambridge, Massachusetts: Harvard University Press, 1985.
Luder, Hope Elizabeth. *Women and Quakerism*. Wallingsford, Pennsylvania: Pendle Hill, 1974.
Ludlow, Dorothy Paula. *"Arise and be doing": English "Preaching" Women, 1640–1660*. Ph.D. diss., University of Indiana, 1978.
Macfarlane, Alan. *The Family Life of Ralph Josselin*. New York: W. W. Norton, 1970.
———. *Marriage and Love in England, 1300–1840*. Oxford: Basil Blackwell, 1986.
Mack, Phyllis. *Visionary Women: Ecstatic Prophecy in Seventeenth-Century England*. University of California Press, 1992.
Magrath, John Richard, ed. *The Flemings in Oxford: Being Documents Selected from the Rydal Papers*. Oxford: Oxford Historical Society, Clarendon Press, 1904.
Manners, Emily. *Elizabeth Hooten; First Quaker Woman Preacher, 1600–1672*. London: Headley Bros., 1914.
Manning, Brian, ed. *Politics, Religion and the English Civil War*. London: Edward Arnold, 1973.
Marshall, John Duncan, ed. *The Autobiography of William Stout of Lancaster, 1665–1752*. Manchester: 1967.

———. *Lancashire.* London: David and Charles, 1974.
Matthews, A. G. *Calamy Revised, Being a Revision of Edmund Calamy's Account of the Ministers and Others Ejected and Silenced 1660–2.* Oxford: Clarendon Press, 1934.
McGee, J. Sears. *The Godly Man in Stuart England: Anglicans and Puritans and the Two Tables 1620–1670.* New Haven, Connecticut: Yale University Press, 1976.
Mendelson, Sara Heller. *The Mental World of Stuart Women.* Amherst: The University of Massachusetts Press, 1987.
Midgley, J. Herbert. *Margaret Fell.* 1908.
Morgan, Nicholas. "The Social and Political Relations of the Lancaster Quaker Community, 1688–1740." Ph.D. diss., University of Lancaster, 1984.
Morton, A. L. *The World of the Ranters: Religious Radicalism in the English Revolution.* London: Lawrence and Wishart, 1970.
Mounsey, Thomas. *A Brief Account of Th. Fell of Swarthmore Hall.* Manchester: William Irwin; London: Ch. Gilpin, 1846.
Mullett, Michael, ed. *Early Lancaster Friends.* Lancaster: University of Lancaster, Northwest Regional Studies, Paper no. 5, 1978.
———. *Radical Religious Movements in Early Modern Europe.* London: George Allen & Unwin, 1980.
Newman, Josiah. *The Quaker Records: Some Special Studies of Genealogy.* 1908.
Nichols, John Gough, ed. *The Autobiography of Anne Lady Halkett.* New York: Johnson Reprint Corp., 1965.
Nicolson, Marjorie Hope. *Conway Letters.* New Haven, Connecticut: Yale University Press, 1930.
Nightingale, B. *Lancashire Nonconformity, or, sketches, historical and descriptive, of the Congregational and Old Presbyterian churches in the county.* 6 vols. Manchester: J. Heywood 1890–93.
———. *The Early Stages of the Quaker Movement in Lancashire.* Edinburgh: 1921.
Nuttall, Geoffrey F. *Early Quaker Letters from the Swarthmore MSS to 1660: calendared, indexed and annotated.* London: Friends' House Library, 1952.
———. *The Beginnings of Nonconformity.* London: J. Clarke, 1964.
———. *The Holy Spirit in Puritan Faith and Experience.* Oxford: Basil Blackwell, 1947.
———. *Studies in Christian Enthusiasm: Illustrated from Early Quakerism.* Wallingsford, Pennsylvania: Pendle Hill, 1948.
———. *Visible Saints.* Oxford: Basil Blackwell, 1957.
Ogg, David. *England in the Reign of Charles II.* Vol. I and II. Oxford: Clarendon Press, 1934, 2d ed. 1956.
Outhwaite, R. B., ed. *Marriage and Society: Studies in the Social History of Marriage.* New York: St. Martin's Press, 1981.
Peare, Catherine O. *William Penn.* Philadelphia: Lippincott, 1957.

Pease, Alfred E. *A Dictionary of the Dialect of the North Riding of Yorkshire*. London: Horne and Son, 1928.

Penney, Norman, ed. *"The First Publishers of Truth:" Being early records (now first printed) of the introduction of Quakerism into the counties of England and Wales*. London: Headley Bros., 1907.

Pepys, Samuel. *Diary and Correspondence of Samuel Pepys*. London: H. Colburn, 1854.

Phillips, Roderick. *Untying the Knot: A Short History of Divorce*. Cambridge: Cambridge University Press, 1991.

Pinchbeck, Ivy. *Women Workers and the Industrial Revolution*. London: Virago, 1930; reprints, 1969, 1981.

Prior, Mary, ed. *Women in English Society 1500–1800*. London: Methuen, 1985.

Punshon, John. *Portrait in Grey: A Short History of the Quakers*. London: Quaker Home Service, 1984.

Reay, Barry. *The Quakers and the English Revolution*. New York: St. Martin's Press, 1985.

———. "Early Quaker Activity and Reactions to It, 1652–1664," Ph.D. diss., Oxford University, 1979.

Reay, Barry, and J. F. McGregor, eds. *Radical Religion in the English Revolution*. Oxford: Oxford University Press, 1984.

Richardson, Joseph. *Furness, Past and Present, Its History and Antiquities*. Barrow In Furness: J. Richardson, 1880.

Richardson, R. C. *Purtianism in North-West England: a regional study of the Diocese of Chester*. Manchester: Manchester University Press, 1972.

Robinson, Cedric. *Sand Pilot of Morecambe Bay*. London: David and Charles, 1979.

Roper. *Churches, Castles and Old Halls of North Lancashire*.

Ross, Isabel. *Margaret Fell, Mother of Quakerism*. London: Longman, 1949, reprint, 1984.

Roth, Cecil. *A History of the Jews in England*. Oxford: Clarendon Press, 3d ed., 1964.

———. "The Origins of Hebrew Typography in England." *Journal of Jewish Bibliography*. 1938.

———. *A Life of Menassah ben Israel*. Philadelphia: 1934, 1945.

Ruether, Rosemary, ed. *Religion and Sexism*. New York: Simon and Schuster, 1974.

Ruether, Rosemary, and Eleanor McLaughlin, eds. *Women of Spirit*. New York: Simon and Schuster, 1979.

Russell, Elbert. *History of Quakerism*. New York: 1943.

Schlatter, Richard B. *The Social Ideas of Religious Leaders 1660–1688*. London: Oxford University Press, 1940.

Schochet, Gordon J., ed. *The Authoritarian Family and Political Attitudes in Seventeenth Century England: Patriarchalism in Political Thought*. New Brunswick: Rutgers University Press, 1988.

———. *Religious Resistance and Civil War: Proceedings of the Folger Institute*.

Washington, DC: The Folger-Shakespeare Library, 1990.
Schubert, H. R. *History of the British Iron and Steel Industry*. London: 1957.
Schwoerer, Lois G. *Lady Rachel Russell*. Baltimore: The Johns Hopkins University Press, 1988.
Searle, Arthur. *Barrington Family Letters*. Vol 28 of Camden Fourth Series. London: RHS, 1983.
Seaver, Paul S. *Seventeenth Century England: Society in an Age of Revolution*. New York: New Viewpoints, 1976.
——. *Wallington's World: A Puritan Artisan in the Seventeenth Century*. Stanford, California: Stanford University Press, 1985.
Seitz, Don C., ed. *The Tryal of William Penn and William Mead*. 1719.
Sharpless, Isaac, and Rufus M Jones. *Quakers in the American Colonies*. London: Macmillan, 1923.
Shaw, Wm. A., ed. *Calendar of Treasury Books and Papers, 1729–[1745]*. Nendeln, Liechtenstein: Kraus Reprint, 1974.
Simpson, A. W. B. *An Introduction to the History of the Land Law*. Oxford: Oxford University Press, 1971.
Slack, Paul. *Poverty and Policy in Tudor and Stuart England*. London: Longman, 1988.
Slater, Miriam. *Family Life in the Seventeenth Century*. London: Routledge & Kegan Paul, 1984.
Smith, Hilda L. *Reason's Disciples*. Urbana, Illinois: University of Illinois Press, 1982.
Smith, Hilda L., and Susan Cardinale, eds. *Women and the Literature of the Seventeenth Century*. Westport, Connecticut: Greenwood Press, 1990.
Smith, J. P., compiler. *The Genealogist's Atlas of Lancashire*. Liverpool: Henry Young and Sons, 1930.
Smith, Joseph. *A Descriptive Catalogue of Friends' Books, or Books Written by Members of the Society of Friends*. 2 vols. London: 1867; with supplement, 1893.
——. *Bibliotheca Anti-Quakeriana; or a Catalogue of Books Adverse to the Society of Friends*. London: 1873.
Smith, Nigel. *Perfection Proclaimed: Language and Literature in English Radical Religion 1640–1660*. Oxford: Oxford University Press, 1989.
Smith, Richard M., ed. *Land, Kinship and Life-Cycle*. Cambridge: Cambridge University Press, 1984.
Springer, Keith L. *The Learned Doctor William Ames*. Urbana, Illinois: University of Illinois Press, 1972.
Spufford, Margaret. *Contrasting Communities: English Villagers in the 16–17th Centuries*. Cambridge: Cambridge University Press, 1974.
Spurrier, William W. "The Persecution of the Quakers in England, 1650–1714." Ph.D. diss., University of North Carolina at Chapel Hill, 1976.
Staves, Susan. *Married Women's Separate Property in England, 1660–1833*. Cambridge, Mass.: Harvard University Press, 1990.

Stone, Lawrence. *The Family, Sex and Marriage in England, 1500–1800.* Abridged edition. New York: Harper & Row, 1977.
——. *Road to Divorce: England 1530–1987.* Oxford: Oxford University Press, 1990.
Stoughton, John. *Religion in England.* 1881.
Tawney, Richard H. *The Agrarian Problem in the Sixteenth Century.* London: Longman, Green & Co., 1912.
Tawney, Richard H., and Eileen Power, eds. *Tudor Economic Documents.* London: Longman, 1924.
Taylor, Ernest E. *The Valiant Sixty.* London: Bannisdale Press, 19–1.
Thirsk, Joan, ed. *Agrarian History of England and Wales.* Vol. IV, 1500–1640. Cambridge: Cambridge University Press, 1967.
——, ed. *Agricultural regions and agrarian history in England, 1500–1750.* Atlantic Highlands, New Jersey: Humanities Press, 1987.
Thrisk, Joan, and J.P Cooper, eds. *Seventeenth Century Economic Documents.* Oxford: Clarendon Press, 1972.
Tilly, Louise A. and Joan W. Scott. *Women, Work and Family.* New York: Holt, Reinhart and Winston, 1978.
Tolles, Frederick B. *Quakers and the Atlantic Culture.* New York: Octagon Books, 1960.
Toom, Peter, ed. *Puritans, the Millennium and the Future of Israel: Puritan Eschatology, 1600 to 1660.* Cambridge: James Clarke & Co., 1970.
Trevett, Christine. *Women and Quakerism in the Seventeenth Century.* York, England: The Ebor Press, 1991.
Trueblood, D. Elton. *Robert Barclay 1648–1690.* 1967.
Turner, G. Lyon. *Original Records of Early Nonconfomrity Under Persecution and Indulgence.* London: T. F. Unwin, 1914.
Tuveson, Ernest L. *Millennium and Utopia.* Berkeley and Los Angeles: University of California Press, 1949.
Underdown, David. *Revel, Riot, and Religion: Popular Politics and Culture in England, 1603–1660.* Oxford: Oxford University Press, 1985.
Vann, Richard. *The Social Development of English Quakerism, 1655–1755.* Cambridge, Mass.: Harvard University Press, 1967.
Vann, Richard and David Eversley. *Friends in Life and Death.* Cambridge: Cambridge University Press, 1992.
Venn, Albert. *Alumni Cantabrigirnsis.* Cambridge: 1922.
Victoria History of the County of Lancaster, The. William Farrar and J. Brownhill, eds. 8 vols. London: University of London Institute of Historical Research, 1908; reprinted, London: Dawsons of Pall Mall, 1966.
Watkins, Owen. *The Puritan Experience.* London: Routledge and K. Paul, 1972.
Watts, Michael R. *The Dissenters: From the Reformation to the French Revolution.* Oxford: Clarendon Press, 1978.
Webb, Maria. *The Fells of Swarthmoor Hall and their Friends; with an account of their ancestor, Anne Askew, the martyr.* London: F. B. Kitto,

1867.
West, Thomas. *Antiquities of Furness*. London: T. Spilsbury, 1774.
Westfall, Richard S. *Science and Religion in 17th Century England*. New Haven, Connecticut: Yale University Press, 1958.
White, Barrington R. *The English Separatist Tradition*. Oxford: Oxford University Press, 1971.
Whiting, Charles E. *Studies in English Puritanism from the Restoration to the Revolution – 1660–1688*. New York: A. M. Kelley, 1968.
Whitwell, John. *Kendal During the Civil Wars of the 17th Century*. Kendal, England: 1764.
Williams, George H. *The Radical Reformation*. Philadelphia: Westminster Press, 1962.
Wilson, Bryan R. *Patterns of Sectarianism*. London: Heinemann, 1967.
———. *Religious Sects*. London: Weidenfeld and Nicolson, 1974.
Wing, Donald. *Short Title Catalogue . . . 1641–1700*. New York: Index Society, 1945–51.
Woolrych, Austin. *Commonwealth to Protectorate*. Oxford: Clarendon Press, 1982.
Woodbridge, Linda. *Women and the English Renaissance: Literature and the Nature of Womanland, 1540–1620*. Urbana, Illinois: University of Illinois Press, 1984.
Wright, Luella. *The Literary Life of the Early Friends, 1650–1725*. New York: Columbia University Press, 1932.
Wrightson, Keith. *English Society, 1580–1680*. New Brunswick, New Jersey: Rutgers University Press, 1982.
Wrightson, Keith and David Levine. *Poverty and Piety in an English Village: Terling, 1525–1700*. New York: Academic Press, 1979.
Wrigley, E. A. *An Introduction to English Historical Demography*. London: 1966.
Wrigley, E. A., and R. S. Schofield. *The Population History of England, 1541–1871. A Reconstruction*. London: Edward Arnold, 1982.
Yoshioka, Barbara. "Imaginal Worlds: Woman as Witch and Preacher in 17th Century England." Ph.D. diss., Syracuse University, 1977.
Zagorin, Perez, ed. *Culture and Politics from Puritanism to the Enlightenment*. Berkeley and Los Angeles: University of California Press, 1980.
———. *History of Political Thought in the English Revolution*. London: Routledge & Kegan Paul, 1954.
———. *Rebels and Rulers, 1500–1660*. 2 vols. Cambridge: Cambridge University Press, 1982.

Journal Articles

———. "Thirnbeck MSS." *JFHS* reprints IX nos. 1, 2, and 4. (1912); 53– 65; 95–107; 175–187.
Anderson, Alan B. "A Study in the Sociology of Religious Persecution: The First Quakers." *Journal of Religious History* 9 (1977): 247–262.

Arkell, Tom. "The Incidence of Poverty in England in the Later Seventeenth Century." *Social History* 12 (January 1987): 23–47.
Avis, P. D. L. "Moses and the Magistrate: A Study in the Rise of Protestant Legalism." *Journal of Ecclesiastical History* 26 (1975): 149–72.
Awty, Brian G. "Force Forge in the Seventeenth Century." *Transactions of the Cumberland and Westmorland Antiquarian and Archeological Society* N.S. 77 (1977): 97–112.
Barbour, Hugh. "William Penn: Model of Protestant Liberalism." *Church History* 48 (1979): 156–73.
———. "A Review Essay: Ranters, Diggers and Quakers Reform." *Quaker History* 64 no. 1 (1975): 60–65.
———. "The Young Controversialist." Ina R. Dunn and M. M. Dunn, eds, WWP.
Bitterman, M. G. F. "The Early Quaker Literature of Defense." *Church History* 42 (1973): 203–228.
Blackwood, Bruce G. "Agrarian Unrest and the Early Lancashire Quakers." *JFHS* 51, no. 1 (1965): 72–76.
———. "The Catholic and Protestant Gentry of Lancashire During the Civil War Period." *Transactions of Lancashire and Cheshire Historical Society* 126 (1976): 1–29.
———. "The Economic State of the Lancashire Gentry on the Eve of the Civil War." *Northern History* 12 (1976): 53–83.
———. "The Lancashire Cavaliers and Their Tenants." *Transactions of Lancashire and Cheshire Historical Society* 117 (1965).
———. "The Lancashire Gentry." *Transactions of the Lancashire and Cheshire Historical Society*, date unknown.
———. "The Marriages of the Lancashire Gentry on the Eve of the Civil War." *The Genealogists' Magazine* 16 (1970).
———. "The Cavalier and Roundhead Gentry of Lancashire." *Transactions of Lancashire and Cheshire Antiquarian Society* 77 (1967).
Braithwaite, Alfred. "The Mystery of Swarthmoor Hall." *JFHS* 51 (1965): 22–29.
Braithwaite, W. C. "Westmorland and the Swaledale Seekers." *JFHS* (January 1908): 3–10.
Brockbank, Elizabeth. "The Story of Quakerism in the Lancaster District." *JFHS* 36 (1939): 3–20.
Bronner, Edwin B. "George Fox and William Penn." *JFHS* 56 (1991:78–95.
Brown, Irene Q. "Domesticity, Feminism and Friendship: Female Aristocratic Culture and Marriage in England, 1660–1760." *Journal of Family History* (Winter 1982): 406–24.
Brownbill, and Gaythorpe. "The Askews of Marsh Grange." *Transactions of Cumberland and Westmorland Antiquarian and Archeological Society*, n.s. 10.
Cadbury, Henry J. "From Margaret Fox's Library." *JFHS* 34 (1937): 27–28.
———. "Intercepted Correspondence of William Penn." *Pennsylvania Magazine of History and Biography* 70 (1946): 349–56.

———. "An Obscure Chapter in Quaker History." *Journal of Religion* 24 (1944): 201–14.
Capp, B. S. "Godly Rule and English Millenarianism." *Past and Present* 52 (1971): 106–117.
Cole, Alan. "The Quakers and the English Revolution." *Past and Present* 10: (1956): 39–54.
———. "The Social Origins of the Early Friends." *JFHS* 48 (1957): 117.
Cope, Jackson I. "Seventeenth-Century Quaker Style." *Modern Language Association Publications* 71 (1956): 725–754.
Cowper, Henry S. "Kirkbys of Kirkby Hall, Furness." *Transactions of the Cumberland and Westmorland Antiquarian and Archaeological Society*, n.s. 6.
Cressey, David. "Literacy in Pre-Industrial England." *Societas* 4 (1974): 229–40.
Davis, J. C. "Radicalism in and Traditional Society: The Evaluation of Radical Thought in the English Commonwealth 1640–1660." *History of Political Thought* 3 (Summer 1982): 193–213.
Dunn, Mary Maples. "The Personality of William Penn." In R. S. Dunn and M. M. Dunn, eds, *WWP*.
Durnbaugh, Donald. "Baptists and Quakers: Left-Wing Puritans?" *Quaker History* 62 (1973): 67.
Edwards, Irene. "Women Friends of London." *JFHS* 47 (1955): 3–21.
Endy, Melvin B. Jr. "Puritanism, Spiritualism, and Quakerism: An Historiographical Essay." In R. S. Dunn and M. M. Dunn, eds, *WWP*.
Farrar, William. "North Lonsdale after the Restoration." *Transactions of the Cumberland and Westmorland Antiquarian and Archaeological Society* 12 New Series (1912): 202–215.
Fell, John. "Home Life in North Lonsdale." *Transactions of Cumberland and Westmorland Antiquarian and Archaelological Society* (1891).
———. "The Fells of Dalton Gate." *Transactions of Cumberland and Westmorland Antiquarian and Westmorland Antiquarian and Archaeological Society*, 11 (1882).
Fogelklou, Emilia. "Quakerism and Democracy." *JFHS* 42 (1950): 18.
Ford, Linda. "William Penn's Views on Women." *Quaker History* 72 (Fall 1983): 75–102.
Forster, Ann M. C. "The Oath Tendered." *Recusant History* 14 (1977): 86.
France, R. Sharpe. "A History of the Plague in Lancashire." *Transactions Lancashire and Cheshire Historical Society* 90 (1938).
Galgans, Michael J. "Out of the Mainstream: Catholic and Quaker Women in the Restoration Northwest." In R. S. Dunn and M. M. Dunn, eds, *WWP*.
Greaves, Richard L. "The Puritan-Nonconformist Tradition of England. 1560–1700: Historigraphical Reflections." *Albion* (Winter 1985): 449–86.
———. "Shattered Expectations? George Fox, the Quakers, and the Restoration State, 1660–1685." *Albion* (Summer 1992): 237–59.

Healey, Robert M. "The Jew in Seventeenth Century Protestant Thought." *Church History* 46 (March 1977): 63-79.

Holderness, B. A. "Widows in pre-industrial society." In Richard M. Smith, ed., *Land, Kinship and Life-Cycle*. Cambridge: Cambridge University Press, 1984.

Howell, Roger, Jr. "The Newcastle Clergy and the Quakers." *Archaeological Aeliana* 7 (1979): 191-206.

Howson, W. G. "Plague, Poverty and Population in Parts of North-west England, 1580-1720." *Transactions Lancashire and Cheshire Historical Society* 112 (1960).

Horle, Craig. "Changing Quaker Attitudes Toward Legal Defense: The George Fox Case, 1673-5, and the Establishment of Meeting for Sufferings." In J. William Frost, and John M. Moore, eds, *Seeking the Light*. Wallingford, Pennsylvania: Pendle Hill Publications, 1986.

Hudson, Winthrop S. "A Suppressed Chapter in Quaker History." *Journal of Religion* 24 (April 1944): 108-18.

Hurwick, Judith. "The Social Origins of the Early Quakers." *Past and Present* 48 (1970): 156-62.

Ingle, H. Larry. "A Letter from Richard Farnworth, 1652." *Quaker History* 79 (Spring 1990): 35-38.

——. "A Quaker Woman on Women's Roles: Mary Penington to Friends, 1678." Forthcoming.

—— and Jaan Ingle. "The Excommunication of George Fox." *JFHS* 56 (1991):71-77.

——. "George Fox, Millenarian." *Albion* (Summer 1992): 261-78.

James, Margaret. "The Political Importance of the Tithes Controversy in the English Revolution, 1640-1660." *History* 26 (June 1941): 1-18.

Keirn, Tim and Frank T. Melton. "Thomas Manley and the Rate-of-Interest Debate, 1668-1673." *Journal of British Studies* 29 (April 1880): 147-173.

Kent, Stephen A. "The Papist Charges Against the Interregnum Quakers." *Journal Religious History* 12 (December 1982): 180-90.

——. "Seven Thousand Handmaids of the Lord: English Quaker Women's Political Protest in 1659," unpublished paper, University of Alberta, Edmonton, Canada.

Kim, Chrysostom. "The Diggers, the Ranters and the Early Quakers." *American Benedictine Review* 25 no. 4 (1974): 460-75.

Kirby, E. Williams. "The Quakers' Efforts to Secure Civil and Religious Liberty 1660-96." *Journal of Modern History* 7 (1935): 405.

Kunze, Bonnelyn Young. "An Unpublished Work of Margaret Fell." *Proceedings of the American Philosophical Society* (December 1986): 424-52.

Lamont, William. "Biography and the Puritan Revolution." *Journal of British Studies* 26 (July 1987): 347-530.

——. "The Religious Origins of the English Civil War." In Gordon J. Schochet, ed., *Religion, Resistance and Civil War: Proceedings of the Folger Institute* vol. 3, Washington, DC: The Folger-Shakespeare Library, 1990.

Levine, David. "'For Their Own Reasons': Individual Marriage Decisions and Family Life: Shephed, Leicestershire." *Journal of Family History* 7 (Fall 1982): 255–64.
Lindert, Peter H. "English Occupations, 1670–1811." *Journal of Economic History* 40 (1980): 685–712.
MacArthur, Ellen. "Women Petitioners and the Long Parliament." *Economic History Review* 24 (1909): 698–709.
Macfarlane, Alan. "Review of Stone." *History and Theory* 18 (1979): 103–126.
McIntosh, Marjorie K. "Servants and the Household Unit in an Elizabethan English Community." *Journal of Family History* 9 (1984): 3–23.
McMullen, Norman. "The Education of the English Gentlewoman 1540–1640." *History of Education* 6 no. 2 (1977): 87–101.
MacGregor, J. F. "Ranterism and the Development of Early Quakerism." *Journal of Religious History* 9 (1977): 349–63.
Mack, Phyllis. "Women as Prophets during the English Civil War." *Feminist Studies* (Spring 1982).
Maclear, James F. "Quakerism and the End of the Interregnum: A Chapter in the Domestication of Radical Puritanism." *Church History* 19 (1950): 240–71.
Masson, Margaret W. "The Typology of the Female as a Model for the Regenerate: Purtian Preaching, 1690–1730." *Signs* 2 (Winter 1976): 304–15.
Morgan, Nicholas J. "Lancashire Quakers and the Oath, 1660–1730." *JFHS* 54 (1980): 235–54.
———. "Lancashire Quakers and the Tithe, 1660–1730." *Bulletin of the John Rylands University Library of Manchester* 70 (Autumn 1988): 61–75.
Morrill, J. S. "The Northern Gentry and the Great Rebellion." *Northern History* 15 (1979): 66–87.
Mortimer, Russell. "Quakerism in 17th Century Bristol." Master's thesis, Bristol University, 1946.
Mulligan, Lotte, and Judith Richards. "A 'Radical' Problem: The Poor and the English Reformers in the Mid-Seventeenth Century." *Journal of British Studies* 29 (April 1990): 188–46.
Nicolson, Marjorie. "George Keith and the Cambridge Platonists." *Philosophical Review* 39, no. 1 (January 1930): 36–55.
O'Malley, Thomas. "Defying the Powers and Tempering the Spriit: A Review of Quaker Control over the Publications." *Journal of Ecclesiastical History* 33 (January 1982): 72–78.
Penney, Norman. "George Fox's Writings and the Morning Meeting." *Friends' Quarterly Examiner* 36 (1902): 63–72.
———. "George Fell and the Story of Swarthmoor Hall." *JFHS* 29–31 (1932): 51–61; (1933) 28–39; (1934); 27–35.
Phillips, C. B. "Iron-Mining in Restoration Furness: The Case of Thomas Preston." *Recusant History* 14, (1977–78).
Popkin, Richard H. "Spinoza's Relations with the Quakers in Amsterdam." *Quaker History* 73 (Spring 1984): 14–29.

———. "Jewish Messianism and Christian Millenarianism." In Perez Zagorin ed. *Culture and Politics from Puritanism to the Enlightenment*. Berkeley and Los Angeles: University of California Press, 1980.

Reay, Barry. "Social Origins of Early Quakerism." *Journal of Interdisciplinary History* 11 (1980): 55–72.

———. "Popular Hostility Towards Quakers in Mid Seventeenth Century England." *Social History* 5 (1980): 387–407.

———. "Quaker Opposition to Tithes." *Past and Present* 86 (1980): 98–120.

———. "The Quakers 1659, and the Restoration of the Monarchy." *History* 63 (1978).

Riegel, Stanley K. "Jewish Christianity: Definitions and Terminology." *New Testament Studies* 24 (1978): 410–15.

Robbins, Caroline. "William Penn, 1689–1702: Eclipse, Frustration and Achievement." In R. S. Dunn and M. M. Dunn, eds, *The World of William Penn*. Philadelphia: University of Pennsylvania Press, 1986.

Roberts, Michael. "Sickles and Scythes: Women's Work and Men's Work at Harvest Time." *History Workshop* 7 (1979): 3–29.

Routledge, R. A. "The Legal Status of the Jews in England, 1190–1790." *Journal of Legal History* 3 (September 1982): 91–124.

Schofield, Mary Ann. "Women Speaking Justified: The Feminine Quaker Voice, 1662–1797." NEASECS conference paper, November 1985.

Scott, Jonathan. "Radicalism and Restoration: The Shape of the Stuart Experience." *Historical Journal* 31, no. 2 (1988): 453–467.

Shammas, Carole. "The World Women Knew: Women Workers in the North of England During the Late Seventeenth Century." In R. S. Dunn and M. M. Dunn, eds, *The World of William Penn*. Philadelphia: University of Philadelphia Press, 1986.

———. "The Domestic Environment in Early Modern England and America." *Journal of Social History* 14 (1980): 1–24.

Shanley, Mary Loydon. "Marriage Contract and Social Contract in Seventeenth Century English Political Thought." *Western Political Quarterly* (1979): 79–91.

Sharman, Cecil W. "George Fox and his Family." *Quaker History* 74 (Fall 1985) 1–19; *Quaker History* 75 (Spring 1986): 1–11.

Sharp, Isaac. "Summary of Pastor Theodor Sippell's articles on Origin of Quakerism." *Friends Quarterly Examiner* (1910): 298–302.

Skidelsky, Robert. "Exemplary Lives." *Times Literacy Supplement* (13–19 November 1987): 1250.

Smith, Bonnie. "The Contributions of Women to Modern Historiography." *American Historical Review* (June 1984): 709–32.

Smith, Robert M. "Christian Judayers in Early Stuart England." *Historical Magazine of Protestant Episcopal Church* (June 1983): 123–39.

Speizman, Milton D., and Jane C. Kronick. "A Seventeenth-Century Quaker Women's Declaration." *Signs* 1 no. 1 (1975): 231–45.

Stone, Lawrence. "Literacy and Education in England 1640–1900." *Past and Present* 42 (1969): 69–139.

———. "Illusions of a Changeless Family." *Times Literary Supplement* (May 1986): 525–26.
Stroud, L. John. "The History of Quaker Education in England (1647–1903)." Ph.D. diss. University of Leeds, 1944.
Taylor, Ernest E. "The First Publishers of Truth: A Study." *JFHS* 19 (1922): 66–81.
Thomas, Keith. "Women and the Civil War Sects." In Trevor H. Aston, ed., *Crisis in Europe 1560–1660*. London: Routledge & K. Paul, 1975.
Trinterud, Leonard. "A.D. 1689: The End of the Clerical World." Paper presented at the William Andrews Clark Library Seminar on Theology in Sixteenth and Seventeenth Century England. University of California at Los Angeles, 1971.
Vann, R. T. "Diggers and Quakers – A Further Note." *JFHS* 50, no. 2 (1962): 67–68.
———. "From Radicalism to Quakerism: G. Winstanley and Friends." *JFHS* 49 no. 1 (1959): 41–46.
Wolf, L. "The Jewry of the Restoration." *Transactions Jewish Historical Society of England* 5 (1908).
Whiteside, Joseph. "Some Accounts of Anne, Countesse of Pembroke." *Transactions of the Cumberland and Westmorland Archeological and Antiquarian Society* n.s. 5 (1905): 188–201.
Willen, Diane. "Women and Religion in Early Modern England." In Sherrin Marshall, ed., *Women in Reformation and Counter-Reformation Europe*. Bloomington, Indiana: Indiana University Press, 1989.
———. "Women in the Public Sphere in Early Modern England: The Case of the Urban Working Poor." *Sixteenth Century Journal* 19 (1988): 559–75.
Williams, E. M. "Women Preachers in the Civil War." *Journal of Modern History* 1 (1929): 561–69.
Wright, Luella M. "Literature and Education in Early Quakerism." *University of Iowa Humanistic Studies* 5 (1933).
Wrigley E. "Family Limitation in Pre-Industrial England." *Economic Historical Review*, 2d s. 19 (1966).
Wrigley, E. A., and R. S. Schofield. "English Population History from Family Reconstitution: Summary Results, 1600–1799." *Population Studies* 37 (July 1903): 157–84.
Youd, G. "The Common Fields of Lancashire." *Transactions* 113 (1961): 1–40.

Index

adoptionism, 284
Ames, William, 212–13, 214, 215
Amussen, Susan Dwyer, 238
Anderson, Alan B., 278
Anglican church, 16, 132, 140, 161, 198, 204–5
Anglicanism 9, 20–1
anti-Christ, 20
antinomians, 20
apocalypticism, 194, 197
Askew family, 29
assize court, 18, 33
Assize, North Wales and Cheshire circuit, 29, 33
Atkinson, Alice, 90
Audland, John, 68
autobiography (Fell), 11–26
Awty, Brian, 104–6

Ball, T. Frederick, 271, 146–47
Baptists, 19, 126, 133, 198
Barbour, Hugh, 6, 237, 278, 284
Barclay, Robert, 181–82, 199, 200, 231
Barrington, Ladies Joan and Judith, 31
Bauman, Richard, 278, 171
Beatty, Edward, 277
Beck, Dorothy, 91, 94
Beck, William, 271, 146–47
Bell, Maureen *et al.*, 237
Benson, Lewis, xviii
Benson, Gervace, 79, 112
Benson, Mary, 95
Benson, Sarah, 74
Besse, Joseph, 140–41, 270
Birch, Thomas, 30, 39
Birkett, Elizabeth, 95
Birkett, Ann, 91
bishops, 20, 204
Blackbury, Sarah, 141, 145, 147
Blackwood, Bruce, 27, 29
Blaugdon, Barbara, 6, 141
Blaykling, John, 149

Bradshaw, John, 79
Brailsford, Mabel, 6, 270
Braithwaite, Ellen, 91–2
Braithwaite, Edward, 90
Braithwaite, John, 90
Braithwaite, William C., 5, 102, 143–44, 157, 237, 275 *et passim*
Brayshaw, Alfred N., 227, 288
Bristol, 21, 28, 133, 136
  Nayler's entry into, 133
Brittaine, Isabele, 93
Bronner, Edwin B., 278
Brown, Irene Q., 280
Bunyan, John, 275, 282, 283
Burroughs, Edward, 134, 139, 199
business expenses, 291

Cadbury, Henry J., 213, 237, 277
Camm, Thomas, 191
Camm, John, 68
Camm, Ann (of Westmorland), 166
Capp, Bernard S., 284
Cardinale, Susan, 268
Cartmel, 91, 93, 102, 103
Caton, William, 35, 247, 49–50, 68, 188, 214, 215–16
charity, 83–100, 290–91, *see* Margaret Fell
Charles II, 16, 17, 18, 37–8, 131, 133, 135, 136, 137, 139, 140, 189
Christianson, Paul, 282
Church of England, *see* Anglican Church
circumcision, 217, 218, 220
Civil War, 30, 140, 169
Clarendon Code, 140
Clark, Alice, 73, 77
Clifford, Ann, 97
Collins, John, 282
Colton, Jenny, 74
Colton, Jane, 90
Comfort, William, 277
Committee for Compounding, 29

318

Index

Commonwealth, 131, 140
Condert, Allison, 283, 284
conventicle worship, 134, 140, 141, *see under* Quakers
Conventicle Acts, 140
Cooper, Hester, 160–1
Corporation Act, 140
Court of King's Bench, 21
Coward, Barry, 270
Cowell, Jane, 93
Cowell, Bridgett, 74
Crawford, Patricia, 237
Creasey, Maurice, 283
Cromwell, Oliver, 16, 68, 132–3, 209, 211
Cromwell, Richard, 133
Crosfield, Helen, 213, 235
Crown, The, 20, 45, 140, 189, 200
  her letters and visits to, 21, 23, *see* Margaret Fell
Curtis, Thomas, 151–2, 174
Curtis, Ann, 136, 151–2, 174
Curwen, Alice, 141
Curwen, Ann, 60
Curwen, Thomas, 78

Dalton-in-Furness, 27
Davis, J. C., 235
death and dying, 47–8, 248
Declaration of Breda, 140
Dixon, Ann, 73
Dobree, Bonamy, 277
Dockra, Thomas, 141
Dodgeson, Peggy, 75
domestic expenses, 290–1, *see* Swarthmoor
domestic economy, 8, *see* Swarthmoor
Duchy of Lancaster, 29
Dunn, Mary Maples, 230, 273, 277, 280, 283, 289
Dunn, Richard S., 273, 277, 283
Durston, Christopher, 238
Dutch Republic, *see* Netherlands
Dutch War, 140
Dyer, Mary, 141

ecclesiastical courts, 207
egalitarianism, 169–70, 181, 183
Elson, Mary, 141, 145–6

Endy, Melvin, 6, 148, 181–2, 197–8, 199, 207, 236, 237, 278, 282–3
epistles written by Fell, *see* Margaret Fell
Eversley, David, 250
Ezell, Margaret, 238

family life, *see also* patriarchy
  early modern, 7
  of Margaret Fell, *see* Fell
"fanatics," 1, 133, 139, *see* radical sectarianism
Farnsworth, Richard, 154, 275
Fell children, ix–xviii, 37–53, 47–8, 245 *et passim*
  Fell, Bridget, 37–41, 45
  Fell, George, 38–9, 43, 49–53, 120–21, 191, 248
    his wife, Hannah (Potter), 39, 246, 52, 72
    his will, 52, 255
  Fell, Isabel, 45, 52
  Fell, Margaret Jr, 37–8, 41, 45
  Fell, Mary (Lower), 43, 45, 47–8, 72, 195, 247
  Fell, Rachel (Abraham), 44, 48, 58, 181
  Fell, Sarah (Meade), 43–4, 45–6, 159: clerk for the SWMM, 91–100; her accounting methods, 70; her banking activities, 89–100; her epistle on order and discipline, 159–65; her philanthropy, 90–1
  Fell, Susannah, 43–44
  legacies, 43
Fell, Christopher, 108
Fell, Henry, 269
Fell, Margaret (1614–1702), Margaret Askew Fell Fox:
  A Call to the Seed of Israel (1668)
  A Loving Salutation to the Seed of Abraham (1656)
  and marriage procedure, 157–8
  and Swarthmoor, 8, 23, 32, 39, 40
  apologist, 203–5, 211–24, 233–4
  a study in contrasts, 229–34

A Testimonie of the Touch-stone . . . Some of the Ranters Principles Answered (1656)
autobiography, 11–26; and the oath, 18
business activities, 8, 66, 71–9, 101–28, 123–8, 122
Certain Queries to the Teachers and Rabbis Among the Jews (1656 or 57)
charity, 290–91, 83–100, esp. 87–100, see charity
co-foundership, 135, 137–8, 139, 142, 147, 148, 149–52, 155–7, 158, 159–65, 166–8, 175, 181–3, 198, 211, 215, 219, 223–4, 229–34, 275, 279
co-leadership, 134, 135, 137–8, 139, 142, 147, 148, 149–52, 155–7, 158, 159–65, 166–8, 170–2, 173, 175, 181–3, 198, 211, 215, 219, 223–4, 229–34, 270, 275, 279
Daughters of Zion Awakened (1677)
domestic economy, 37–41, 65–82
establishment of women's meetings, 143–68
family and domestic life, 7, 8, 37–53, 47–8
family background, 29, 31, 37–53, 61, 72, 81–2
feud with Thomas Rawlinson, 101–28
financial independence, 101
fines, 23
For Manasseth-Ben-Israel: The Call of the Jews out of Babylon (1656)
friendship with William and Gulielma Penn, 169–83, 230, 231
gentry lifestyle and status, 123–8
her appearance, 68–9, 107
her authority in the church, 101–2, 106–14, 115, 116–19, 122, 123–28
her biblical imagery, 192
her biographies, 2, 235
her business ethics, 101–28, esp. 123–8
her conversion or "convincement" to Quakerism, 13
her death, 24–5, 48, 189
her dispute with son, 49–53, 191
her epistles, 21, 25, 34–5, 37, 40, 132, 134, 142, 154, 168, 190, 195, 206
her estate income, 39, 190–1, *see* business activities, above
her grandchildren, 37, 45, 49, 53, 72, 255
her imprisonments (at Lancaster), 47, 51, 56, 271
her legal rights to Swarthmoor, 8, 31, 39–40, 50–54, 248, 249
her meetings with the King, 136, 137, 139, 173, 270
her mental world, 187, 196
her neighbors, 66–7
her nonconformity, 18, *see* nonconformity
her peace testimony, 5–6, 34, 131, 137–8
her religious outlook, 187, 190–91, 196, 223–24, 229–34
her sister, 7, 67, 255
her theology, 127–8, 190, 191, 197–210, 219, 222–4, 229–34, *see* religious outlook *above*: apocalypticism, 194–5, 208–10, 216–17, 222–4, 226; christology, 192–3, 200–4; faith, 192, 196, 200, 202, 225–6; restorationism, 154–5, 225; millennialism, 134, 135, 193, 194–5, 208–10, 211, 212–28, 217, 221–2, 225, 226–8, 286; satan, 193–4; sin, 127–8, 191, 193, 208, 216–17, 220, 227; the light, 192, 194, 216, 220; spiritualism, 197–210, 220, 225–8, 282–3; revelation, 196, 197–210, 220, 282–3; perfectionism, 126–8, 196, 203, 208; universalism, 217, 221–2
her travels, 68–69, 156, 291: with her daughters, 156

her trial (1664), 18
her will, 81, 255
her writings, 131–2, 137–8, 236, 197–8, 225–6: to the Jews, 211–28, see epistles *above*
personal correspondence, 141, 170–83, 269, 270
Kendal Fund, 87, 153–4
living expenses, 290–91
London women's meetings, 145–7, 223–4
M[argaret] F[ell]s Answer to Allan Smallwood Dr. priest of Grastock in Cumberland (1668)
oath trial, *see* oaths
on death and dying, 187–9
persecutions, 23, 141, 178–9, 192, *see* Quaker persecutions
political issues, 131, 135, 137–8, 139, 140–2, 170–2, 223–4, 269,
Quaker meetings, 18
spiritual authority, 181–3, 208, 225, 229–34
see spiritualism *above*
SWMM, 83–100
To the Generall Council and Officers of the English Army (1659)
The Citie of London Reproved (1660)
This was given to Major Generall Harrison and the Rest (1660)
translation of her writings into Latin, Hebrew, etc., 211–24
visits to London, 37–41, 46–7
vs. the Ranters, 190 *et passim*
women's separate meetings, 155, 159–65, 229
Women's Speaking Justified (1666), 19
Fell, Margaret and Thomas Fell, Judge,
assize court circuit, 14
barrister, 29
family history of Judge Fell, 1–2, 29–30
family, Margaret Fell's, 31, 37–53, 44–5, 61, 72, 81–2
financial situation
her children, ix–xviii, *see above under* Margaret Fell

spousal relationship, 24, 32, 34–7, 55, 60–1
Fell, Sarah of Ulverston (no relation to Margaret Fell), 93, 163–4
Fell, Thomas
assize circuit court judge, 14, 33
gentry status, 30, 79
his death, 15
his friends and associates, 29–30, 35, 66, 79
his will, 81
judge, 33
offices held, 29–30, 241–42, 79
reaction to his wife's conversion, 14–15, 35, 60
Fell women, 31, 81, 106
their dress, 68–9
Fifth Monarchy Uprising, 42
Fifth Monarchists, 199, 226
Firth, Katherine R., 282, 284
Fisher, Samuel, 134, 212, 213, 215, 216
Fisher, Jane, 93
Fisher, Mary, 6
Fleming, Daniel of Rydal Hall, 80, 97, 170
Fletcher, Elizabeth, 141
flock, 59, 159, 179–80
Force Forge, 76, 78, 101, 103–14, 119–28
Ford, Linda, 278
Fox, George, 134, 137–8, 139, *see* Quakers
his theology, 15, 126, 199, 209, 212, 216, 219, 220, 222, 223–8
charged with blasphemy, 132
death, 22, 23–4, 180
ministry, 24, 103–4, 177
his testimony, 24 *et passim*
persecution, 24
imprisonments, 16, 21–2, 38, 42, 133, 136, 172
re Judge Fell, 33
itinerant ministry, 34, 54
Journal, 47, 133, 139, 143, 146, 167, 176, 180, 195, 223, 226, 231, 232
his epistles, 139, 144, 271, 148, 149, 272–3, 152–3, 275, 155
peace testimony, 138–9

Swarthmoor, 143, 232, *see*
    Swarthmoor Hall
  meetings, men's and women's, 143,
    168 *see also under* Fell
  on women's ministry, 152–168, 275
  re spiritual authority, 153, 167–8
  voyage to America, 155, 172
  on marriage procedure, 158
  his leadership *vis-à-vis* Fell, 166–8,
    230–4, *see* Fell, co-leadership
  his travels, 167, 179
  friendship with William Penn,
    170–83, 231
  printing of books, 175
  founder status, 176, 229–34, *see also*
    Fell, co-founder
  his relationship with his stepchildren,
    52, 53–4
  his marriage, 53–4, 55
  his agreement not to take over his
    wife's estate, 54, 249
  his correspondence, 72 *et passim*
  on charity, 86, 87–8, 89–90
  on settling disputes, 111, 122, 125
  spiritual authority, 177, 183, 230–4
  spirit, inner light, 199
  the Jewish mission, 215, 286–7
  depressions, 226–7: at Enfield, 227–8
  expenses while at Swarthmoor, 291
Fox, George and Margaret Fell Fox,
    and Swarthmoor, 8, 143, 149
  *A Paper Concerning Such as are
    made Ministers* (1659)
  co-founder, 6–7, 83, 87, 101, *see*
    Margaret Fell
  co-leadership, 83, 87, 99–100,
    101–2, 106–8, 123–8, *see*
    Margaret Fell
  establishment of women's meetings,
    143–7, 166–8, *see* Margaret Fell
  establishment of men's meetings,
    143–4, 145
  Fox, Penn, and Fell, 169–83
  her position *vis-à-vis* George Fox
    and his coming to Swarthmoor
    Hall, 13–15
  his first meeting with Margaret
    Fell and her household,
    13

imprisonments, *see* prison, Fell,
    and Fox
  London's women's meetings, 145–7
  marriage, 21, 53–4, 55
  persecutions, 23
  their oath trials, 170–2
  their relationship, 22–4, 33, 42, 52,
    53–7, 166–8, 224–8, 230–4,
    291 *et passim*
  their theology, 212, 228, *see also* Fell,
    and Fox
  Wilkinson-Story Schism, 147–52
Fox, Margaret Askew Fell (1614–1702),
    see Margaret Fell
Fraser, Antonia, 271
Freiday, Dean, 283–4
Friends, The Society of, 167, 176, 232,
    232, *see* Quakers
friendship, 170, 176, 183
Frost, J. William, 237, 271, 278, 279
Furness, 27, 65
  economy, 27
  population, 28

Gardener, Isabel, 108
Geldart, Margaret, 93–4
Geldart, Ann, 74
gender, 26, 100, 168, 169–83, 179,
    229–34, *see* patriarchy
  and class, 169–83, 179, 183, 231–4
  and Quaker nonconformity, 6, 10, *see*
    nonconformity
  female preaching and ministry, 19–20
  feminism in seventeenth century, 7, 10
gentry status, 230–4
gentry status and lifestyle, 68–70, 79
Gersham, Gerard Scholem, 286
gospel, 25
"gospel order," 163, 164
Great Fire of London, 140
Greaves, Richard, 1, 235, 237, 278
Gregg, Pauline, 279
Gwyn, Douglas, 224, 225, 288

Halhead, Miles, 112
Harrison, Major General, 135
Haselrig, Arthur (Sir), 279
Hawkshead, 96, 102, 112, 118
Hawkswell, 48, 49

Index

Healey, Robert M., 286
Hebrew translations of Fell's works, 211, 216
Hebrew prophecy, 219, 220–1
Higgins, Leslie, 286, 287
Hill, Christopher, 9, 139, 226, 235, 270, 277, 281, 285
Holme, Elizabeth Leavens, 107–8
Holme, Isabel, 94
Holmes, Reginald, 104
Holme, Thomas, 107–8
Hookes, Ellis, 138, 173
Hooten, Elizabeth, 6, 136
Horle, Craig, 261, 279, 280
Houlbrooke, Ralph, 238
Howgill, Francis, 68, 132, 137, 188–89
Hubberthorne, Richard, 87, 133, 134, 138, 139
Hudson, Jane, 91
Hudson, Winthrop, xviii, 167, 168, 236, 270, 274, 277
Hull, William, 237, 277, 213
Hutchinson, Julius, 278
Hutchinson, Col., 133
Hutchinson, Lucy, 133, 169, 275
Hutton, Thomas, 106

Ingle, H. Larry, 272, 282
Inner Light or light of Christ, 6, 15, 19, 81, 135, 144, 148, 181, 196, 199, 212–13, 218, 220–1, 225, 227
Interregnum, 6, 9, 16, 133
Irish rebellion, 133
iron business, *see* Force Forge
Irwin, Joyce, 7, 238

James II, 23
Jerusalem, 217, 218, 222, 223, 226, *see* Zion
Jews, *see* Menasseh-ben-Israel
  philo-semitism, 211–24
  Quaker mission to, 203, 211–24
  re-entry into England, 131, 209, 211, 213, 221–2
Josselin, Ralph, Essex Anglican priest, 1–2, 36, 77–8, 81, 97–8, 187, 189, 190–2
justices of the peace, 23, 29, 97, 102, 141, 58, 170–2, 179

Kanner, Barbara, 7
Katz, David, 286
Keeble, N. H., 270
Keith, George, 275
Kelley, Donald R., xviii, 59–61
Kendal, 91, 97, 141, 148
Kendal Fund, 87
King's Bench Court, 21
Kirkby family, 66–7
Kirkby, Richard Col., 66–7, 251
Knype, Thomas, 79
Kroll, Richard, 283
Kronick, Jane C., 6
Kussmaul, Ann, 252

Lambs' War, 174, 216–28, 225, 227, *see* Fell's and Fox's theology
Lamont, William, 268, 284
Lampitt, William, 15, 195, 242
Lancashire, 27, 95, 134
  population, 28
Lancaster, 65, 171
  Duchy, 29
  Quarter Sessions, 33
Lately, Gilbert, 146–7
Latin translations of Fell's works, 213
law, the, 219
Lawson, Thomas, 46, 83
letters written by Fell, *see* epistles *and under Margaret Fell's writings and correspondence*
Levellers, 226, 279
Levy, Barry, 27, 277
Light, *see* Inner Light
Lilburne, John, 279
Lloyd, Arnold, 144, 271, 275
London, 131, 135, 137, 145
  London women's meetings, 145–7, *see* Fell and Fox, and women's meetings
Lovejoy, David S., 285–6
Lower, Mary, 72
Lower, Thomas, 47, 112, 151–2
LQM, 101, 104, 112, 118, 122
LWQM, 112, 115, 159–65

Macfarlane, Alan, 1–2, 98, 187–8, 281
Mack, Phyllis, 7, 237, 271, 275
marriage, 162–3

Marsh Grange, 27, 29, 32, 51, 72
matriarch, 8, 22, 31, 42, 81, 101, 179, 182, 229
McGregor, J. F., 278
Meade, William, 43–4, 52, 180, 279
men's monthly meetings, 160, *see* SMMM
Menasseh ben Israel, 209, 211, 222, *see* Jews in England
Mendelson, Sara Heller, 237
millennialism, 134
Miller, Perry, 197, 282
Milton, John, 226, 227
ministry
  female, 7, 19–20
  ordained, 20, 24, 32, 132, 193, 198, 204–5, 206–7
  Quaker, 143–68, 201
Mistress of Swarthmoor, 23, 32, 71, *see* Margaret Fell *and* Swarthmoor
Monk, General, 137, 269
Moore, John M., 271, 279
Moses, Peter, 108
mothers of Israel, 165
Mucklow, William, 277
Mullett, Michael, 94–5, 257, 260, 283

Nayler, James, 24, 79, 86, 132, 133, 168, 215, 216, 230
Netherlands, 211–24, *see* Jews, Quaker mission to
Netherlands mission, 34
nonconformity, 6, 18, 59, 82, 133, 139–42, 170–2, 233, *see* sectarianism
  political implications, 133, 229–34

Oaths, 18, 20, 111, 138, 140, 278
Oath trial, 142, 170–2
Oxford, 141

parliament, 37, 83, 84–5, 131, 133, 134, 140, 269
  Long Parliament, 132
Parker, Alexander, 137
patriarchy, 1, 7–8, 35–6, 101, 151, 166, 169, 179, 232–3, 238, 277
patrimony, 70
patronage system, 142, 168,

169–72, 173, 175–6, 177, 183, 232, 279
  deference, 132, 136, 142, 166, 173, 232, 274
Peace Testimony, 5–6, 17, 34, 131, 137
Pearson, Anthony, 79, 254
Pearson, John, 149
Pease, Mary, 57–8
Penn, Gulielma, 80, 156, 172, 174, 177–9
Penn, William, 9, 24, 248, 55–6, 72, 80, 97, 169–83, 230, 231
  his connections with the Crown, 173, 180
  his conversion, 172
  his trial at the Old-Baily, 171–2
  leadership, 180
  see Margaret Fell and friendship
Penney, Norman, ed., 236
The Household Account Book of Sarah Fell (1920), 65, 251
Pennsylvania, 28, 177–8, 182
Pepys, Samuel, 43, 135–6, 175
perfectionism, 126, 267–8
persecution, 132, 133, 137, 138, 139–42, 148, 170, 178–9, 189–90, 192–3, 200, 233, 271
  of Fell, *see* Fell
  of Quakers, *see* Quakers
Phillips, Roderick, 238
political-religious issues, 170–2
politics and religion, 226, 227
Pollard, Ellen, 74–5
poor relief, 83–100
  almshouse, 97
  state system, 84–5, 260
  SWMM activities, 83
  the Fell's activities, 83, *see under* Margaret Fell
Popkin, Richard, 213, 215, 285, 286
praemunire, 171, 247, 278
Presbyterian, 140, 198
Presbytery of Furness, 29
priests, *see* ministry
Prior, Mary, 7, 237
prison and imprisonment, 19, 21, 23, 38, 42
  Fell and Fox, *see under Fell and Fox, Fox and Quakerism*

writings, 19–20
prophets, prophecy, 7, 155, 229, 233
protestant, 187, 189, 194, 198, 203, 207, 209–10, 233
Publishers of Truth, 34, 50, 148
Puritan, Puritanism, 140, 182, 187, 189, 191, 196, 197–210, 219, 221, 222, 223
  Quaker opposition to, 9 see also above

Quaker nonconformity, 6, see sectarianism
Quaker(s), 133
  Act, 278
  books and publications, 175, 280
  children, 162–8
  church hierarchy, 181–3
  church order and discipline, 106–28, esp. 122–8, 148, 149, 150, 151, 153, 159–68, 169, 181, 272–3
  church organization, 143–45, 148, 272–3, 158–68, 181
  conventicle worship, 17–18, 23, 42, 134, 269, 159–68
  demography, 28
  dress, 25
  in northwest England, 28
  equality of the sexes, 144, 153–4, 170, 208, 225
  historiography, 1–10, 139, 145–7, 166–8, 230–4
  imprisonments, 137, 139
  in prison, 17
  internal disputes, 106–28, 147–52
  in the Netherlands, 211–24
  itinerant ministers, 57, 102, 144
  its appeal to women, 233–4
  London women's meetings, 145–7, see Fell and Fox
  marriages, 149, 157–8, 162–3
  orphans, 95–6
  persecutions, 17–18, 240, 23, 40, see persecutions
  poor relief, (1561–2), 84, 86–8, 90–100, 145–7, 155, 156, see Margaret Fell's charity
  reaction to the Restoration, 16
  refusal to do hat honor, 132, 138, 169–70

resisting civil authority, 17–8
religious ideas, 57
spirit, see Inner Light
spiritual authority, 148, 152–3, 154, 177, 181–2, 272, see theology under restorationism, and under Fell
testimonies, 23, 24
theology, 25, 159–68, 194, 196, 197–9, 200, 211–28, 233: apocalypticism, 194, 208–10, 216–17, 226; christology, 198, 199, 201–2, 284; millennialism, 212–28; perfectionism, 126–8, 196, 203; Quaker vs Puritan, 197–210; restorationism, 153, 154, 159, 274, 275; sin, 198, 126–8; spiritualism, 197–210, 206, 282–3
voluntarism, 181–2
women, 155–68, 229, 271: women's meetings, see Fell and Fox, and women's meetings; women's petition to Parliament, 134; women writers, 131–2, 159
Quaker women's meetings, see women's meetings and Margaret Fell

radical, radicalism, 170, 223–4, 229, 288
Ranters, 16, 19, 190, 193, 199, 281
Rawlinson, Thomas, 8, 50, 101–28, 191, see Margaret Fell
Reay, Barry, 6, 133–34, 268, 269, 270, 277, 278
regicides, 38
regratresses (female peddlers), 73
religious toleration, 17, 23, 132, 134, 138, 140, 180
Restoration, The, 6, 16, 37, 135, 140
Richardson, Matthew (Fell's brother-in-law), 52, 67, 263, 111–12
Robbins, Caroline, 278, 280
Roberts, Gerard, 35
Roberts Arthur, 271
Rogers, William, 149, 275
Roman Catholics, 133, 205
Roper, Richard, 68

Index

Ross, Isabel, 144, 213, 235, 275, 277, 286 *et passim*
Rous, Margaret, 155 *see also under* Margaret Fell
Rous, John, 36, 45, 51–2, 111, 187–8
Rusland Hall (home of Thomas Rawlinson), 102
Russell, Rachel, duchess of Devonshire, 3

sacraments, 20, 198, 205
Salthouse, Thomas, 68, 70, 104, 116, 118
Salthouse, William, 90
salvation, 25, 135
Satterthwaite, William, 116
Savoy Conference, 140
Schochet, Gordon J., 238
Schwoerer, Lois G., 3, 4, 236
scripture, 20, 193, 194, 198, 199, 206, 216, 229
Seaver, Paul, 21, 127
sectarians and sectarianism, 6, 10, 42, 139
  radical, 133
seed, Seed of the Promise, 19, 159, 203, 211, 217, 218, 221
Seekers, 199
Serrarius, Peter, 213
servants, 8, 57–9, 73–5
Shammas, Carol, 251, 253, 259
Sharpe, Joseph (bailiff of Marsh Grange), 90
sheriffs, 18
Simcock, Jean, 156
Slack, Paul, 84–6, 99, 260
Slater, Miriam, 35–6, 41, 246, 249, 280
Smallwood, Allen, 20, 201, 205–7, 230
Smith, Bonnie, xviii, 236
Smith, Nigel, 235
Smith, Hilda, 7, 237, 268
SMMM, 96, 104, 114–19, 122, 125
social status, 169–70, 173, 175–76, 177, 182–3, 279
soldiers, 134–5, 141, 164, 193, 269
Speizman, Milton D. and Jane C. Kronick, 6, 159–66
Spinoza, Baruch de, 212, 214–15
spirit, 25, *see inner light*

spiritualism and Quakers, 9 *see also under* Quakers *and* Fell
St. Mary's Anglican Church (Ulverston), 32, 37
Standish, Ann, 75
stock, *see* SWMMM *and* SMMMM, 156, 260
Stone, Lawrence, 238, 280
Story, John, 24, 147, 149–50, 230
Strickland, Jane, 93
Stubbs, John, 35, 212, 213, 215, 269
Swarthmoor Hall, 13, 15, 19, 21, 22, 24, 25, 27, 29, 37, 42, 44–5, 47, 48, 50–3, 71–2, 83, 174, 178–9, 192, 210, 234, 246, 290–1
  domestic economy, 31, 65–82, 290–1
  income, 77–9, 290–1
  labor force, 73–75, 77, 90
  meetings at, 23
  rents, 76, 253, *see* Margaret Fell
SWMM, 6, 60, 74, 83, 91–100, 101, 112, 115, 156, 158–68
  stock or treasury, 91–100

Taylor, George, 153
Taylor, James, 104, 118
Taylor, Mary (wife of James Taylor), 91–2, 95
testimony, 160–1
Thomas, Keith, 6, 237
tithes, 20, 28, 48, 134, 138, 140, 161, 207, 276
Todd, Margo, 238
Toleration Act, 23, 200
Tolles, Frederick, 81–2, 268
Travers, Rebecca, 131, 141, 145, 147
Trevett, Christine, 238
Trinterud, Leonard, 284
Truth (Quaker), 32, 52, 123, 132, 141, 143, 156, 160–1, 174, 182, 192, 203

Ulverston, 29–30, 65, 73
Uniformity Act, 140
unity, 181–2, *see* church order
universalism, 217, 221–2, *see* seed *and* Margaret Fell's theology

Vann, Richard, 6, 80, 144, 250, 278

Index

Verney, Lady (of Claydon House), 179
Verney, Sir Ralph and family, 3–6, 41, 246
visitation, 218, 220
Vokins, Joan, 141

Walker, Mary, 94
Walker, Reginald, 263, 119
Waller, Richard, 68
Wallington, Nehemiah, London Puritan artisan, 2, 127–8, 187, 189–92
Watts, Michael, 6, 144, 237
weaker vessel image, 36, 101, 154–5, 159, 164, 166
Webb, Maria, 235
West, William, Col., 35, 79, 111–12
White, Dorothy, 131
Whitehall, 21, 247, 136, 209, 286
Whitehead, Anne Downer, 131, 141, 145–6, 147
Whitrow, Joan, 131
Widders, Robert, 112
Wilkinson, John, 24, 147–230
Wilkinson-Story Schism, 123, 147–52, 156, 161, 173–5, 181–2
Willan, Thomas, 153
Willen, Diane, 316
William and Mary, 23
Williams, George, 274
Williamson, Elizabeth, 98
Wilson, William, 104, 263, 118
Wilson, John, 112

women
  and Civil War, 6
  Countesse of Warwich, 31
  poor, 83
  prophets, 7
  sectarian prophetic ministers, 19–20
  sectarians, 7
  Stuart period, 6
  the Fell women, 31, *see under* Fell
  writers, 7, 131–2, 159–68
Women's Box Meeting, *see under* London
women's meetings, 5, 143–68, esp. 158–68, 229, *see also* Fell *and* Fox
women's ministry, 143–68, 159, 166, 208, 223–4, 229, 288
Women's Speaking Justified (1666), 19
women's work, 73–7
Woodall, Jennett, 96
Woodell, Jane, 91
Woolrych, Austin, 269
Worcester Gaol, 21
Wright, Luella, 6
Wrightson, Keith, 238, 278

Yorkshire, 156
  women's quarterly meeting, 156

Zagorin, Perez, xvii, 239, 268, 274, 279, 285
Zion, 217, 218, 219, 221, 224